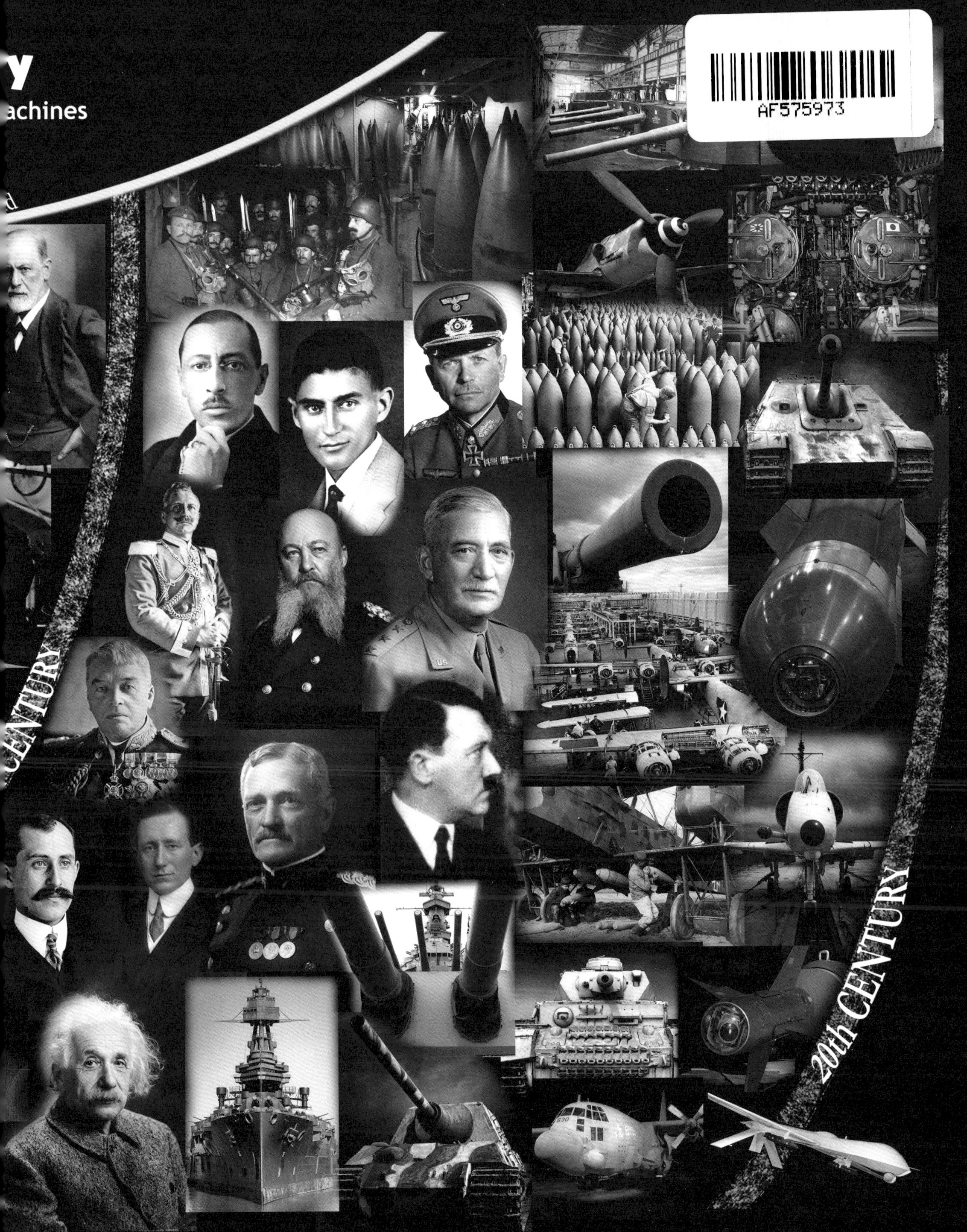

y
achines
d
CENTURY
20th CENTURY
AF575973

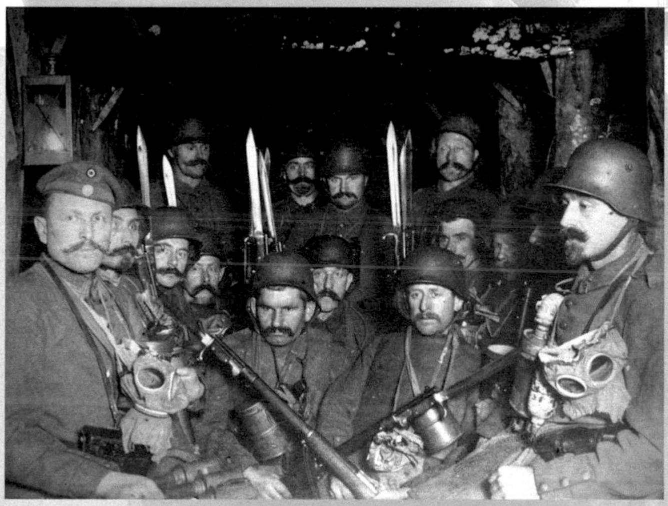

TRIUMPH AND TRAGEDY

The Evolution and Legacy of 20th Century War Machines

Photographs and History by Martin Miller

with Foreword by R. Cargill Hall

Emeritus Chief Historian, National Reconaissance Office

Triumph and Tragedy: The Evolution and Legacy of 20th Century War Machines
by Martin Miller

Library of Congress Control Number: 2020915505

ISBN 978-0-9862127-2-7

The Chelsea Press
Baltimore, Maryland

Printed in the United States of America

ALSO BY MARTIN MILLER

The Neutron's Long Shadow: Legacies of Nuclear Explosives Production in the Manhattan Project, Schiffer Publishing, Ltd., 2017.

Weapons of Mass Destruction: Specters of the Nuclear Age, Schiffer Publishing, Ltd., 2017.

FOR FURTHER INFORMATION PLEASE VISIT

www.WorldWarWorks.com

www.Martin-Miller.us

This book is dedicated to the memory of my beloved uncle, Harry Luck Miller, who served as a Private in Company A, 304th Division Supply Train, American Expeditionary Force in France from May 1918 to July 1919, and for the two million other American men in the AEF who lost their innocence, limbs, and lives in the cataclysm of the First World War.

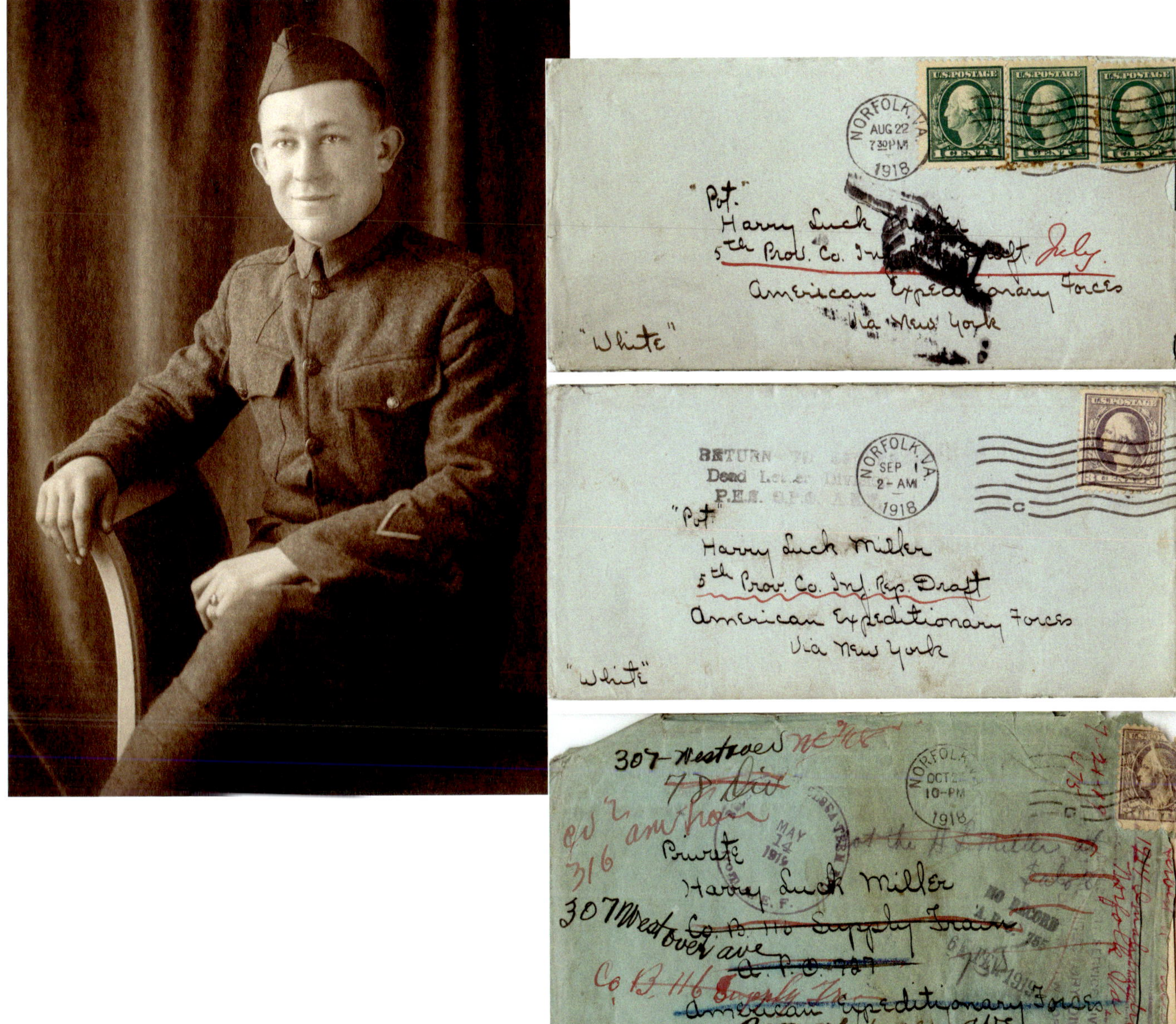

Envelopes of letters to Private Miller in France 1918 from his fiancé. The increasing chaos of the war from August 22 to September 1 is reflected in the Army's inability to deliver the last two letters.

Turning and turning in the widening gyre
The falcon cannot hear the falconer;
Things fall apart; the centre cannot hold;
Mere anarchy is loosed upon the world,
The blood-dimmed tide is loosed, and everywhere
The ceremony of innocence is drowned;
The best lack all conviction, while the worst
Are full of passionate intensity.

Surely some revelation is at hand ...

—from William Butler Yeats'
The Second Coming 1919

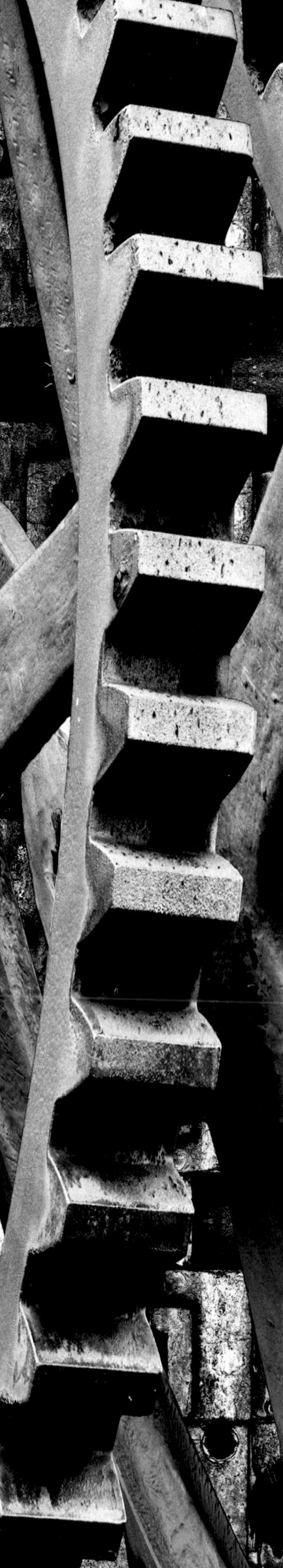

CONTENTS

FOREWORD

Triumph and Tragedy is a history of the pursuit by modern nation states of ever more deadly weapons with which to subjugate enemies in times of war. Its author, Martin Miller, brings a physicist's training to bear in his perceptive prelude, an accounting of the 18th and 19th century scientific, cultural and industrial revolutions that made possible the engines of war that would make the 20th century the most deadly in human history. This history turns on the primary weapons fashioned and introduced in the two world wars, the huge companies that built them, and how governments met the materiel demands of modern warfare. Once underway in 1914, First World War weapons include advances in artillery, airplanes, machine guns, tanks, and naval forces. The range of artillery, for example, increased dramatically, from targets viewed in line of sight to targets attacked far out of sight; that is, from direct fire to indirect fire. One large German gun struck Paris itself from 75 miles away. The advent of airplanes enabled the major powers to use them first for reconnaissance and artillery spotting of indirect fire, then, equipped with machine guns, as fighter and ground attack planes. Finally, loaded with bombs, they served to attack enemy emplacements and installations and, finally, cities. Having bombed Antwerp in late August 1914 with Zeppelins, the Germans followed up with strikes on the British coast and on London—aerial attacks on civilians that continued throughout the conflict.

As evidenced by indiscriminate attacks on civilian targets, the First World War saw Germany embrace "total war," wherein no distinction would be drawn between soldiers and civilians. Germany introduced poison gas on the battlefield, and the Allies responded in kind. (So terrible were the effects of this weapon, no combatant in the Second World War would use it again.) Ignoring the law of the high seas, Germany sank ships from neutral countries without warning. It bombed hospitals and impressed civilians into slave labor in its mines and factories. The Allies imposed a naval blockade that by 1918 brought starvation to the German populace. At war's end on 11 November of that year, a subjugated Germany would be forced to sign the Treaty of Versailles, which imposed on it reparations to the Allies and proscribed its rearmament. But, as Miller observes, in Germany "it would not be long before preparations began for another war with these precedents guiding the way."

The Second World War began on 1 September 1939 when the Nazi Third Reich invaded Poland. Some eight months later, on 10 May 1940, German mechanized forces struck at France through the Ardennes, flanked the Maginot Line, and trapped French and British forces against the English Channel. In six weeks they conquered France and drove the British army into the sea at Dunkirk. In the fall of 1940 an aerial blitz of Britain's air force and then of its major cities followed. Abandoning an invasion of Great Britain, on 22 June 1941 Germany invaded the Soviet Union in a surprise attack termed Operation Barbarossa. This action placed the Third Reich in a war on two fronts, to its east and west, which ultimately would prove its undoing. Japan, meanwhile, had been carving out spheres of influence in Asia during the 1930s. The United States imposed sanctions on shipments of raw materials to that country in early 1941, and on 7 December Japan launched a carrier-based surprise attack on the American fleet at Pearl Harbor, in Hawaii. The United States declared war on Japan and, before the end of December, Germany declared war on the United States. The combatant states, known as the Axis and Allied powers, were now fully engaged in what would be a total war never matched before or since. The Second World War would see more people die in its maw than had died in all the wars in recorded human history.

In Miller's book the "mobilization" phases of both the First and Second World Wars are introduced by illuminating graphs prepared by the author: GDP and crude steel production of the major powers in the lead-up to both WWI and WWII. In economic strength and steel the United States out-classed all combatant states during both wars by a wide margin, but especially so in WWII. In the that war, thousands of tanks, trucks, guns, warships and airplanes would be mass-produced and see combat; most of the materiel was delivered by American-made Liberty Ships to North Africa, England, the Soviet Union and to India, Australia and various islands in the Pacific Theater of war. If Japanese leaders expected the United States to sue for peace after the destruction of much its fleet at Pearl Harbor, they were woefully mistaken. Indeed, the atomic bomb, the most portentous weapon innovation of that war, would ultimately be employed against Japan. In somewhat less than four years, in less than a single American presidential term, the Allies utterly defeated the Axis powers. The unity of purpose, incredible endeavors and sacrifices of the Allied peoples on their respective home fronts during the Second World War can hardly be imagined today by those who did not live through that cataclysm.

The focus of *Triumph and Tragedy* is conventional weaponry: the development of its enabling technologies and its deployment in total war. These machines reached their apotheosis in the First and Second World Wars so weapons that appeared after World War II during the so-called "Cold War" are assayed in much less detail. The Soviet Union, America's postwar antagonist, would acquire atomic weapons and long range bombers. Both the United States and the Soviet Union would eventually mate atomic and thermonuclear warheads with Intercontinental Ballistic Missiles (ICBMs), developments that were treated in Miller's previous volume, *Weapons of Mass Destruction*. For American and Soviet leaders, any thought of total war in the years that followed was tantamount to a suicide pact. The term coined for such an event: "mutually assured destruction" or MAD. Instead, conventional wars were initiated by proxies of the major powers in Korea and Vietnam. These wars ushered in various weapon innovations, most notably, perhaps, Surface-to-Air Missiles (SAM) that could bring down warplanes at altitudes up to 75-80,000 feet. Miller's book touches on the advent of stealth airplanes and advances in space systems that will shape actions and events in any wars that follow in the 21st century. The author's detailed photographs of twentieth-century war machines taken at home and abroad bring his book to a close. They depict those remarkable objects as mute and solemn reminders of a world in extremis. Anyone with an interest in how and why the fearsome twentieth-century war machines came into being and what they looked like will find in this volume a welcome and useful overview.

R. Cargill Hall
Emeritus Chief Historian,National Reconnaissance Office
January 2018

PREFACE

This book is the third and final volume of a photographic odyssey begun fourteen years ago to document and to interpret my responses to war machines of the twentieth century. I first encountered these extraordinary objects in the Army Ordnance Museum collection at Aberdeen Proving Ground where I was posted as a US Army Ordnance officer in 1972–1974. At that time an historical collection of tanks lined the median strip of Maryland Boulevard and proved an eerie sight with the sun rising behind them at dawn. I was already doing large-format photography and, though my subjects at the time were confined to nature, a curious affinity for those stark objects of war was awakened. However, it was not until January 2006, after I had retired from a career as research physicist at the same installation, that I engaged these subjects with my camera. Some eighteen months later I encountered and photographed my first nuclear weapons at Hill Aerospace Museum in Utah. In the early years of my career as a physicist I had done research on the atmospheric response to high altitude nuclear detonations but had never been involved with the hardware agency of these explosions. I was fascinated and sought out more such subjects across the country. In my nuclear weapons effects research I had computed radar blackout contours above the Mickelsen Safeguard anti-ballistic missile site in North Dakota that had operated for only a single day before sinking into obsolescence. In 2008 I made arrangements to photograph its remnants. Also in 2008 I began wondering what was left of the Manhattan Project facilities and obtained permission to photograph at Oak Ridge and Hanford that year.

In 2013 I returned to Hanford for another photoshoot to discover that many of my subjects from 2008 had been demolished and their sites remediated. The same was true of the major Manhattan Project facilities at Oak Ridge. In 2010 the Army Ordnance Museum at Aberdeen Proving Ground was closed and most of its collection relocated to Fort Lee in Virginia. Very few of its objects are on display there as of mid 2020. My collection of photographs began to take on some historical significance. By 2011 I had written an historical narrative to accompany my Manhattan Project photographs and began, with the help of editor-turned-agent Robert Morton, to seek a publisher. The narratives explained the nature and significance of the photographed objects, but they also gave me an opportunity to explore what about these subjects had so captivated my imagination.

The central problem in writing the histories was to fish a coherent account out of an ocean of facts and stories told in a mountain of books. The goal was both to give context to the images and to convey something of my motivations. A recurring preoccupation in all three books has been to understand how these strange objects came into being, rather like an archeologist approaching a relic of a past civilization. Who built them? What motivated their construction? Why were such huge sums of money lavished on them? As a scientist, I was also curious about the chain of discoveries and circumstances that made them possible at all. In this volume, I was able to indulge this last interest by charting the course of modernity's rise, starting with the agricultural and scientific revolutions in the seventeenth century. The suddenness with which attitudes toward change and discovery occurred is quite remarkable, and rapid change has characterized the journey ever since.

Of course, the narrative here has to be highly selective in covering such a long period of history. It is certainly not intended to be a comprehensive history of wars in the twentieth century, rather, a history of the evolution of its mighty war machines—and this at a gallop. I am not a professional historian, and, not knowing what might go unchallenged by such, I have endeavored to document sources of the many facts brought out in the narrative. My hope is that the angle-bracketed <> endnotes will not unduly interrupt the flow of the story yet still provide the interested reader with a starting point for more detailed study. I am well aware that the interpretation of historical facts and even the facts themselves are often open to debate. Where I have been sensitized to such controversies, I have tried to present alternative viewpoints.

In studying this long sweep of history, one cannot help but be struck by the extreme spasms of violence and destruction that occurred in the twentieth century. Walking the streets of cities today, it is almost beyond the powers of imagination to visualize hundreds of heavy bombers in the sky above raining down bombs day and night. As I write this, it has been more than seven decades since the last world war, which ended the day after I was born; there have now been some three generations that have not known total war. It is tempting to think that total war is now a relic of the past, that humankind took to heart the lessons of the great wars of the 20th century, that nuclear war is simply irrational and therefore irrelevant, that the new "smart weapon" paradigm has obsoleted mass civilian casualties, that the same exquisite rationality that informs us of our place in the universe will render total war unthinkable. One can hope that such thinking will prevail, but it is worth considering the thoughts of writer and social critic, Paul Fussell, embittered by his experience as an infantry platoon leader in WWII,

> ***This was the fourth time I'd been involved in a night operation that went awry, and most had ended in near disaster. I was learning from these mortal-farcical events about the eternal presence in human affairs of accident and contingency, as well as the fatuity of optimism at any time or place. All planning was not just likely to recoil ironically: it was almost certain to do so. Human beings were clearly not like machines. They were mysterious congeries of twisted will and error, misapprehension and misrepresentation, and the expected could not be expected of them.***<1>

One would have thought that the industrialized terrors of the First World War would have inoculated the major powers against another, yet barely one generation had passed before the Second World War ushered in an even larger toll of death and destruction. It is also well to remember that the tensions and stakes of the Cold War could have escalated into the worst world war yet in the space of only a day or two. We were lucky then, but the winds of "twisted will and error, misapprehension and misrepresentation" still blow.

Martin Miller

ACKNOWLEDGMENTS

None of my photographs would have been possible without the foresight, dedication, and perseverance of museum directors, curators, and benefactors. These artifacts are large and heavy and for the most part must be kept outside where the weather exacts an unending toll in maintenance. Although museums are credited in the captions, we acknowledge them all together here as well. The largest number of photographs were taken at the US Army Ordnance Museum that was at Aberdeen Proving Ground, MD before its multiyear move to Ft. Lee beginning in 2009. Other museums hosting my photoshoots were the Canadian War Museum in Ottawa; Patriot's Point in Mount Pleasant, SC; Hill Aerospace Museum at Hill Air Force Base, UT; the National Museum of the US Air Force in Dayton, OH; Pima Air and Space Museum in Tucson, AZ; the Air Force Armament Museum at Eglin Air Force Base, FL; Battleship Texas State Historic Site in La Porte, TX; Battleship Cove in Falls River, MA; Battleship Memorial Park in Mobile, AL; and the Liberty Ship SS John W. Brown in Baltimore, MD. Although the historical narrative was constructed primarily to support my images accumulated over a dozen years, a last-minute trip to the British Imperial War Museum in London helped provide images that supported the nearly completed manuscript.

I am very grateful to Dr. R. Cargill Hall for his generosity in writing the foreword. Cargill pointed out to me the importance of the proximity fuze development for artillery in WWII, and I included a section on it in subsequent drafts. He also made a number of other constructive suggestions that stimulated me to revise the manuscript for the better.

I would also like to acknowledge and thank the Baltimore County Public Library staff who procured countless books for me through the inter-library loan system in a most timely and efficient manner.

My brother, Grayson, physician and life-long dedicated historian, graciously consented to reviewing the manuscript for any glaring inaccuracies, and for this as well as many enlightening discussions, I am grateful. Brothers being brothers, there were occasions where I declined his advice so, of course, any surviving errors are on me and not him.

Working with such a grim subject would have taken a greater psychological toll if it had not been for my delightful granddaughter, Aryani. She endured greater absences of her grandpa because of books whose nature was quite incomprehensible to her. On each greeting card occasion she playfully inscribed "No More Books" and she even included drawings of an encircled open book with a diagonal line through it to clinch the message. On a serious note, nothing brings home the absurdity of war more poignantly than trying to explain what it is to a six-year-old child.

Finally, there is the irredeemable debt that I owe my dear wife, Gail. My choice of high-resolution photography, first with large view cameras and then by stitched mosaics using digital cameras, required long hours in the field, in the darkroom, and at the computer. Gail accompanied me on all out-of-town trips and thus had to exercise great patience and forbearance while I was actually photographing. While I was in the darkroom or in the lightroom (i.e., at the computer), I was neither doing chores nor providing companionship. Probably the worst for her was my writing of the narratives, during which I was either totally absorbed at my desk or excitedly engaging in a conversation topic long gone stale. Nevertheless, she gamely provided totally invaluable expert editing of each manuscript and endured countless discussions on syntax and word usage. Whatever improvements my writing style has undergone is due to her. Beyond all this, her love and support has given me the serenity of mind and emotion to undertake what has been a rewarding but arduous experience. As a chemist and physician, she will understand what I mean when I say that she is the first term on the right hand side of all my governing equations.

CHAPTER 1. THE 20th CENTURY CALAMITY

On the eve of the twentieth century, H.G. Wells had imagined a "War of the Worlds"—a Martian invasion that devastated the earth. In the hundred years that followed, men proved that it was quite possible to wreak comparable havoc without the need for alien intervention.

—Niall Ferguson, *The War of the World*<1>

The twentieth century was unique in many ways. It saw the full flowering of the incipient and simmering age of rationality that stretched back to the Renaissance. The widespread harnessing of electricity to illuminate the nighttime world; the perfection of mass production bringing the fruits of technology to everyone; the development and mastery of flight connecting even the most remote parts of the world; the invention of the telephone, radio, and television connecting people and events down the street or across the globe nearly instantaneously; the amazing advances in medicine reducing suffering and extending life; the digital-electronic revolution that has left no aspect of our lives untouched—all these things and more characterize the singular nature of the century just passed. Yet amid this cornucopia of progress, there was a darker kind of flowering, one in which all the same, hard-won scientific, technological, and industrial prowess was lavished on the tools of war.

This other flower's fruit proved bitter in the extreme. In the twentieth century some 110,000,000 human beings lost their lives to governmentally sanctioned violence. Of these, over half were killed during the Second World War, unquestionably the historical zenith of mechanized slaughter. In fact, according to historian Max Boot, more people were killed in WWII than in all previous wars combined.<2> As precious as each of these victims was to someone, their exact number will never be known—lost in the desperate struggles of the survivors and with the destruction in a thrice of records that were carefully kept for generations. Shockingly, the uncertainties in these numbers are not small but run in the millions. Attempts to quantify the toll with a precision greater than about 20 percent, noble and conscientious as they may have been, have proved to be a fool's errand, especially though not exclusively, in the undeveloped world.<3> Imprecision is particularly attached to the counting of civilian deaths.

There is a trend beneath the surface of these numbers that is as appalling as the figures themselves. As the ghastly mid-century zenith was approached, the number of civilian casualties relative to military casualties mushroomed. In the First World War there were around 13 million military deaths with about as many civilian deaths.<4> In the Second World War there were some 20 million military deaths and 46 million civilian deaths.<5> In this narrative we shall not dwell on the necrology of the wars except to point out some of the enormous difficulties in counting methodologies.

Wars are not tidy affairs that lend themselves to unambiguous statistics. Counting war deaths among soldiers and among civilians are each fraught with their own separate sets of problems. Military personnel may be killed outright in battle. They may be wounded and later die of their wounds. They may succumb to illness as a result of contagion, malnutrition, or exposure. They may be listed as missing in action and actually be dead. They may be captured and die in captivity or killed while trying to escape. But in professional armies, at least, soldiers at risk are generally accounted for at some level with records usually kept far from the battlefield. From a statistician's viewpoint, at least the total number of personnel at risk is generally known even if the number killed can be ambiguous. To give a sense of the uncertainties involved, Brzezinski cites military deaths in WWII as 19 million, but Urlanis puts the figure at 22 million.<6> While this disagreement represents only a fifteen percent difference, the uncertainty of 3,000,000 lives is unsettling. And this is the more straightforward of the two cases; counting civilian deaths is much more complicated and subject to many more uncertainties.

Civilians may be killed incidentally during direct clashes between opposing forces. They may be killed by aerial bombing or naval bombardment. They may die as a result of privations caused by naval blockade (British blockade of Germany in WWI). They may be deliberately starved by occupying armies. They may be intentionally killed by invading armies for political or security reasons (Nazi Einsatzgruppen). They may be massacred by their fellow citizens for reasons of ethnic hatred—opportunistically cloaked by the chaos of war (Armenian Massacre). They may be killed by non-state forces engaged in civil war triggered by the wider international war (Russian Civil War). They may be killed by their own government as a part of state terror policies instituted to squelch dissent and consolidate political power (Stalin's Reign of Terror). Civilian population records are usually held locally and, if the local archives are destroyed by bombing or other battle action, the total number at risk as well as the number killed become obscure. Typically, then, civilian casualty counts are widely disparate among sources. For example, Rummel gives 85 million as the total number of civilian dead in WWII whereas Urlanis puts the total civilian deaths at 28 million.<7> They might both be right but include/exclude different groups in their methodology. However, it makes any nuanced arguments based on wartime statistics futile. Moreover, these quoted statistics, disappointing as they are, represent one of the most thoroughly studied periods in all of history. In the end, as historian John Keegan has aptly put it,

> ***It is difficult to be categorical about casualty figures; they are, as any military historian knows, a quagmire into which the scholar sinks ever deeper the more effort he makes to wade his way out.***<8>

A PERFECT STORM

What was it about the twentieth century that enabled this unprecedented calamity? Though complex and intertwined chains of events were involved, to be discussed in succeeding chapters, an inauspicious event occurred in the last half of the previous century. The American Civil War in many respects presaged Modernism's great tragedy that played out in the First World War. In the 1860s the marriage of steam and steel was still young and seeking its potential. The Union dominated the 30,000 miles of North American railroad track (more than all other nations of the world combined) and added to this asset every month of the war. At the same time they made a point of destroying any Confederate track they encountered. With its much smaller industrial capacity, the South was hard pressed to replace its vital track.<9> The result was a more mobile, better fed, and better equipped Union fighting force. Telegraph lines were also laid next to new track providing better communications as well. The war-fighting value of these infrastructure innovations in moving and controlling masses of men and materiel was not lost on European observers.

Apart from railroads and the telegraph, the American Civil War also saw important innovations in weapons, if not during the war, then shortly after. In 1866 the US Ordnance Department adopted the six-barrel, hand-cranked Gatling gun—the first practical machine gun. The Gatling gun did see some limited action during the war at the siege of Petersburg, Virginia, only because Union General Benjamin F. Butler purchased 12 of these new guns from his own pocket.<10> In the 1880s the US and all the European powers adopted the Maxim gun, the first fully automated machine gun. This weapon, coupled with the repeating, breach-loading rifles first appearing in the Civil War, made possible the mass casualties to come in WWI. Advances in artillery, development of tanks and airplanes in WWI and WWII, and, of course, the atomic bomb in in the latter war completed the potent arsenal of death for the new century.

In addition to mass-casualty-producing weapons and efficient transportation systems, a necessary ingredient for the unprecedented loss of life was an almost inexhaustible supply of combatants. Ironically, advances in agriculture coupled with improved public health especially in the last half of the nineteenth century served up this last critical ingredient. From 1800 to 1914 the British population quadrupled, and that of Europe as a whole including Russia more than doubled.<11> Moreover, after being roughly constant since the sixteenth century, life expectancy rose sharply from 36 to about 47 years from 1870 to 1913 in Europe.<12> Apart from contributing to a longer lifespan, advances in medicine also insured that more men were healthy enough to actually fight. Until the twentieth century more deaths of British soldiers were caused by illness than by combat.<13> Mass conscription is often associated with Napoleon Bonaparte in the early years of the 19th century, but he was only able to muster 7 per cent of his population compared with Britain's 13 percent and Germany's and France's 20 per cent in WWI.<14> With larger, healthier populations, universal conscription, newly available methods of mass producing war materiel, and megalomaniacal national leaders, there gathered a perfect storm of factors favorable to the mass slaughter of human beings in the wars of the twentieth century.

THE WAR MACHINES

And then there are the engines of death themselves—the war machines, which are the primary subject of this book. They are the Janus figures of the twentieth-century story, showing two faces to the world. On the one hand, they are marvels of engineering design and manufacturing skill. They are also products of prodigious organizational efforts, especially true in wartime when national production boards must prioritize and ration critical materials among many competing military and civilian demands. Production engineers essentially must redesign prototypes to meet mass production requirements. Factory workers and supervisors must be retrained to accomplish each task efficiently and safely. Materials must be procured in the face of shortages, stored, and fed to the production line at just the right time to make mass production work at all. The factories' issue must be tested and transported to railheads and ports, and shipped overseas if necessary. The surprise is not that shortages occurred in the field, but that the whole complicated clockwork worked at all. These were triumphs of the human mind.

On the other hand, these war machines had a savage function to perform, with grim consequences. In the unforgiving conditions of battle, nuances of their design could well make the difference between prevailing or not, between living or dying. Their aesthetics of form and elegance of function evaporated as abstractions in the presence of their actual use in combat. What remained was their stark brutality. The contrast between the fragility of the human body and the ferocity and scale of twentieth-century war machines is glaring indeed. But, regardless of their size or thickness of their armor plate, the element of human courage remained vital to their success.

The principal story told here is how they came to exist at all. These imposing machines were, perhaps, the inevitable products of agricultural and industrial revolution, competition among industrialized nations in the face of limited natural resources, and the lust for power. No doubt the extent of their deadly harvest was aggravated by a lagging understanding of the full potential of modern industrialized warfare. In succeeding chapters we will trace the developments that transformed feudal communities into modern industrialized states, the emergence of huge private armament companies to fuel the tragedy, and the mechanisms that governments evolved in mobilizing these resources to fight world wars. We will show how these factors combined to create the most awesome war machines the world had ever seen or experienced, enabling the bloodiest century in the history of humankind. Our focus in this book is on conventional weapons; the reader is encouraged to see my previous books on the Manhattan Project to build the atomic bomb, ***The Neutron's Long Shadow***, and on the development and deployment of nuclear weapons during the Cold War, ***Weapons of Mass Destruction***. In the latter book I argued that the apocalyptic nuclear confrontation that occurred in the Cold War could only be understood in light of the existential experiences that its key leaders had endured in the Second World War. In this book I maintain that the intensity and deadliness of wars of the twentieth century can only be understood in light of the war-making patterns that were set in the First World War. Accordingly, the present narrative places its heaviest emphasis on the developing paradigm of total war that unfolded in WWI.

CHAPTER 2. THE LONG PRELUDE

> ***Never before have the conditions of life changed so swiftly and enormously as they have changed for mankind in the last fifty years. ... Quite a small number of people, heedless of the ultimate consequence of what they did, one man here and a group there, have made discoveries and produced and adopted inventions that have changed all the conditions of social life.***
>
> —H.G. Wells, *What are we to do with Our Lives?* [1931]<1>

AGRICULTURAL AND SCIENTIFIC REVOLUTIONS

Medieval Europe was a feudalistic society in which roles were rigidly perpetuated. Land was wealth, and the nobility owned the land. Vassals owed allegiance to their lord, agreed to fight for him and manage his lands in return for the right to occupy and manage a portion of those lands. Peasants worked the land, and a middle class emerged, mostly from the peasantry, of merchants and traders using money instead of barter as a medium of exchange. Success motivated them to increase their wealth, now denominated in money, as well as their social status, and they became the drivers of a new economic, social, and intellectual order. Witnessing the changes that they had wrought encouraged them to challenge old assumptions and modes of thought. New inventions arose, most significantly the movable-type printing press of Gutenberg in the middle of the 15th century. By the early 16th century there were printing houses all over Europe spreading ideas and ending the Church's monopoly of education.<2> But the world was still hostage to uncertain food production.

We have become accustomed to abundance in food and goods, but it was not always so. Food production was the central preoccupation of life before the seventeenth century. Some eighty percent of the people were engaged in agriculture, and periodic famines robbed all but the rich of any real security.<3> In one documented case in the early 17th century, a merchant-class family spent eighty percent of their income on food.<4> Not until agriculture could be made more efficient and productive could there be excess workers for factories or excess money to buy goods. This impasse was first broken by the Dutch who learned that they could rotate types of crops from one season to the next instead of leaving one third of their land fallow each year as was the tradition. Alternating crops of grain, turnips, hay, and clover not only restored nitrogen to the soil but produced fodder for livestock whose grazing further enriched the soil with their manure.<5> Protestants in the Low Countries fleeing religious persecution brought the new practices to Britain, and they were adopted by British farmers. Cycles of innovation followed. By 1800 only about a third of adult labor was engaged in farm work.<6> In another hundred years the figure would drop below ten percent.<7>

Innovation was not limited to agriculture. There was also a change in the way of thinking about the physical world. In 1543 the Pole, Nikolaus Copernicus, published his work, *Dē revolutionibus orbium coelestium* (*On the Revolutions of the Celestial Spheres*), overturning the Ptolemaic world view, which had the earth at its center.<8> Copernicus based his new model of the world, with the sun at its center, on careful astronomical observations that conflicted with the older model, which had held sway for fourteen hundred years. The scientific method, that has proved so valuable ever after, had just taken its first steps into the realm of human affairs. Furthering the dissemination of the new theory (and its methodology) was Gutenberg's recently developed printing press.

William Gilbert, an English physician/scientist, followed in 1600 with his book of seminal research on electricity and magnetism, *De Magnete, Magneticisque Corporibus, et de Magno Magnete Tellure* (*On the Magnet and Magnetic Bodies, and on That Great Magnet the Earth*), in which he revealed his insight that the whole earth acts as a magnet.<9> Close on the heels of Gilbert came Austrian astronomer/mathematician Johannes Kepler, who knew of Gilbert and even drew an analogy between gravity and the earth's magnetism developed in *De Magnete*.<10> The remarkable mathematical formulations of Kepler showed that the planets moved in elliptical and not circular paths. His 1609 *Astronomia Nova* analyzed the precise astronomical observations of the Dane, Tycho Brahe, in Prague to reach his conclusions, a process akin to the modern scientific method. He even benefited from correspondence with the brilliant Italian, Galileo Galilei, who had just invented a novel optical telescope. Kepler's mathematical abstractions, however, were not as threatening as Galileo's new vision enhancer, which was not universally embraced.

> ***Some of Galileo's academic colleagues from the University of Padua refused to look through the telescope. ... The telescope might indeed reveal things which the eye did not see. But it revealed them, said the critics, by the agency of the devil: it was a form of conjuring and therefore at bottom an illusion.***<11>

Kepler subsequently published a paper analyzing Galileo's telescope and proposed an improved design of his own. The double synergisms of scientific cooperation and competition were being unleashed.

Amazingly, in 1572 Tyco Brahe noticed a new star in the night sky where none had been before.<12> This keen observation was the death knell for the then accepted philosophical proposition that the heavens were unchanging. Importantly, it also encouraged the growing suspicion that other received wisdoms might, in fact, be wrong and provided fresh impetus to invest authority in actual observations in preference to theoretical musings. We now know that Brahe had observed a supernova, an old star that explodes in a brilliant display as its nuclear fuel is exhausted. The remnants of this explosion can still be "seen" by the Chandra X-ray Observatory, which orbits the earth.

In a 1662 publication English chemist/physicist Robert Boyle laid some of the groundwork for doing work with gases by experimentally establishing the reciprocal relationship between pressure and volume of a gas at constant temperature (Boyle's Law). His success owed in part to his use of the precepts of the scientific method as articulated in 1620 by his countryman, Francis Bacon, in *Novum Organum Scientiarum* (*New Instrument of Science*). Bacon is known as the father of empiricism, a philosophical forerunner of the scientific method in which evidence of the senses is given primacy over theoretical speculations no matter how persuasive the latter may appear. Fundamental changes were afoot in how human beings viewed the authority of tradition. As polymath writer Peter Watson describes the shift: "In addition to changing ***what*** we think about, science has changed ***how*** we think."<13>

The master practitioner of this new approach to understanding the world, of course, was Isaac Newton, who secured his eminence in mathematical physics by inventing differential calculus and classical gravitational theory, among a host of other accomplishments. Through his propounding of the laws of motion, Newton was able to derive Kepler's planetary orbits from his gravitational theory, dispelling any lingering doubts about the heliocentric nature of the solar system. Newton's work inspired generations of scientists that the natural world could be rationalized and made the object of quantitative prediction. His *Principia*, published in 1687, is a work of genius by the standards of any age. The theoretical fundamentals were being readied for major advances in technology.

The success of this new scientific approach owed as much to an understanding of and resignation about what questions could not be answered as with the questions that could be answered. Early on, all academic fields of knowledge were closely associated with philosophy whose goal was the unveiling of the meaning of reality. Newton's discovery that gravity could be explained with mathematical precision by assuming that bodies attract each other instantaneously at great distances with no apparent material connection begged the question of the agency of this "force." Newton was himself aware of this "shortcoming" of his theory; to a friend he had written:

> ***That gravity should be innate, inherent, and essential to matter, so that one body can act on another at a distance, through a vacuum, without the mediation of anything else, by and through which their action and force may be conveyed from one to another, is to me so great an absurdity that I believe no man who has in philosophical matters a competent faculty of thinking, can ever fall into it.*** <14>

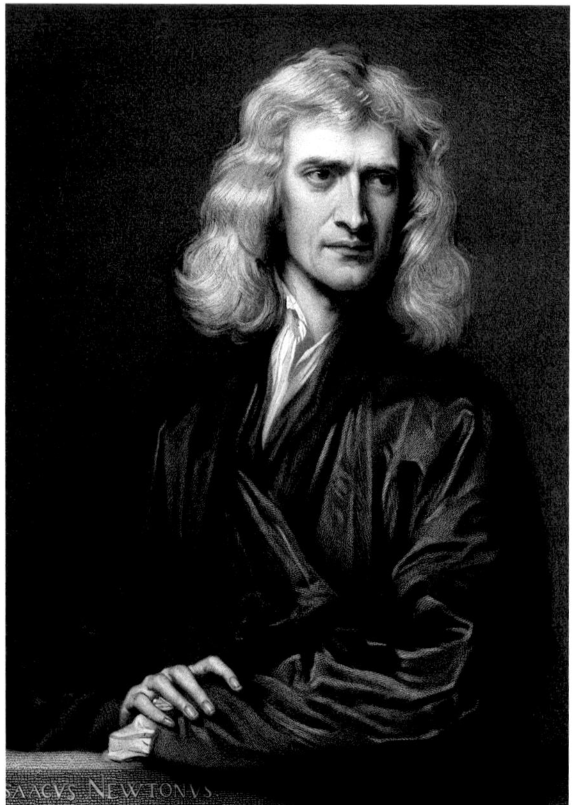

Isaac Newton (1643–1727)
Courtesy Wellcome Library, London

This willing and necessary suspension of the need to understand the ultimate nature of reality would come to characterize science done in the modern era. Though the ideals of empiricism sought to base belief in sensory perception alone, the human senses were altogether inadequate to the task, practically from the start. Hence the need for scientific instruments. Neither would the capacity of human understanding, apart from abstract mathematical representation, be equal to the strangeness of quantum mechanics, quantum electrodynamics, dark matter, and dark energy in the twentieth century and beyond. Humankind is still struggling with the limits of its intellect, remarkable though it is.

After Newton's great triumph in physics, it was chemistry's turn. In a 1756 publication British physician/chemist Joseph Black determined that air was, in fact, a mixture of gases.<15> Since classical Greece, air had been considered one of the four, indivisible "elements" along with water, earth, and fire. Black also discovered the existence of "latent heat" in the conversion of ice to water, i.e., an amount of heat equal to the latent heat (of fusion) must be supplied to the ice before the temperature of the ice/water mixture will begin to rise. He also found an analogous latent heat (of vaporization) associated with the change of phase from liquid water to vapor. This fact proved important to the development of steam engines, a lesson not lost on Black's instrument maker and friend, James Watt.<16> Some thirty years later British chemist Henry Cavendish showed that water was not a classical "element" either, but a compound made from two parts hydrogen to one part oxygen.<17> By 1789 French chemist Antoine-Laurent Lavoisier had published the first table of modern chemical elements, and established the concept of conservation of mass in chemical reactions, thus establishing the basis for the chemical industry to follow.<18>

The scientific revolution was thus launched and, through the gradual refinement of its methodology, steadily improving communications, and growing community of practitioners, it would gather strength and momentum until, in the 20th century, governments would come to rely on it for their wartime survival. The new way of thinking extended to the practical as well as the esoteric. By the middle of the 18th century the industrial revolution had begun its near magical transformation of our world. New inventions in powered machinery, textile manufacture, and iron making led the way.

INDUSTRIAL REVOLUTION

The coming industrial revolution was foreshadowed in 1712 when Thomas Newcomen devised a crude steam engine that was used to pump water from coal mines. A court judged Newcomen's engine to be inessentially different from a patent granted to Thomas Savery in 1698, though it involved a moving piston and Savery's did not. The patent system itself would greatly incentivize individual inventors to create and profit from their inventions while at the same time providing encouragement to others to find small improvements that would warrant a new patent. In this case, anticipating a land rush of patent-control wrangling and political meddling, Savery's patent was extended beyond the normal 14 years by 21 years by a special act of Parliament and at his death in 1715 was acquired by a consortium, *The Proprietors of the Invention for Raising Water by Fire*. The consortium's name reflected the conceptual breadth of Savery's original patent and this sweeping claim may well have stifled progress until its expiration in 1733.

Newcomen's steam-engine pump relied on atmospheric pressure to lift pump rods from a mine and low-pressure steam to reset the system. It was also very inefficient, as a 27-year-old Glasgow University instrument maker named James Watt discovered when he was asked to repair a model of Newcomen's pump.<19> In Newcomen's engine, steam was introduced into a cylinder fitted with a movable piston and pushed the piston to the top of its stroke. Then the cylinder was cooled rapidly, condensing the steam and effectively producing a vacuum. The power stroke was provided by atmospheric pressure, doing work by pushing down on the top of the piston and returning it to the bottom of its stroke. More steam was introduced below the piston to start the next cycle. Watt realized that an enormous amount of heat was wasted in reheating the entire cylinder in each cycle, so he attached a second cylinder, called the condenser, which was kept cool by a water bath. When steam had pushed his piston up on the power stroke, a valve connected the working cylinder to the condenser at the end of the stroke created a vacuum below the piston by condensing the steam in the condenser cylinder. The power stroke was now performed by the steam and no longer limited to atmospheric pressure. He also provided a steam jacket for the working cylinder so that heat from the steam would not be lost in reheating the cylinder on every cycle. A later improvement alternately introduced steam on each side of the piston to provide a double power stroke. Further improvements followed in quick succession.

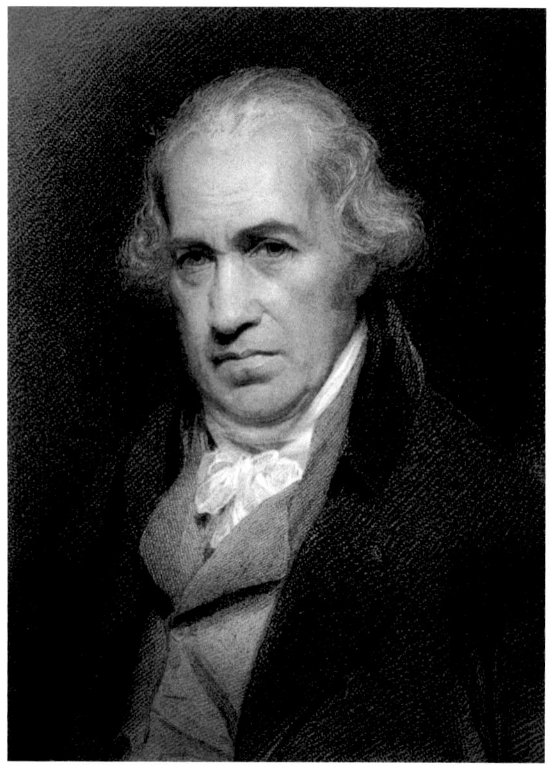

James Watt (1736–1819)
By Permission National Portrait Gallery, London

Creating a commercially viable engine proved challenging for Watt because of the close machining tolerances required of the piston parts, and Watt received financial backing first from Prof. Joseph Black and ultimately from Matthew Boulton with whom he founded the company of Boulton and Watt. Their company prospered, and by 1825 they had produced more than a thousand steam engines.<20> The role of investment proved a pivotal factor in the budding industrial revolution and only increased in significance the more ambitious the projects became. Capitalism, though older than the industrial revolution, certainly played a critical role in its development. This was particularly true for the colossal projects taken on by the iron and steel industry.

Great Britain was not always preeminent in ironmaking. In 1700 it did export iron finished goods but bought much of its pig iron and wrought iron from Russia and Sweden, which together produced half of Europe's iron. Pig iron is the product of smelting iron ore in a blast furnace and collecting the liquid metal. Originally this was done by allowing the liquid iron to spill into depressions in sand with furrows connecting the flow to other depressions. The resulting castings resembled pigs at suckle. At this stage the iron (actually an alloy with about 3.5-4.5 percent carbon) is brittle and of limited use, but its properties are improved by reheating and working by hammer in a forge, a process that drives out impurity inclusions and reduces the carbon content to 0.05-0.25 percent carbon.<21> The material that emerges is wrought iron. By 1700 this was old but skilled-labor-intensive technology, having been imported from France before 1500.<22>

In the early 18th-century English blast furnaces were vertical masonry structures in which intense fires were maintained with iron ore, charcoal, and limestone as the feed materials. Water wheels powered bellows to provide the necessary air "blast." The concept is that combustion of the charcoal creates carbon monoxide which, through a series of reactions, removes oxygen atoms from the iron oxide in the ore. At the same time, reaction of the carbon monoxide with the calcium carbonate (limestone) scavenges sulfur from the iron sulfide in the ore. Liquid iron emerges from all these reactions and trickles down through the stack to be collected.

These same basic processes are still used today.<23> As production increased, the expense of charcoal became an important factor constraining output. Also, because of the dependence on full streams to operate the water-wheel bellows, production was seasonal. By the 1750s Abraham Darby II successfully substituted coke for charcoal in a blast furnace. Coke is produced by cooking coal at high temperatures in the absence of air; this process drives off tar, water, and other volatiles leaving nearly pure carbon. The water wheel limitation was removed by the pioneering work of John Wilkinson in using steam engines to "blow" the blast furnace.<24> In 18th-century Britain coal was becoming plentiful and cheap enough to supply both purposes.<25>

The efficient mining of coal in Britain proved to be the midwife of the industrial revolution. Britain was blessed with a practically unlimited source of the black rock, fueling multiple agents of change. The steam engine transformed the productiveness of the mines. The mines returned the favor by powering the steam engines with their fiery heat and boosting production of iron with coke. Iron and steel became the structural materials of choice for building railroads and countless manufacturing machines and factories, all of which ultimately depended on cheap energy for their success. All in all—a tangle of synergisms that irresistibly propelled us into the modern age at a breathtaking pace. British coal output went from 3 million tons per year in 1700 to 115 million tons by 1871.<26>

Iron production in Britain followed a similar but even more impressive progression, increasing from 48 thousand tons in 1788 to 6,743 thousand tons in 1872, a factor of 140.<27> The production of steel, however, followed a still more abrupt path. Before 1856, which marks the introduction of the Bessemer process, the making of steel was almost a cottage industry. Wrought iron (whose carbon content had literally been hammered down to 0.05-0.25 percent) was stacked with charcoal in alternating layers in a "cementation" furnace and heated for days to increase the carbon content of the alloy to 0.07 -1.3 percent carbon. The resulting product was called "blister" steel.<28> It then was rolled or hammered to produce "shear" steel, suitable for sharp tools and blades. A variation was called "crucible" steel and was produced by heating the blister steel to high temperatures in a graphite crucible and casting into an ingot mold. This steel was more suited to clock springs but also worked for edged tools.<29>

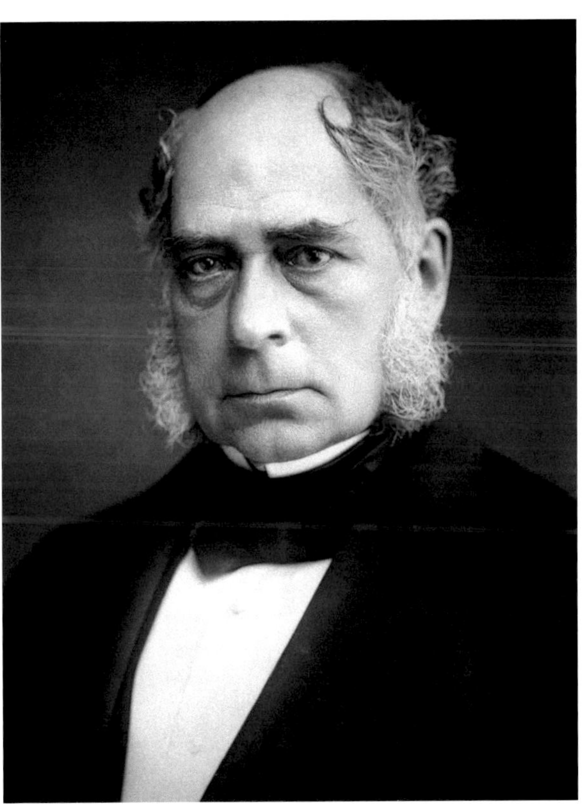

Henry Bessemer (1813–1898)
By Permission National Portrait Gallery London

Henry Bessemer found a way to short-circuit this tedious several-step process of first decreasing carbon content, only then to increase it. By blowing air through molten pig iron, oxygen combined with the carbon and carried it off as carbon monoxide gas, making steel directly. Batches of steel by the ton could be made in this way very quickly (about 15 minutes).<30> This very speed, however, made it difficult to control the carbon content with precision. In 1867 the open-hearth process bearing the names Siemens-Martin was patented. This new, slower process (10 or more hours)<31> employed a large rectangular furnace to increase the surface area across which heated air and gaseous fuel were forced, again with the result of lowering the carbon content of the molten pig iron. Both processes proved useful, depending on the kind of ore available and the desired steel properties, and were often found in the same mill. The ability to produce steel in bulk came just in time to supply and spur the development of revolutionary means of transport: railroads and steamships.

TRANSPORTATION AND COMMUNICATION REVOLUTIONS

Railways predated the steam engine. By the early 18th century in Britain and Germany there were wagonways that used wooden rails with flanges either on the track or on the wheels to keep the wagon on the track. They found use in and around mines to move coal and ore but also in moving loads from mine or factory to nearby rivers and ports for movement by barge or ship. The gauge, or distance between rails, was about five feet determined by trial and error as most suited for loads that could be pulled by horse. At the close of the 18th century iron rails were beginning to replace wood rails, and, as the new century dawned, Britain found itself with a hodgepodge of iron and wood rails plied by flesh-and-blood horses as well as experimental iron horses and even rail cars pulled by cables that were powered by steam engines. The time was right for a powerful personality to fashion order from the chaos.

George Stephenson, now often called the "Father of Railways," was born of poor working parents in Northumberland. Though working at a coal mine twelve hours a day, the determined young man managed to learn to read at eighteen years of age.<32> As a brakesman in charge of the steam-powered winding engine, he would spend his precious Saturday-night free time disassembling and cleaning his engine to better understand its operation.<33> He put this aptitude and knowledge to good use in designing a succession of steam engines beginning in 1814.<34,35> His inventiveness did not stop there. Improvements in rail joints, "steam springs" to distribute the great weight of the locomotive evenly on all of its wheels,<36> and rail-bed grading all drew his attention. Pulling all of his ideas and skills together, he built and operated his first complete railway, the Stockton and Darlington line, in 1825. In doing so Stephenson settled on what would become the "standard" gauge, used ultimately by most railroads in the world, as 4 feet 8 1/2 inches. His engine, Locomotion No. 1, pulled a train of 34 cars with 600 passengers and freight.<37> His next project, the 31-mile Liverpool and Manchester line, opened in 1830 and is considered the first modern railway, with separate tracks in each direction. In a competition against other designers, Stephenson's son, Robert, designed the locomotive Rocket, which was selected for the line.<38>

Though Stephenson's gauge eventually become the world standard, at the time it had competition. As chief engineer, Isambard Kingdom Brunel completed his Great Western Railway from London to Bristol in 1844.<39> It was as noteworthy for the out-sized character of Brunel himself as for its gauge of 7 feet. His railway pioneered the construction of a nearly two-mile long tunnel through Box Hill and a long suspension bridge over the Avon River.<40> It ultimately was extended from Bristol to the western tip of Wales and the southern tip of Cornwall, incorporating numerous branch lines along the way. Not one to rest on his achievements, Brunel proposed extending his railway to America—by means of steamships. The Great Western Steamship Company was formed to realize this ambition. Its first ship, the SS Great Western, was the first steamship to make regular transatlantic voyages beginning in 1838, the same year that the first leg of Brunel's railway opened.<41> At 236 feet long and displacing 2,300 tons, the SS Great Western, constructed mostly of wood and propelled by a side paddle wheel, became the longest passenger ship in the world.<42> It was followed by the SS Great Britain, the first iron hulled steamship to be driven by a screw propeller. It completed its first trip to New York in 1845.<43> Again setting the record for the longest passenger ship in the world, the SS Great Britain stretched for 322 feet and displaced 2,984 tons.<44> It has been restored and is currently a museum ship at Bristol. Brunel's last project, the SS Great Eastern, was more than twice as long at 693 feet, displaced 27,419 tons,<45> and had both a paddle wheel (powered by a 1000 horsepower engine) and a screw propeller (powered by 1600 horsepower engine). He supervised its launch in 1858 but did not live to see its first transatlantic voyage in 1860.<46> Dying of illness at age 53, Isambard Kingdom Brunel was the quintessential 19th-century hard-driving industrialist, and Britain had been blessed with many. But the scientific and technological revolutions were not over. Much of the 19th century was still a world lit by oil lamps and messages sent by horseback. The experiments of Michael Faraday and other notables would change all that.

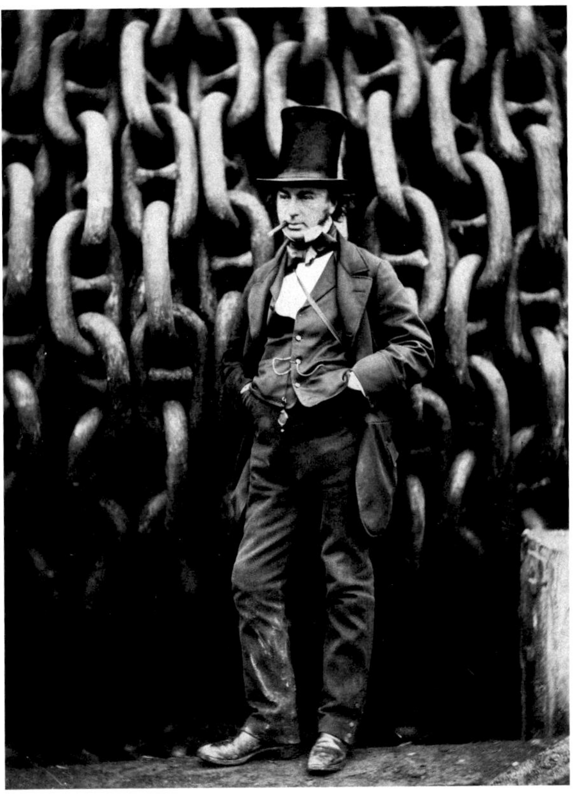

Isambard Kingdom Brunel (1806–1859)
By Permission National Portrait Gallery, London

The 19th century was to witness yet another triumph of the new discipline of science—the discovery and mathematical formalization of the coupled actions of electricity and magnetism. In 1813 the self-educated British researcher Michael Faraday (1791-1867) began his long career investigating electromagnetic, electrochemical, and chemical phenomena. His many contributions included the discovery of electromagnetic induction, which is the basis for electrical generators and motors. He studied electrolysis, which had implications for the corrosion of metallic ship hulls. He learned how to liquefy gases, including chlorine that would later be used as a weapon in WWI. Faraday's fellow Brit, James Clerk Maxwell (1831-1879), was inspired by Faraday's studies and by 1864 had developed an elaborate set of equations rigorously describing electrostatics (the forces generated by fixed electric-charge distributions) and electrodynamics (the generation both of magnetic fields by moving electric charges and of electric fields by changing magnetic fields). Furthermore, Maxwell surmised that light consisted

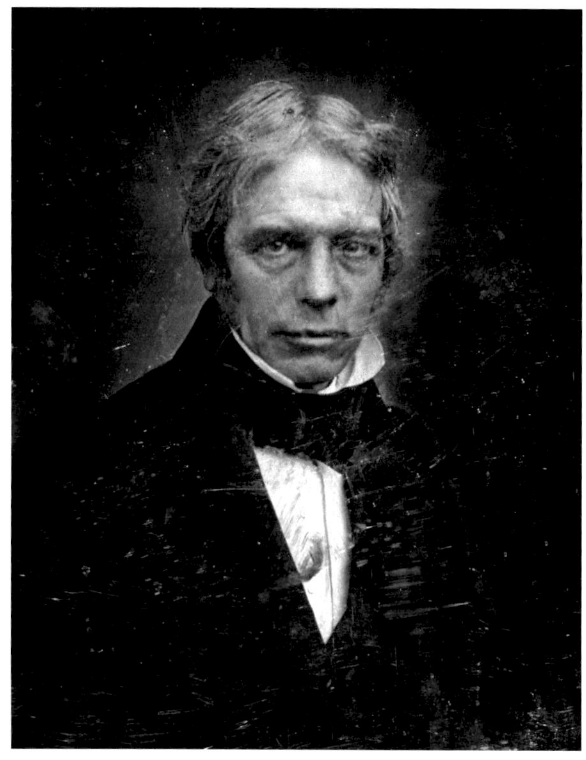

Michael Faraday (1791–1867)
Courtesy Library of Congress. Damaged daguerreotype produced by Mathew Brady's studio and partially restored by the author.

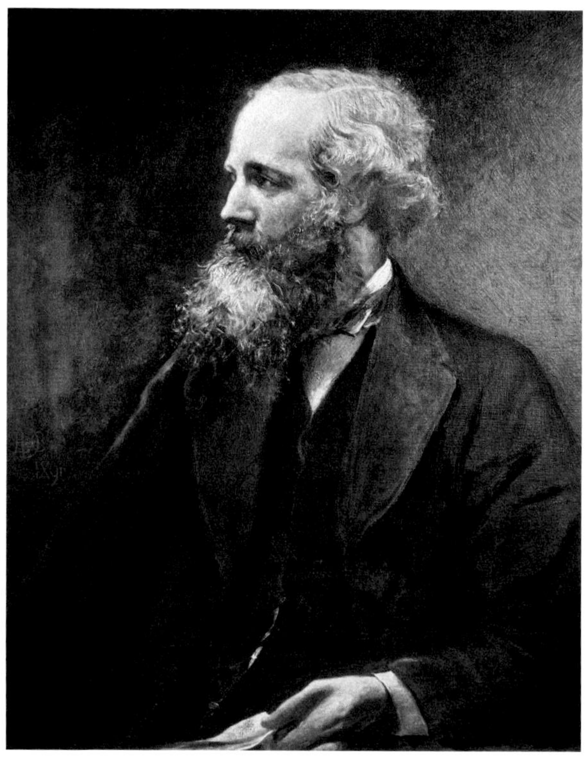

James Clerk Maxwell (1831–1879)
By Permission National Portrait Gallery, London

of mutually inductive electric and magnetic waves—electromagnetic waves—a truly stunning achievement of the human intellect, and one that would have a powerful consequences for the wars of the twentieth century. As Isaac Newton had found with his gravitational action at a distance two hundred years before, Maxwell in his grand theory had arrived at an almost total abstraction, describing nature with mathematical precision but leaving the inadequate human mind with little intuitive grasp of its meaning.<47> As was true for Newton's generation and generations to this day, scientists would have to get used to this unsatisfying reality.

The technological implications of Faraday's researches did not materialize overnight and devices demonstrating his principles remained in the realm of parlor tricks until the middle of the century. The telegraph developed by Charles Wheatstone and William Fothergill Cooke in Britain and Samuel Morse in the United States in the 1840s was immediately recognized as a great boon and soon connected major cities in both countries.<48> Alexander Graham Bell received a patent in 1876 for the electrical transmission of the human voice using Faraday's concepts. But efforts to bring electricity to the masses in the late 19th century created chaos as Thomas Edison's company (later General Electric) promoted a direct-current (DC) power-distribution system while Westinghouse and Nikola Tesla pushed for alternating current (AC). In an epic commercial battle known as the current wars, Edison tried to discredit the safety of AC by using it to electrocute dogs and even a horse. Edison is also thought to have instigated the use of Westinghouse dynamos for the first electric chair used in criminal executions by the State of

New York. But alternating current, making use of Faraday's principles of electromagnetism, prevailed because of the ease of changing voltages using transformers, a direct application of Faraday's work. Edison, however, found solace in the widespread adoption of his electric light bulb patented in 1880.

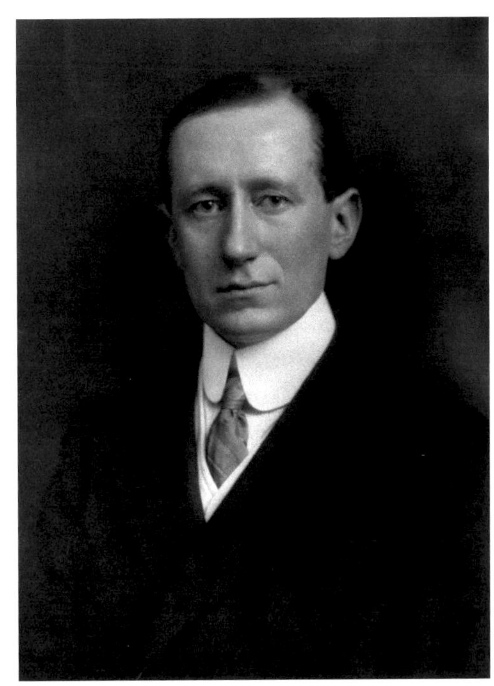

Guglielmo Marconi (1874–1937) c1908
Courtesy Library of Congress, ID 3a40043

Electromagnetic waves as a carrier for wireless telegraphy was pioneered by Guglielmo Marconi in the closing years of the century. In a fitting herald of the new century he managed to make the first transatlantic transmission in 1901. The first wireless broadcast of speech was on Christmas Eve of 1906 by the Canadian inventor, Reginald Fessenden.<49> By this time so many inventors had joined the fray to exploit electromagnetic technology that heated disputes often attended claims of priority. Fortunes were also being made in other technologies.

The last quarter of the 19th century saw major developments in alternatives to the steam engine. In 1876 self-taught German inventor Nikolaus Otto developed an internal combustion engine that achieved considerable commercial success. Otto's engine used illuminating gas (a mixture of coal-generated gases used for street lamps before electric lamps) for fuel and employed the four-stroke cycle still used today. The first modern engine (operating on the Otto cycle) with carburetor and gasoline as fuel was developed by Gottlieb Daimler in 1883. Three years later he adapted it to power a four-wheeled vehicle, and in 1890 the Daimler Motor Company opened for business. BMW was founded by Otto's son, Gustav. Whereas Daimler's engine used a spark plug to ignite the fuel/air mixture, in 1895 Rudolf Diesel patented an engine that achieved ignition simply by compression heating of the fuel-air mixture. Diesel's engine could be run on more crudely refined (and therefore cheaper) fuel. These powerful and relatively lightweight engines enabled automotive and truck transportation ... and later tanks.<50>

Perhaps the most momentous pre-WWI invention of them all was made by American bicycle manufacturers Orville and Wilbur Wright in the early years of the 20th century. The Wright brothers achieved sustained, powered flight in 1903 near Kitty Hawk, N.C. Their patent described a three-axis control system that is still used today. Weight of the aircraft, of course, was crucial and power for their airplane came from a home-built internal combustion engine with a light-weight aluminum block. They tested their ideas about control, wing shape, even propeller shape in a subscale wind tunnel of their own design. Neither had even a high school diploma, yet their meticulous and methodical approach was a model of the modern approach to engineering development. Their contribution obviously changed the world—not only for the better. An airplane that could hardly have been envisioned by the Wrights, yet using their principles, was to deliver an even more unimaginable device a half a century later.

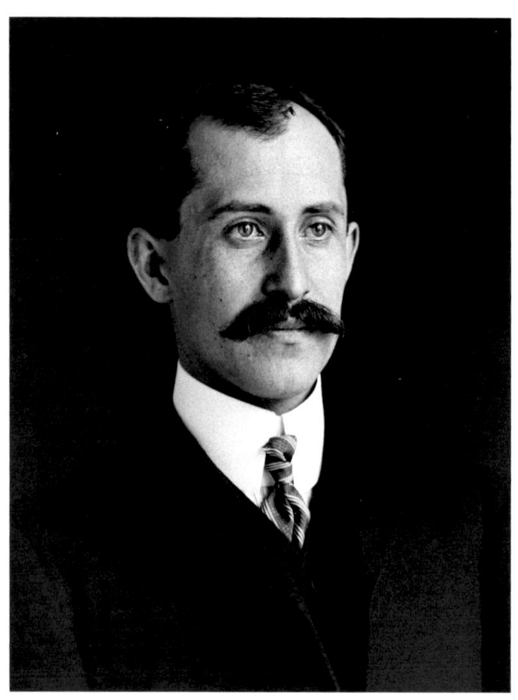

Wilbur Wright (1867–1912) in 1905
Courtesy Library of Congress, ID 0683

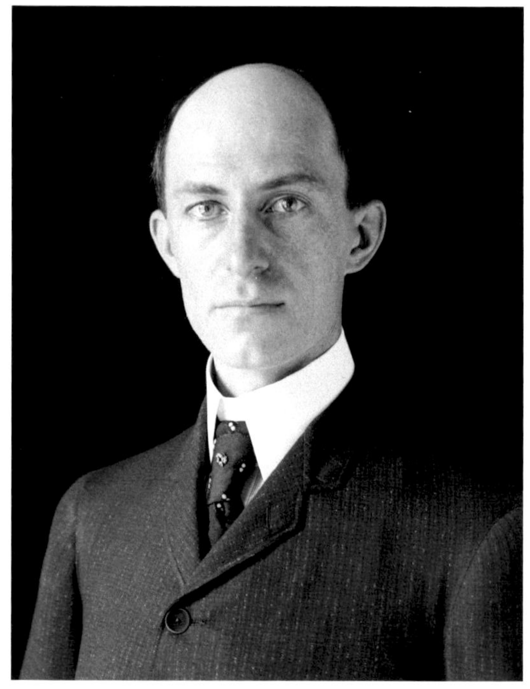

Orville Wright (1871–1948) in 1905
Courtesy Library of Congress, ID 0680

That device, the atomic bomb, was based on the portentous discovery of the Special Theory of Relativity in 1905 by a 26-year-old physicist working in the Swiss patent-office, Albert Einstein.<51> Defying the established methodology of the scientific revolution of giving primacy to experiment, Einstein's thinking was guided by *Gedanken* (thought) experiments and painstakingly working through the possibilities theoretically. Though he conceived his theory without experimental guidance, Einstein ingeniously pointed the way to its verification by the experiments of others. The theory was an extraordinary feat of rigorous imaginings, one consequence of which was the astonishing proposition that mass could be equated with energy and *vice versa*. He even derived a simple relation for computing just how much energy could be released by this conversion in his famous formula, $E=mc^2$, where the proportionality constant between mass m and energy E is the square of the speed of light *c*. The later

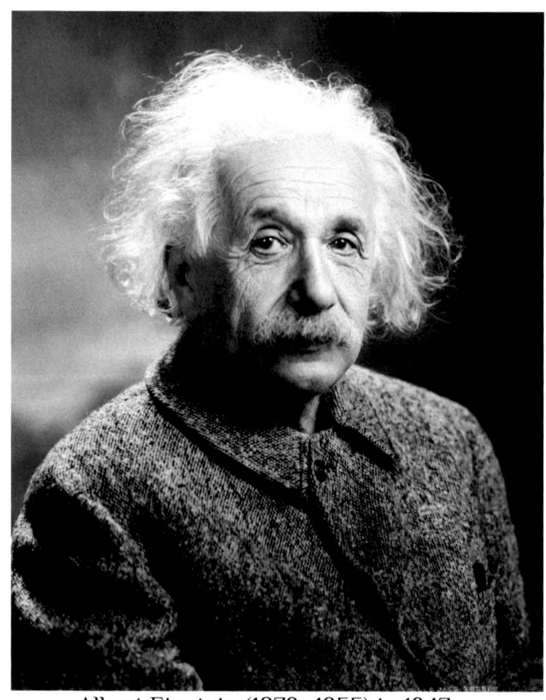

Albert Einstein (1879–1955) in 1947
Courtesy Library of Congress, ID 3b46036

implications of this short formula would spell a fiery end to the Second World War and menace civilization in the second half of the twentieth century and beyond.

WAR MAKING REVOLUTION

This survey of innovations prior to the First World War would be incomplete for our purpose without mention of the revolutions in war making enabled by all the other advances. Some of these were briefly mentioned in the introduction. These 19th-century inventions contributed importantly to the unprecedented level of violence experienced in WWI. A foretaste of this new world was given in the experience of the American Civil War in which 620,000 service members died, more than the combined deaths in all other American wars before or after. Though most deaths were due to disease, the majority of the battle wounds were from small arms fire.<52> It had been known since the early 16th century that inscribing helical grooves on the interior of a barrel, an operation called rifling, imparts a spin to the bullet that dramatically increases the weapon's accuracy and range. Before breech-loading mechanisms were invented, however, ramming the bullet down the muzzle of a rifle was too time consuming for a battlefield use. Consequently, before the Civil War, smooth-bore muzzle-loading muskets were used in battle because of their fast reloading attributes. In 1843 a French army captain, Claude-Etienne Minié, invented an elongated bullet with a hollow base that could be made slightly smaller in diameter than the weapon's bore for easy loading but expand by action of the combustion gases to engage the rifling grooves of the barrel when fired. This simple invention resulted in musket barrels being replaced with rifled barrels, making the so-called rifle musket. The benefits were greater accuracy, higher velocities hence greater range, and easy loading—but also more devastating wounds than those made by round musket bullets. This accounted for the large number of limb amputations that characterized the Civil War casualties. The introduction of breech-loading repeating rifles toward the end of the war raised the soldier's firepower still another notch.

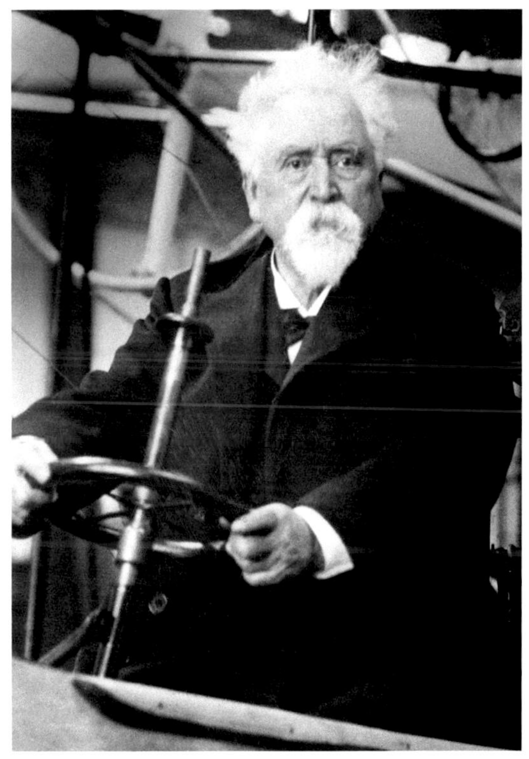

Sir Hiram S. Maxim (1840–1916), prolific inventor about age 75 still inventing within a year of his death. *Courtesy Library of Congress, ID 23307*

Firepower was boosted again by the introduction of the hand-cranked Gatling gun during the Civil War—and yet again with the invention of the Maxim gun in the 1880s. Hiram Maxim succeeded in making a belt-fed, recoil-operated, self-loading, rapid-fire machine gun that, in partnership with the

Swedish company Nordenfelt, he marketed in 1888 to all the European powers.<53>

The flurry of inventions continued unabated with the development by the French chemist Paul Vieille in 1886 of nitrocellulose-based smokeless gun propellants that replaced traditional black powder.<54> The higher energy content of the new powder increased bullet velocities at the same time that stronger nickel-steel alloys were appearing to make the stronger barrels required. This new combination revolutionized naval guns, now rifled, using smokeless powder and explosive projectiles.<55> Smokeless powder burns more slowly than black powder so barrels grew longer to take advantage of the longer-duration push that the smokeless powders could impart to the projectile.

Though the Brits had pioneered the factory system, the Yanks had an important contribution to make to the rapidly evolving concept of mass production—uniformity in manufacture enabling interchangeable parts. In what became known as the American System of manufacture, Springfield Armory and Harpers Ferry Armory introduced the concept of interchangeable parts into small arms manufacture in the first half of the 19th century. At the time Britain was still making its firearms in small workshops with highly skilled gunsmiths. If a part on a musket broke, it would have to be hand-fitted by a gunsmith with a new part. The new American System utilized automated machine tools to make the parts to such a tolerance that any one part could substitute for another with no additional fitting, i.e., they were interchangeable. Pioneering work on interchangeability in the manufacture of muskets at the US armories was taken up by Samuel Colt to fulfill a contract to supply his revolvers to the US Army just prior to the Civil War. This proved a more exacting task than for muskets. Colt hired Elisha Root, a prominent production engineer, to make it happen. The system was based on the extensive use of measuring gauges and fixtures for insuring uniformity of fit between parts. Though Colt never quite achieved complete interchangeability (but claimed that he had!), the path to success in mass producing anything had been established.<56>

So far in our narrative it has been established that improvements in agriculture in Britain meant fewer people were needed to produce food. The rise of capitalism and science together stimulated invention. Inventions were exploited by mines and factories, which in turn provided work for the displaced farm workers. Fortunate as they were to have alternative employment, their living and working conditions worsened as they crowded into squalid cities. For some it was much worse—slavery was an old practice but the industrial revolution in textiles gave it new impetus. The inventions of the spinning jenny in 1764, the water frame for making yarn in 1769, the spinning mule in 1779, and the cotton gin in 1793 mushroomed the demand for cotton in the British textile industry. This in turn boosted the slave trade from Africa to the New World where the cotton was grown.<57> The law of unintended consequences had sneaked into the maelstrom of change, and those individuals swept up into the slave trade would pay the price, all with their freedom and many with their lives. The explosion of knowledge, and more importantly mindset, over the three and a half centuries prior to the twentieth was indeed revolutionary if marked by fits and starts. Its suddenness relative to millennia of comparatively static thought is remarkable, and at the end of the nineteenth century, the Fin de Siècle, the acceleration of change so characteristic of the modern era was having an effect on the psychological stability of society and culture.

SOCIAL AND CULTURAL REVOLUTION

In a macroeconomic sense the Industrial Revolution was a godsend, providing new employment for those losing their old jobs to the Agricultural Revolution. Some, however, began to feel that it was not an even trade. The new factory work was monotonous, dangerous, and frequently denied workers the satisfaction of participating in the whole of a project. In the middle decades of the nineteenth century, German philosopher Karl Marx developed a view of the modern worker as alienated from the product of his labor, from other individuals, and from his very nature.<58> Marx saw workers as being defenseless before the exploitative behavior of industry bosses. Profits, or "surplus value," rightfully were the results of labor's efforts and therefore belonged to the workers; instead they went to the business owners, a process abetted by the laws and customs of the capitalist state. But according to Marx's logic, capitalism bore the seeds of its own destruction. An inevitable decline in the rate of profit would lead to higher competition, which in turn would lead to ever greater monopolization, which ultimately would lead to a collapse of the capitalist state.<59> Later, many would see the fulfillment of Marx's vision in the cataclysm of WWI.

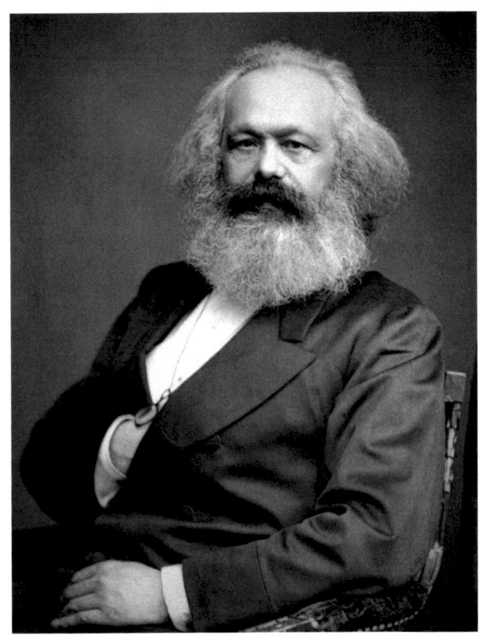

Karl Marx (1818-1883)
By Permission International Institute of Social History (Amsterdam), photographer John Mayall

Though with less direct impact on the changing economic realities, the biological and social sciences were also reexamining old assumptions and trying to make sense of the rapidly changing world. The age of exploration had provided much grist for the mills of biologists puzzling over the enormous diversity of species across the globe. Charles Darwin's 1859 book *On the Origin of Species* shook the foundations of human beings' understanding of their place in the world—with reverberations lasting even to the present. Struggles to understand the implications of Darwin's radical proposal led

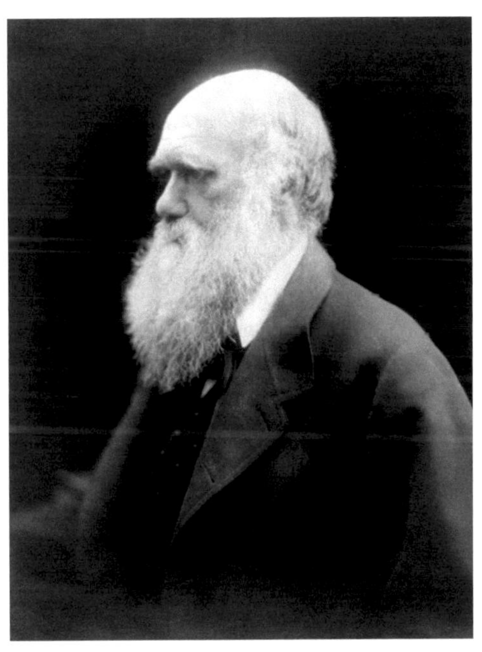

Charles Darwin (1809–1882) c1870, about ten years after his magnum opus, On the Origin of Species. Portrait by Julia Margaret Cameron
Courtesy Library of Congress, ID 3b00370

to a social application of the biological theory, "Social Darwinism," which was to have such catastrophic consequences when taken up as government policy in the next century. Of course, it was many decades before the weight of evidence made Darwin's theory inescapable, but a key idea implied by it was a refutation of the old philosophical notion that change was illusory, that history simply ran in cycles. If the changes over the previous two centuries had left any doubt that this was not true, Darwin's theory attested to the fact that the arrow of time is real and does not double back on itself.

Inevitably perhaps, in the probing spirit of the times, a more aggressive exploration of social Darwinism appeared. German philosopher Friedrich Nietzsche was aware of Darwin's theory but objected to the "struggle for existence" as the driver of change. Seeing no particular scarcity of resources about him in mid nineteenth-century Europe, he doubted that people were motivated by mere survival. Instead, he saw that the "will to power" was the stronger motivating impulse. He decried the shackling of high animal spirits in the most bold and dominating individuals by the timid, bourgeois values of society. For him evolution would lead to the emergence of the Übermensch, or superman, who would achieve his goals by acquiring power with an iron will. By 1889 his own will was extinguished by a descent into madness and he died in 1900. Nietzsche's sister, Elizabeth, began editing, interpreting, and publishing his notebooks and successfully promoted her version of his ideas in Germany. One acolyte was impressed, in fact, believed himself to be the incarnation of the Übermensch—Adolf Hitler. Elizabeth welcomed the Nazi era and even met with Hitler in 1934. That same year the German filmmaker Leni Riefenstahl created her chilling propaganda film, *Triumph of the Will*, elevating Hitler to demigod status. Through Elizabeth's efforts, Nietzsche's ideas permeated German culture of the 1930s.

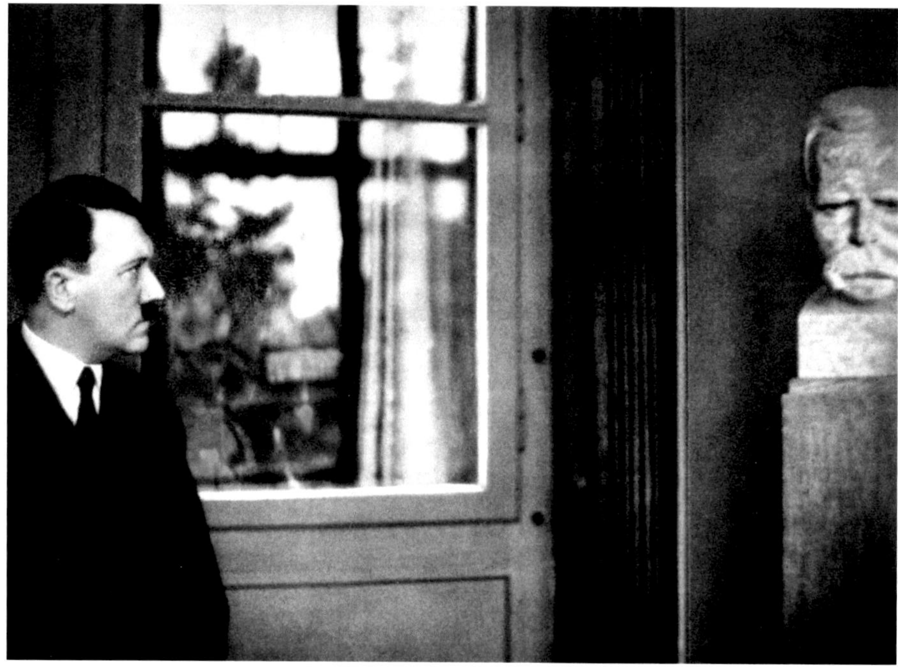

Adolf Hitler captured by his photographer Hans Hoffman pondering a bust of Friedrich Nietzsche.
By Permission bpk Bildagentur / Nietzsche-Archive in Weimar / photographer Hans Hoffman / Art Resource, NY

With his revolutionary and controversial book, *Die Traumdeutung (The Interpretation of Dreams)*, Sigmund Freud sparked a century (and counting) of preoccupation with the self. He suggested that our actions and emotions are governed by competition among three components of our mind: the id, expressing our unconscious primal impulses, the super ego, our moral conscience, and the ego, which mediates between the two. Though Freud's ideas were never actually proved, his systematic approach seemed to fall in line, at least superficially, with the new scientific approach that had proved so successful in the hard sciences. In any case, the age of individuality and psychology was launched, and had considerable influence on the modern mind, particularly in the arts.

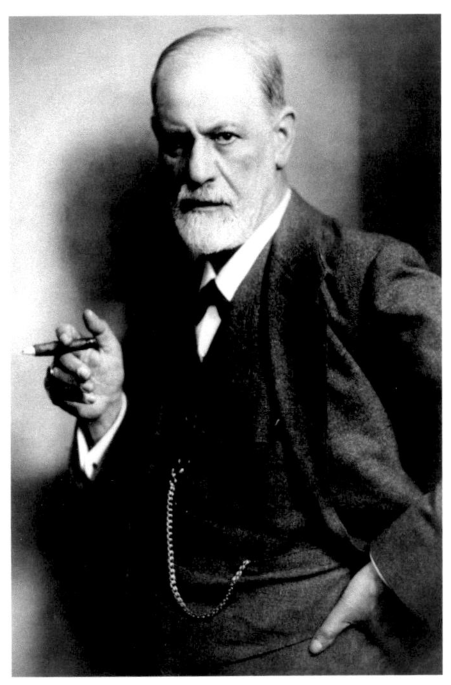

Sigmund Freud (1856–1939)
By Permission Adoc-photos / Art Resource, NY

Not surprisingly, the upheaval in intellectual, economic, and social life brought about by the scientific and industrial revolutions was most poignantly expressed in works of literature, music, and art. Perhaps no literary work better captured the psychological effect of these multiply intertwined revolutions on the human consciousness than that by German-Czech writer Franz Kafka. His 1912 short story Die Verwandlung (The Metamorphosis) begins thus:

> ***One morning, when Gregor Samsa woke from troubled dreams, he found himself transformed in his bed into a horrible vermin. He lay on his armour-like back, and if he lifted his head a little he could see his brown belly, slightly domed and divided by arches into stiff sections. The bedding was hardly able to cover it and seemed ready to slide off any moment. His many legs, pitifully thin compared with the size of the rest of him, waved about helplessly as he looked.***<60>

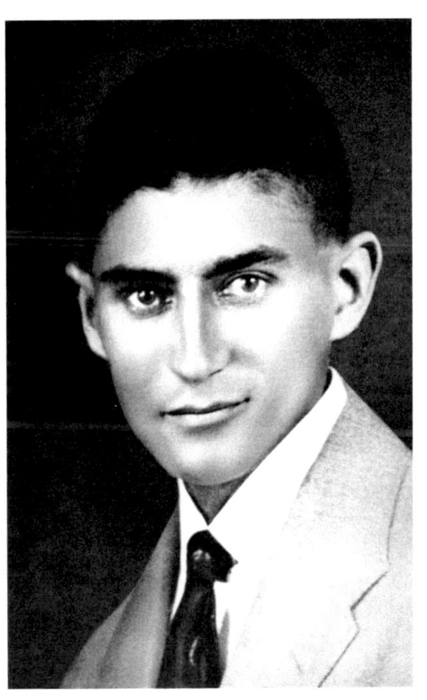

Franz Kafka (1883–1924) in 1914
By Permission Snark / Art Resources, NY

The story is a humorous but sad tale of Gregor and his family coming to grips with his overnight transformation. What is striking is

Gregor's resignation to adjust to his new circumstances. What choice does one have but to adapt, no matter the difficulty? His family, on the other hand, looked at it very differently. The story, of course, lends itself to many interpretations, and it is doubtful that the industrial revolution figured into Kafka's thinking directly. His work is rife with Freudian interpretations, however, and *The Metamorphosis* stands as an apt creative metaphor of the social dislocations of the time.

In music Igor Stravinsky's 1913 ballet, *Le Sacre du Printemps* (*The Rite of Spring*), well expressed the social chaos brought about by the industrial revolution and hauntingly presaged the tragic and pointless debacles that were to commence the very next year. *The Rite of Spring* in music and dance tells the story of a pagan Russian ritual, celebrating the renewal of spring, in which a young maiden is chosen to dance herself to death in a sacrifice sanctified by the elders. Having previously electrified Parisians with his musically and choreographically innovative *Firebird* (1910) and *Petrushka* (1911) ballets, there was high anticipation over the newest work. Advance publicity stoked interest to the point that ticket prices doubled. However, this time the assault against convention was too much for the times; hoots and jeers nearly drowned out the performance, which went on despite the pandemonium.<61> The shock of the new against the sensibilities of the old had aroused anger—and not for the last time.

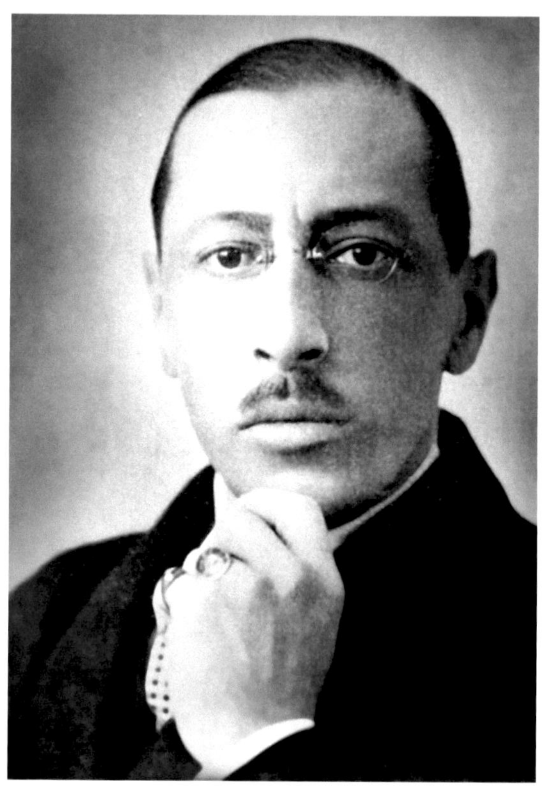

Igor Stravinsky (1882–1971)
Courtesy Library of Congress, ID hec 23698

Pablo Picasso's inventive new painting style that came to be known as Cubism was developed around 1909-1910. In it the comfortable figurative conventions, already under mild attack by Cezanne and others, were now transformed. Subject planes were broken into shards and reassembled on the canvas. The effect, though fascinating, was of a fragmented reality that many found disturbing and that eerily resonated with the new social realities. Taking the trend even further, Wassily Kandinsky abandoned subject matter altogether, making paintings that expressed emotions directly through color and massed forms as early as 1910.<62> The 20th century would see every conceivable canon transgressed in an orgiastic frenzy of creative expression and innovation. In every field of thought, the tempestuous swirl of new ideas, some brilliant some half-baked, engulfed the human mind of the late nineteenth and early twentieth centuries. The disruption of the old economic, social, and cultural orders was now virtually complete. What remained was political convulsion.

CHAPTER 3. THE GATHERING STORM

> ***The war of 1870 was the most splendid moment in Germany's military history, and the subjects of the new Kaiser, not knowing that it was also Germany's last triumphant conflict, were variously awed, reverent, and increasingly haughty.***
> —William Manchester, *The Arms of Krupp*<1>

GERMAN UNIFICATION & FRENCH ANIMOSITY

The seeds of the First World War were sown in a number of prior conflicts. Two, in particular, were instrumental in both strengthening Prussia and leading to her overconfidence. In the 1866 Austro-Prussian War (also known as the Seven Weeks War) Prussia challenged Austrian control of the northern Germanic kingdoms, most of which sided with Austria.<2,3> The Austro-Hungarian Hapsburg Empire boasted some 35 million people compared with Prussia's 19 million and the general opinion in Europe was that Prussia would be overwhelmed if war were to break out.<4> Prussia, however, had been secretly preparing for the fight. Otto von Bismarck, later known as Germany's Iron Chancellor, had accepted the position of Minister President of Prussia in 1862 and proclaimed (with German unification in mind):

> ***The great questions of the Day will not be decided by speeches and the resolutions of majorities—that was the great mistake of 1848 to 1849—but by iron and blood.***<5>

Prussia had then embarked on a program of increased military spending and enlarging their army. Their exceptional Chief of General Staff, Helmuth von Molke, had early on seen the possibilities of the steam railways transporting troops to battle, and they had been used in just this way to quash the liberal uprisings of 1848. Prussia had also been an early adopter of the "needle gun," a breech-loading rifle (with a forerunner of the firing pin) that could be reloaded from any position and tripled the rate of fire of an infantryman armed with a Minié rifle. By 1866 every Prussian soldier carried a needle gun. By comparison the Austrian infantryman was compelled by his government to rely on the Minié rifle and massed bayonet charges.<6> Using superior weapons, railways, telegraph, training, leadership, and strategy, Prussia so whittled the odds back in its favor

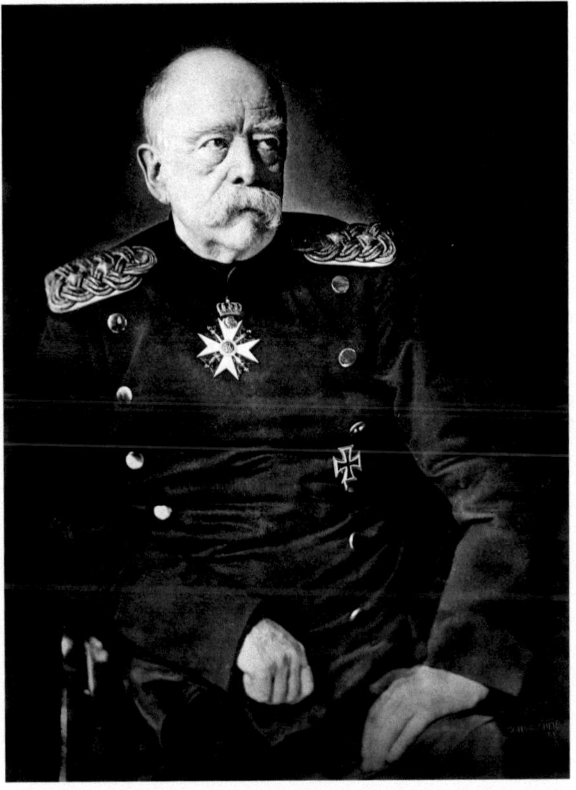

German Chancellor Otto von Bismarck (1815–1898)
By Permission bpk Bildagentur / photographer Karl Hahn / Art Resource, NY

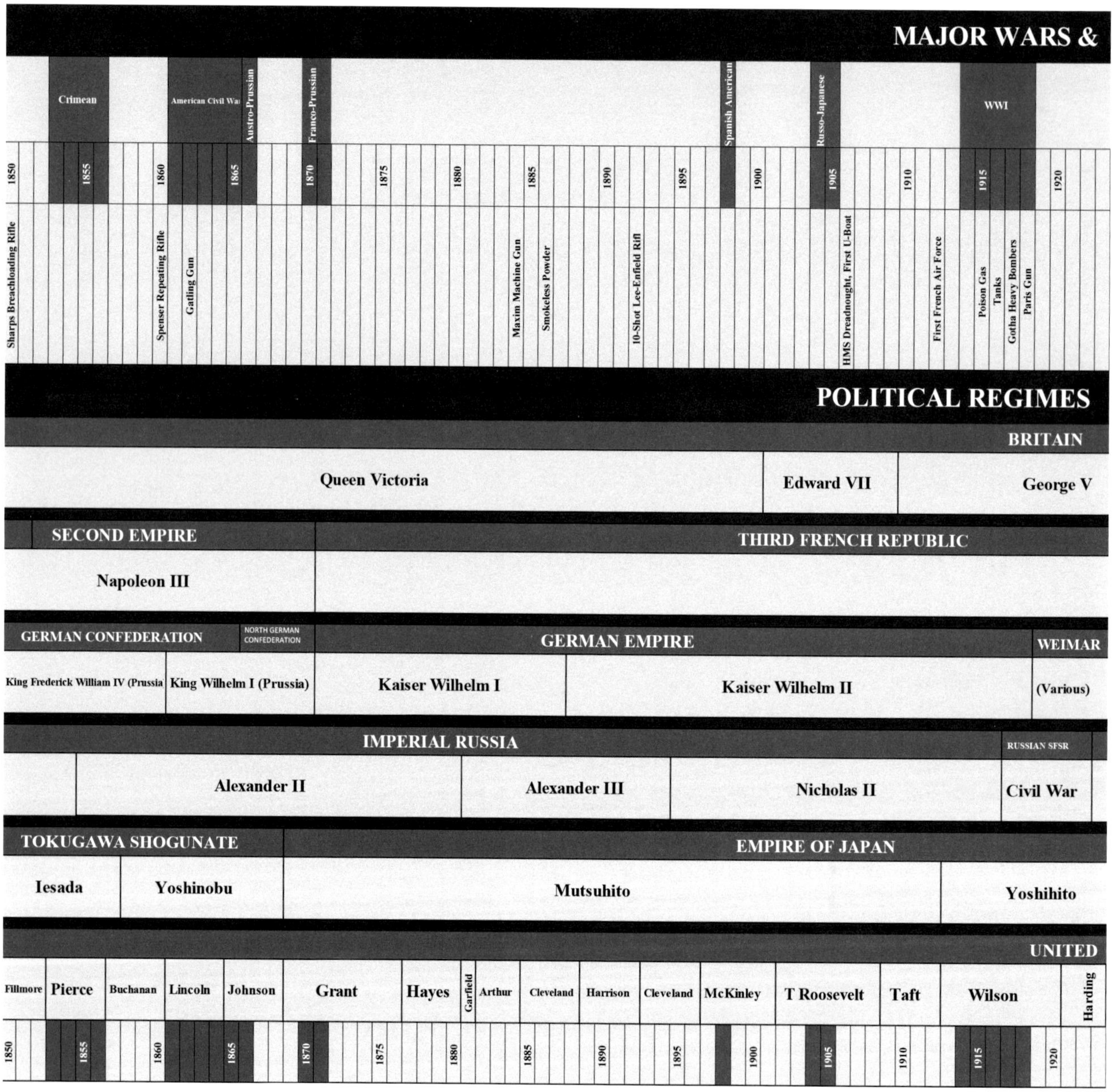

MAJOR WARS &

Crimean

American Civil War

Austro-Prussian

Franco-Prussian

Spanish American

Russo-Japanese

WWI

1850 1855 1860 1865 1870 1875 1880 1885 1890 1895 1900 1905 1910 1915 1920

Sharps Breachloading Rifle

Spenser Repeating Rifle

Gatling Gun

Maxim Machine Gun

Smokeless Powder

10-Shot Lee-Enfield Rifl

HMS Dreadnought, First U-Boat

First French Air Force

Poison Gas

Tanks

Gotha Heavy Bombers

Paris Gun

POLITICAL REGIMES

BRITAIN

Queen Victoria | Edward VII | George V

SECOND EMPIRE | **THIRD FRENCH REPUBLIC**

Napoleon III

GERMAN CONFEDERATION | NORTH GERMAN CONFEDERATION | **GERMAN EMPIRE** | **WEIMAR**

King Frederick William IV (Prussia | King Wilhelm I (Prussia) | Kaiser Wilhelm I | Kaiser Wilhelm II | (Various)

IMPERIAL RUSSIA | **RUSSIAN SFSR**

Alexander II | Alexander III | Nicholas II | Civil War

TOKUGAWA SHOGUNATE | **EMPIRE OF JAPAN**

Iesada | Yoshinobu | Mutsuhito | Yoshihito

UNITED

Fillmore | Pierce | Buchanan | Lincoln | Johnson | Grant | Hayes | Garfield | Arthur | Cleveland | Harrison | Cleveland | McKinley | T Roosevelt | Taft | Wilson | Harding

1850 1855 1860 1865 1870 1875 1880 1885 1890 1895 1900 1905 1910 1915 1920

A timeline of key wars, weapon developments, and political regimes of the major powers. *Chart by the author. Note that dates of weapons introduction are often ambiguous due to a number of factors and, therefore, should be considered approximate. Political regimes are not identified with completeness, only the most relevant to the narrative.*

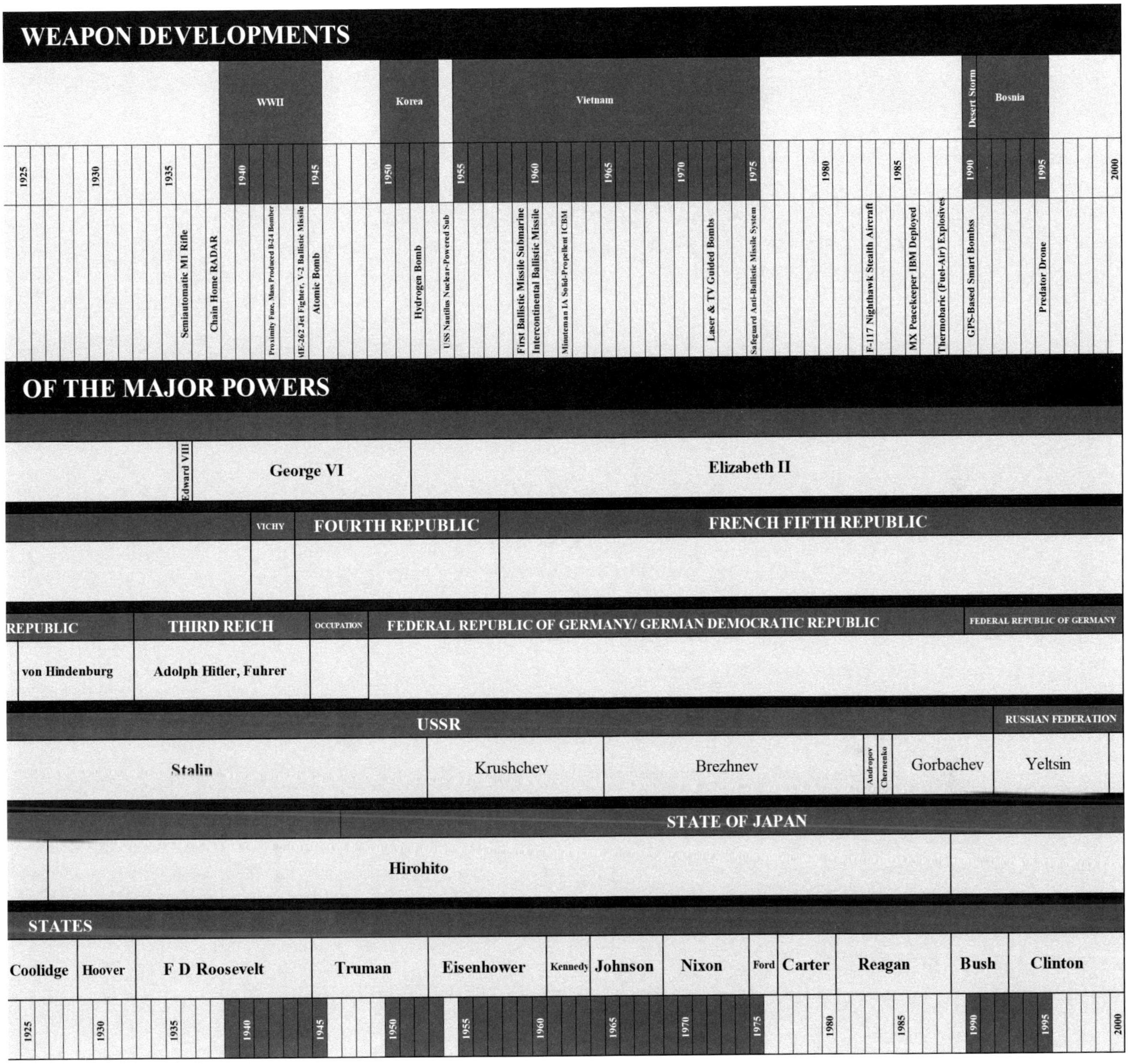

WEAPON DEVELOPMENTS

Wars: WWII | Korea | Vietnam | Desert Storm | Bosnia

Years: 1925 | 1930 | 1935 | 1940 | 1945 | 1950 | 1955 | 1960 | 1965 | 1970 | 1975 | 1980 | 1985 | 1990 | 1995 | 2000

- Semiautomatic M1 Rifle
- Chain Home RADAR
- Proximity Fuze, Mass Produced B-24 Bomber
- ME-262 Jet Fighter, V-2 Ballistic Missile
- Atomic Bomb
- Hydrogen Bomb
- USS Nautilus Nuclear-Powered Sub
- First Ballistic Missile Submarine
- Intercontinental Ballistic Missile
- Minuteman IA Solid-Propellent ICBM
- Laser & TV Guided Bombs
- Safeguard Anti-Ballistic Missile System
- F-117 Nighthawk Stealth Aircraft
- MX Peacekeeper IBM Deployed
- Thermobaric (Fuel-Air) Explosives
- GPS-Based Smart Bombss
- Predator Drone

OF THE MAJOR POWERS

Edward VIII | George VI | Elizabeth II

VICHY | FOURTH REPUBLIC | FRENCH FIFTH REPUBLIC

REPUBLIC | THIRD REICH | OCCUPATION | FEDERAL REPUBLIC OF GERMANY/ GERMAN DEMOCRATIC REPUBLIC | FEDERAL REPUBLIC OF GERMANY

von Hindenburg | Adolph Hitler, Fuhrer

USSR | RUSSIAN FEDERATION

Stalin | Krushchev | Brezhnev | Andropov | Chernenko | Gorbachev | Yeltsin

STATE OF JAPAN

Hirohito

STATES

Coolidge | Hoover | F D Roosevelt | Truman | Eisenhower | Kennedy | Johnson | Nixon | Ford | Carter | Reagan | Bush | Clinton

1925 | 1930 | 1935 | 1940 | 1945 | 1950 | 1955 | 1960 | 1965 | 1970 | 1975 | 1980 | 1985 | 1990 | 1995 | 2000

that it won the war in less than two months. The political result was a unification of the northern kingdoms with Prussia under the banner of the North German Confederation.

Four years later the wily Bismarck maneuvered France's overconfident Napoleon III into declaring war on Prussia on July 19, 1870. The Franco-Prussian War ensued. The French were filled with enthusiasm for the war and visions of national glory, having been told by their minister of war that the French Army was ready "down to the last button on the last gaiter of the last soldier."<7> It was a dramatic but empty boast. Cannon ammunition, clothing, food, and other essentials were in short supply. Louis Napoleon also had counted on help from Austria, but Bismarck had made an alliance with Russia, thereby effectively keeping Austria out of the conflict. France found itself alone. Molke invaded France with an army of 400,000 using plans previously laid. By September 1 Louis Napoleon, himself, had surrendered along with his army of 104,000 at Sedan, 130 miles northeast of Paris. While laying siege to Paris, Molke, Bismarck, and Prussian monarch Wilhelm I took up residence at Versailles. In this conflict the southern German states found common cause with the North Confederation and agreed to band together, forming the German Empire with Wilhelm as *Kaiser*, or Emperor. His coronation in the Versailles Hall of Mirrors might have been thought to be the crowning humiliation of the French nation, but worse was to come. On January 28, 1871 Paris surrendered and, by the peace imposed, France agreed to pay the new German Empire 5 billion marks and cede Alsace and Lorraine, a large, coal-and-iron-rich chunk of eastern France including the cities of Strasbourg and Metz.<8> The resulting injury to French pride engendered bitterness, resentment, and fear in regard to the German nation for many years to come. Compounding their plight was the fact that Germany was now stronger and more confident than ever.

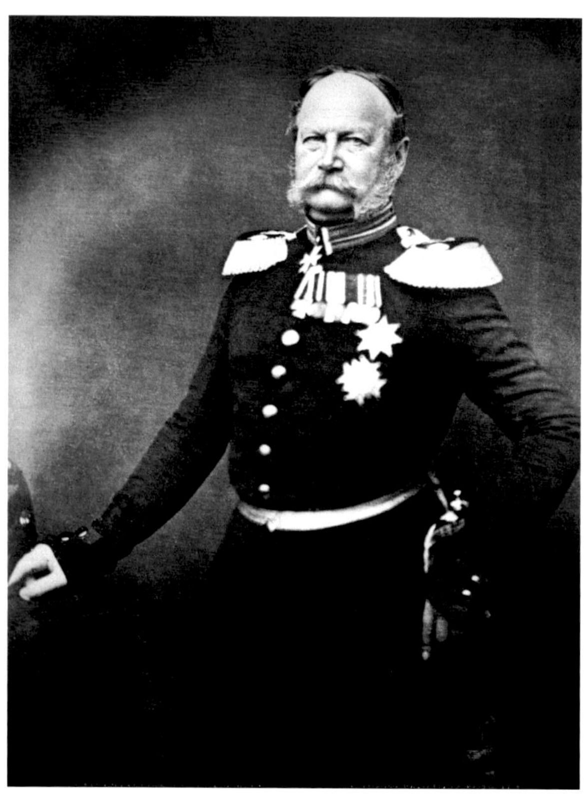

Prussian King Wilhelm I (1797–1888) in 1858, 13 years before being crowned *Kaiser* of the German Empire in the Versailles Palace
By Permission bpk Bildagentur/Art Resource, NY

LATE 19TH CENTURY IMPERIALISM

Unification of the German principalities into the German Empire boosted the German economic growth rate before the First World War to the highest in Europe.<9> Before the end of the century, its population ranked among the top three countries in the world, just behind Russia and that waking leviathan, the United States of America. Britain and France were being left behind though Britain still derived great leverage from its vast colonial empire. Likewise, Germany's Gross Domestic Product (GDP) was surging, by

1900 surpassing Russia and even Britain's by the eve of the Great War. Shortly after 1870 the United States roared past them all, more than twice the size of its nearest competitor by 1913. France was the laggard in this race for wealth, with Russia, under *Tsarist* rule, nearly keeping up with Germany in GDP growth and exceeding them in population growth.

In 1897, heady with its success, Germany adopted a policy called *Weltmacht*, or world power. It thus became the policy of the government to pursue an expansionist colonial agenda designed to assure both raw materials and new markets without which they feared their economy would dwindle.<10> The fact that most of the desirable colonizable parts of the world had already been taken by other countries did not dampen their enthusiasm, but it did mine the future with potential conflict. Its recent rise into the upper ranks of population and industrial output had infused the new German Empire with a sense of national destiny, and even entitlement. Sentiment was growing that, once again, "iron and blood" would be required. Chillingly foreshadowing both the coming war and Hitler's motivating concept of *Lebensraum* (living space) in the subsequent war, the following words appeared in an 1896 letter from Admiral Georg von Mueller, Chief of the Reich's Naval Cabinet, to Kaiser Wilhelm I's brother:

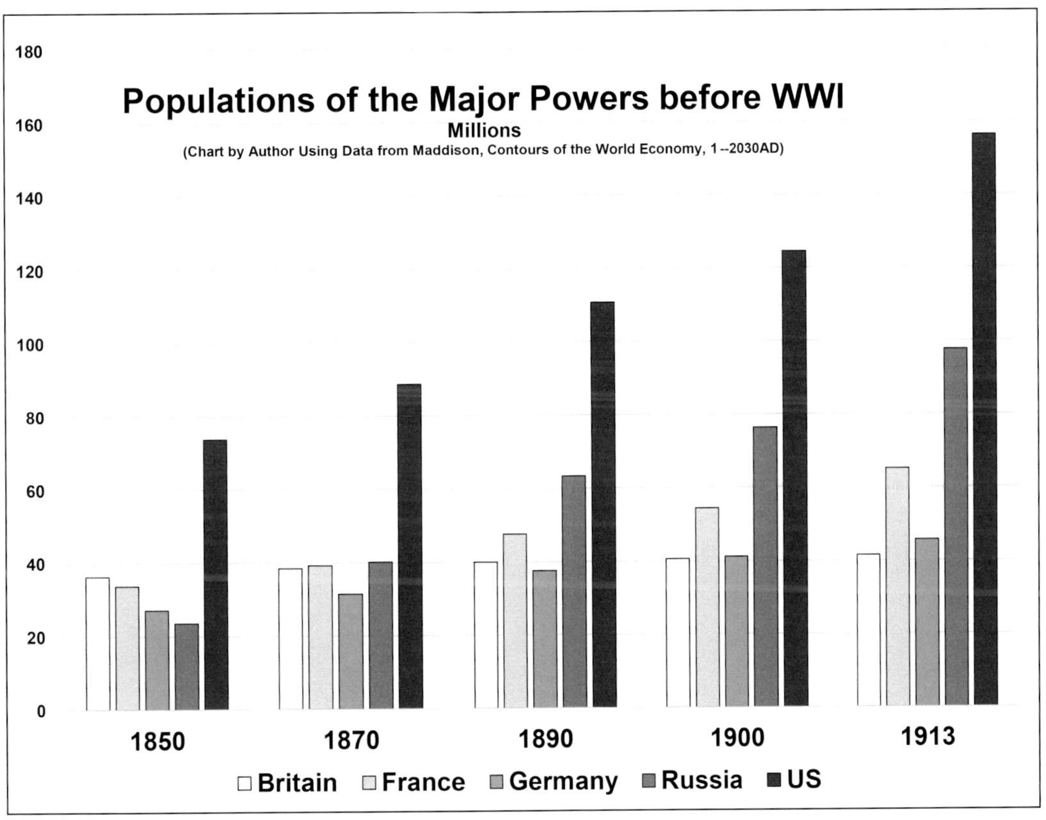

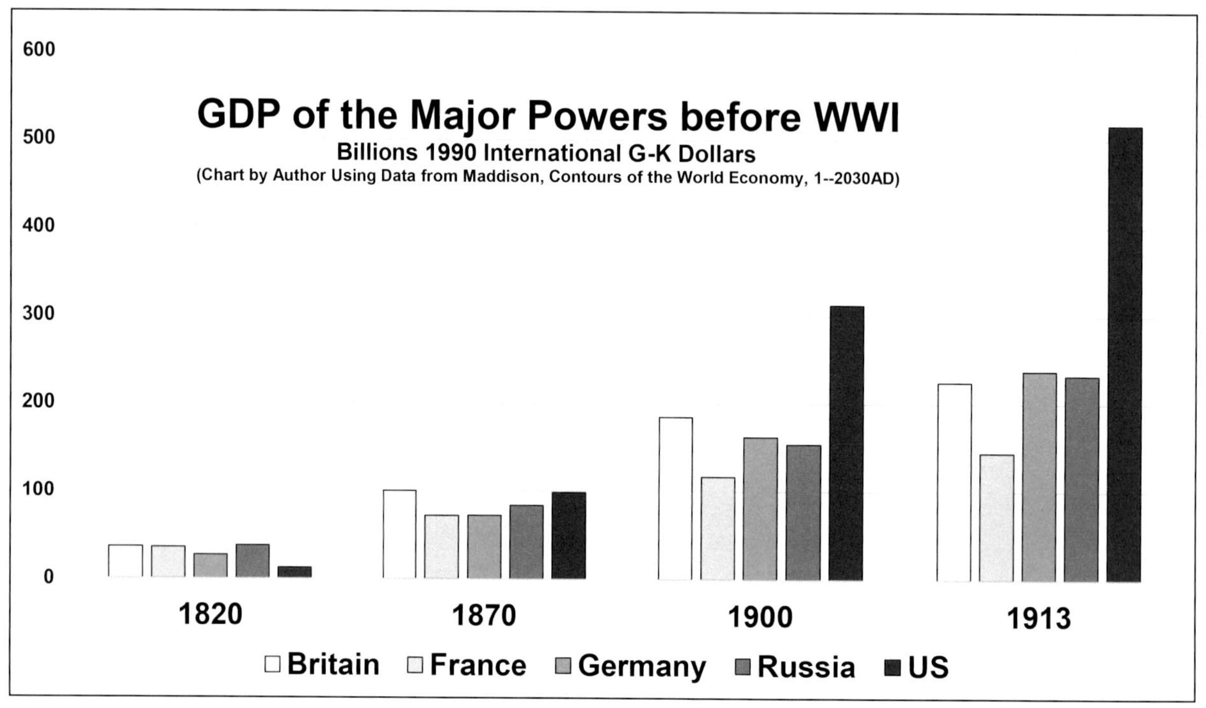

> ***... world history is now dominated by the economic struggle, that Central Europe is getting too small and that the free expansion of the peoples who live here is restricted ... by the world domination of England ... The war which could—and many say, must—result ... would have the aim of breaking England's world domination so as to lay free the necessary colonial possessions for the Central European states who need to expand.*** <11>

If iron was what it took to secure Germany's "place in the sun," then it was well placed to compete. By 1903 German steel production surpassed that of every other European power; by 1913 it had more than twice the steel output of its nearest European competitor, though it failed to catch the American juggernaut. France and Russia lagged significantly behind Britain, Germany, and the United States. The extremely rapid rise of German steel production at the turn of the century may well have stimulated a sense of urgency and even panic that, without increasing access to raw materials and new markets, their country might not continue to prosper.

The fear of losing out on the critical access to natural resources and fresh outlets for their manufactured goods was very real, not only for Germany, and led to widespread colony envy. Each country began looking over its shoulder at potential enemies and potentially threatening alignments of enemies. A frenzy of protective alliances, some public, some secret, ensued in which guarantees were given of military support in the event of attack. The causes of the First World War have since been extensively studied and debated in detail and are many and complex. Not surprisingly, however, these entangling alliances coupled with a concomitant arms race fueled by new wealth, new technology, and fear are important factors given by historians for a world poised for war and awaiting some triggering event. The tension and ambitions were great

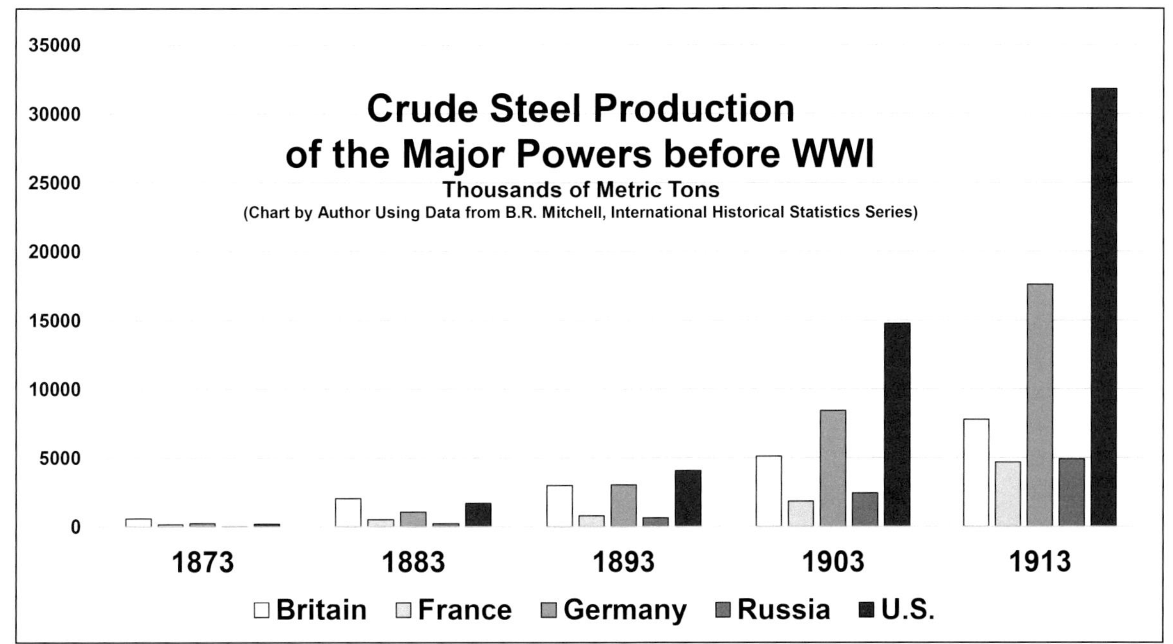

enough for the *Kaiser* (then Wilhelm II, grandson of Wilhelm I) to go to war both against the country of his own grandmother, Queen Victoria of England, a country even then led by his first cousin, George V, and against a country led by his third cousin, *Tsar* Nicholas II of Russia.

With the five major powers growing wealthy from their newborn industrialization, mass production, and colonial exploitation, their populations booming as never before, and their economies at once stimulated and squeezed by increasing global competition, each nation sought to protect its interests by developing portable armed strength. At the end of the 19th century, that meant naval power. Their exploding capacity to make steel would soon be put to use. At the same time, the lessons of Bismarckian diplomacy and the man himself began to fade from influence. In 1890 the 29-year-old Wilhelm II, who had become *Kaiser* in 1888, forced the Iron Chancellor into retirement and then, critically, proceeded to neglect Germany's alliance with Russia that had proved so vital in the Franco-Prussian War. In doing so the insecure but arrogant young emperor had unwittingly doomed his country to a future fight on two fronts when France filled the void by sealing an alliance with Russia in 1894.<12>

WWI NAVAL ARMS RACE

Although not all factions of the German government bought into *Weltmacht*, most notably the aging Bismarck thought it imprudent, a wave of nationalism and business support swept it into being. Its realization, of course, depended on Germany's having a credible naval presence. In 1897 Germany had the smallest navy of the European powers: 12 armored ships over five

HEART OF DARKNESS

Even tiny Belgium got a piece of the imperialist action with unforeseeable consequences for the end of WWII. In the 1840s under the sponsorship of the London Missionary Society, Dr. David Livingstone explored and mapped parts of the African interior including the Zambesi River and Victoria Falls. After years with no word from Livingstone, adventurer/correspondent Henry Morton Stanley was sent by the New York Herald to find him. In 1871 Stanley famously did locate him in a village on Lake Tanganyika. Livingston would stay in Africa until his death less than two years later. Stanley continued Livingstone's explorations particularly in the Congo River basin. He foretold great potential riches from copper, copal (for varnish), and rubber but could not interest the English government in backing development of the area. Belgian King Leopold II, however, took up the cause and personally funded Stanley's efforts to establish the Congo Free State with Leopold as its ruler. By various and devious subterfuges Leopold was able to get an international commission led by Otto von Bismarck to recognize the legitimacy of Leopold's personal colony. This he led with brutal oppression and exploitation that would cost the lives of an estimated ten million natives.<13> In 1908 the Belgium government took over the colony. As fate would have it, most of the uranium used in the atomic bomb delivered to Hiroshima in August 1945 was refined from the uniquely high-grade ore of the Shinkolobwe mine in the Congo. Infrastructure built during the colonial era helped to get it out. The ore had been shipped to New York and was being stored in a Staten-Island warehouse to keep it out of the hands of the Nazis.<14> The Congo Free State experience also served as inspiration for Joseph Conrad's 1899 novel, *Heart of Darkness*. Soon Europe would enter its own heart of darkness.

thousand tons compared with Russia's 18, France's 36, and Britain's 62.<15> The answer was the 1898 First Navy Law, brilliantly coaxed through the rebellious *Reichstag* by Wilhelm's new Navy Minister, Admiral Alfred von Tirpitz. The law provided for the construction of 19 battleships and 8 armored cruisers over the following six years.<16,17> After perceived provocations by British war vessels during the Boer War in 1899, the *Reichstag* passed the Second Navy Law, which was to double the size of the German fleet by 1920.<18> These were followed by Supplementary Navy Laws in 1906, 1908, and 1912, each after some real or perceived incident. The ultimate effect was to raise the target to 41 battleships.

If the intention had been to challenge the British fleet, Admiral Tirpitz had chosen a moving target. In 1889 Britain had adopted the policy of adjusting its naval forces to exceed the combined navies of the next two strongest powers. At the time these were France and Russia. With the commissioning of ships prescribed by the German Second Navy Law, Britain realized that the German naval expansion could only have one plausible target—the British fleet—as the other credible opponents, France and Russia, were only land threats to Germany. Between 1893 and 1901, Britain either commissioned or began construction of 29 new battleships. By 1904 the Brits had committed to 35 new armored cruisers with displacements between 10,000 and 15,000 tons.<19> Moreover, they also began recalling some of their ships from foreign stations. The great naval arms race was under way.<20>

In 1904 Admiral Jacky Fisher became British First Sea Lord, responsible for navigating the Royal Navy through the troubled waters ahead. He wasted no time in overhauling both existing deployments and ship-building strategy. The result of the latter was a new class, indeed a new genre that virtually redefined the battleship ever after. The prototype of the new thinking, the HMS *Dreadnought*, was commissioned in December 1906.<21> It redefined the genre in two important ways. Rather than a variety of large caliber guns (The previous Majestic class and its variants typically carried four 12-inch guns and sometimes as many as ten 9.2-inch guns.), the *Dreadnought* was armed with ten 12-inch guns. This all-big-gun strategy reduced the complexity of managing several different large-caliber shells and associated gun parts. It also helped standardize the gunners training and interchangeability in the event of wounded gun crews and reduced the complexity of multiple fire-control computations and corrections. Furthermore, though the 12-inch gun could be effective at 11 miles,<22> firing at long range had a better chance of success by firing simultaneous salvos and observing the patterns of splash. With different calibers firing at the same time, it was difficult or impossible to determine which caliber shell made which splash. The second innovation characterizing the *Dreadnought* was replacement of the reciprocating steam engine by a steam turbine. Lighter and more efficient, the turbines propelled the dreadnoughts several knots faster than their forerunners and could endure long stretches of running at top speed without breakdowns, unlike their piston counterparts.<23> Astonishingly, by the time

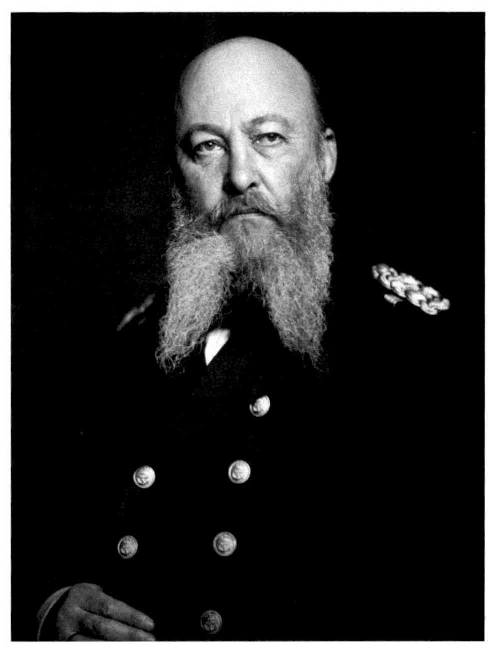

German Admiral Alfred von Tirpitz (1849–1930)
By Permission bpk Bildagentur / photographer Nicola Perscheid / Art Resource, NY

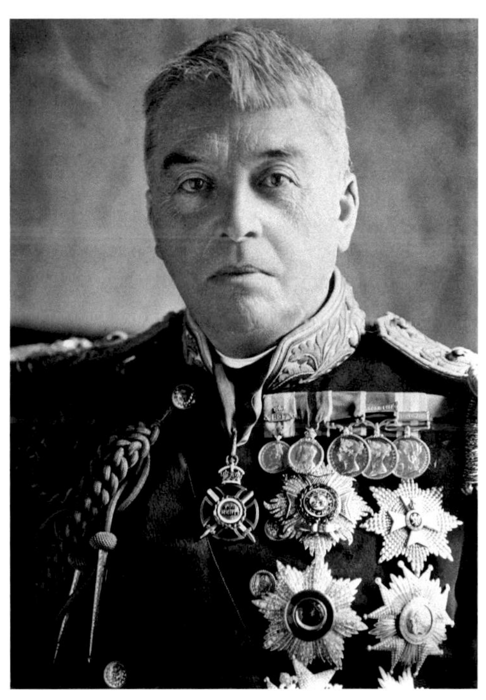

British Admiral John "Jacky" Fisher (1841–1920)
Courtesy Library of Congress, ID ggbain 18059

the *Dreadnought* age arrived, the long chain of British industrial advances had matured to the point that the first of its kind was built and ready for sea trials in exactly one year. The previous record for battleship construction had been 31 months.<24>

Despite Britain's veil of secrecy surrounding the *Dreadnought* project, Germany became aware of the new all-big-gun concept in December of 1904. The Imperial Navy was then faced with a formidable obstacle to building an answer to the HMS *Dreadnought*. The Kiel Canal, connecting the Baltic Sea to the North Sea, was not deep enough for ships the size of the *Dreadnought* to pass. There arose opposition to deepening the canal, but Admiral Tirpitz pushed the alterations through anyway. Work on the canal would not be completed until July 1914.<25> (Admiral Fisher had actually predicted that war would begin with the completion of the canal modifications. He was only a month off.) Though Tirpitz had by July 1906 begun construction of his own all-big-gun battleship, SMS (Seiner Majestät Schiff—His Majesty's Ship) *Nassau*, the sea trials of the HMS *Dreadnought* brought construction to a halt pending analysis of her capabilities compared with the *Nassau*. A year was lost to this task before construction resumed. The 18,900-ton *Nassau's* 12 eleven-inch guns achieved a rough equivalence to the 17,900-ton *Dreadnought's* 10 twelve-inch guns. The Germans reckoned that their eleven-inch guns were the equal of the British twelve-inch guns, but the *Nassau's* top speed of 19 knots fell short of the *Dreadnought's* 21 knots.

In the ensuing naval arms race between December 1906 and October 1917, the British Royal Navy managed to commission 51 capital ships (battleships + battlecruisers) in a succession of 21 classes.<26> The German Imperial Navy commissioned 26 capital ships in 10 successive classes over the same period of time.<27> By the time of the *Nassau's* commissioning in October 1909, the Royal Navy had already commissioned five dreadnoughts in three class modifications. In their final classes of battleships before war's end, the dreadnoughts of both navies bristled with eight fifteen-inch guns. In the secondary category of capital ships, that of battlecruisers—as heavily armed as battleships but faster and more lightly armored, the Royal Navy also had the advantage at ten ships to the *Kaiser's* seven by the end of the war.

The Americans, having had one of the world's most powerful fleets during the Civil War, including the pioneering iron-hulled warships and guns mounted in revolving turrets, were exhausted and demoralized by the war and had allowed their capabilities to wither. By 1881, however, it was widely recognized that the fleet needed modernizing. Importantly, it was not only the fleet that had languished; navy yards and ship-building skills had also declined. By mid-decade designs for the latest warships, armor, large-caliber guns, smokeless powder, large propeller shafts, and steam engines were being purchased in Europe.<28> This approach, though expedient, was obviously not in the long-term interests of national security. It was essential that a vibrant, competitive home-grown naval shipbuilding industry be nurtured. In 1883 Congress mandated a study of how best to proceed in updating the nation's military. The Gun Foundry Board, composed of Naval and Army officers, was established, and it sent members to Britain, France and Germany to assess the capabilities of the prominent arms makers and to make recommendations for the United States. The Board judged that the best barrel forgings were made by Joseph Whitworth of Manchester, England and the best armor plate by Henri Schneider of Le Creusot, France.<29>

To provide America with its own capability to produce large caliber naval guns,

Schneider constructed, tested, and shipped to the US an entire armor-plate factory to duplicate Schneider's revolutionary nickel-steel process. With the capacity to build, arm, and armor their own modern warships in place and with new overseas responsibilities arising from the accession of the Philippine Islands following the Spanish-American War in 1898, it became the policy of the US government to build a navy second only to Britain's.<30> In world ranking the US moved from 6th to 4th place between 1900 and 1902 and, after a major build-up during Theodore Roosevelt's administration, moved close to second place by 1909.<31> Under Woodrow Wilson's presidency America set as its goal a navy second to none.<32> But just as Germany had discovered, British hegemony of the seas would not be easily relinquished. As can be seen from the accompanying chart, by 1921 the US still had a long way to go to catch, let alone exceed, Britain's dominance.

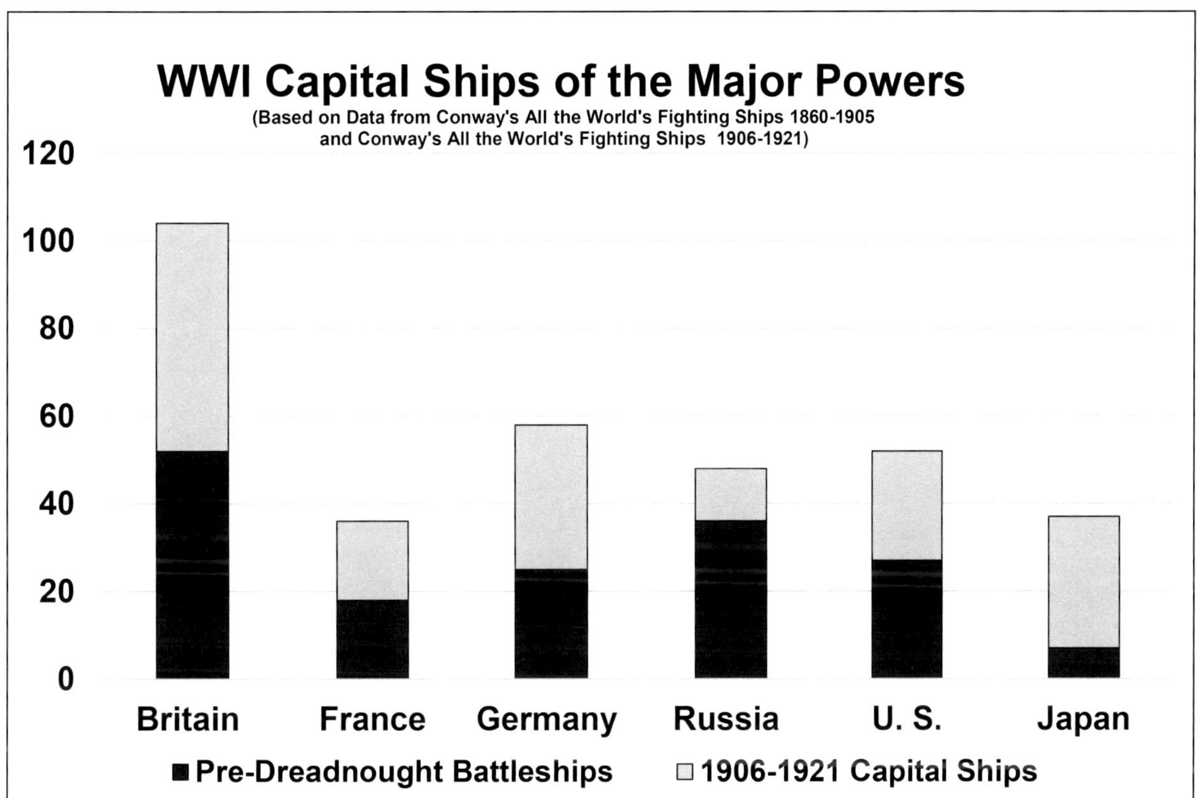

"MERCHANTS OF DEATH"

Of course, these massive new armaments did not spring from nowhere. They were built by the successors of the early iron and steel producers who had learned how to handle large, heavy objects and manage complex construction projects. Many of these firms grew prosperous and politically powerful as a result of the conflicts leading up to and including the First World War.

The heading of this section is taken from the title of a 1934 book by American academic H. C. Engelbrecht and American journalist F. C. Hanighen.<33> The book was a Utopian polemic against the world-wide armaments industry, which they perceived had profiteered during WWI. While they recognized that the arms industry

> ***is an essential element in the chaos and anarchy which characterize our international politics ... [eliminating] it requires the creation of a world which can get along without war by settling its differences and disputes by peaceful means. And that involves remaking our entire civilization.***<34>

(There is a sad irony in this lofty proposal in light of the events that would take place a scant few years later. In the September 1938 Munich Agreement, British Prime Minister Neville Chamberlain believed that he was averting war by "peaceful means" in allowing Adolf Hitler to annex more than a third of Czechoslovakia.<35> Instead, it merely whetted the *Fuehrer's* appetite for more and bolstered his confidence in the weakness of his opponents.) The authors' motive had been to expose the role of the armaments manufacturers as fomenters in The Great War. In the US, at least, charges of profiteering and even instigating US participation in the war by US companies were investigated by a Senate committee over an 18-month period beginning in September 1934, but no evidence was found in support of it.<36> Despite the findings, some companies were so stung by the allegations that they dramatically scaled back their post-war armaments operations—DuPont, for example, to less than 2 percent of their business.<37> And this action was taken only half a dozen years away from another world war that would require much greater production than the first. However, whatever one's view of the profiteering allegations, the book proved to be a fascinating glimpse into the largely hidden operations of the major world-wide producers of war materiel. Those players are pertinent to the story told here.

Despite the cynical thesis of Engelbrecht and Hanighen, the truth was that in the US, before the First World War, some companies had to be coaxed into the armaments business. American iron production had flourished along the anthracite-rich Lehigh Valley in Pennsylvania in the 1850's and 1860's in support of railroad building. The first American Bessemer-process steel plant was built in 1867,<38> but no large-scale forging capability existed to produce large-caliber gun barrels and armor plate. Forging, i.e., working the steel by hammering or pressing, had been found to be necessary to increase the strength of the steel for ordnance applications. In 1887 Navy Secretary William C. Whitney, a Harvard-trained lawyer who had helped break up the Boss Tweed Ring of political corruption in New York, made Bethlehem Steel an offer they couldn't refuse—a $4.5 million contract to build a steel-forging plant. Secretary Whitney initiated secret negotiations among Bethlehem, Whitworth (England), Schneider (France), and the US government. The liaisons for these agreements were active-duty American naval officers, Lt. Comdr. F. M. Barber assigned to Schneider and Lt. William H. Jaques to Whitworth. Even as Secretary Whitney was rigging the contracts for armor and ordnance in Bethlehem Steel's favor, he was also soliciting bids from other steel companies. The two naval officers meanwhile were acting as sole U.S. representatives of their respective foreign steel firms.

All this, of course, was illegal but ultimately gave the Unites States the industrial capacity to produce state-of-the-art warships in the shortest period of time.<39> The dilemma presented by the government's shaping of a civilian free-enterprise economy to generate technologically advanced war materiel was intrinsic to the operation of the arms industry in a capitalist democracy. In fact, Bethlehem Steel learned its armor-plate trade so well that Russia bought its product for their battleships. In 1894, however, a new US Navy secretary discovered that Bethlehem was charging Russia $250 per ton while charging the US Navy $625 per ton. The difference was due to former Secretary Whitney's guarantee of Bethlehem's ability to recoup the costs of their plant, but politically both the government and the company were left with egg on their faces.<40>

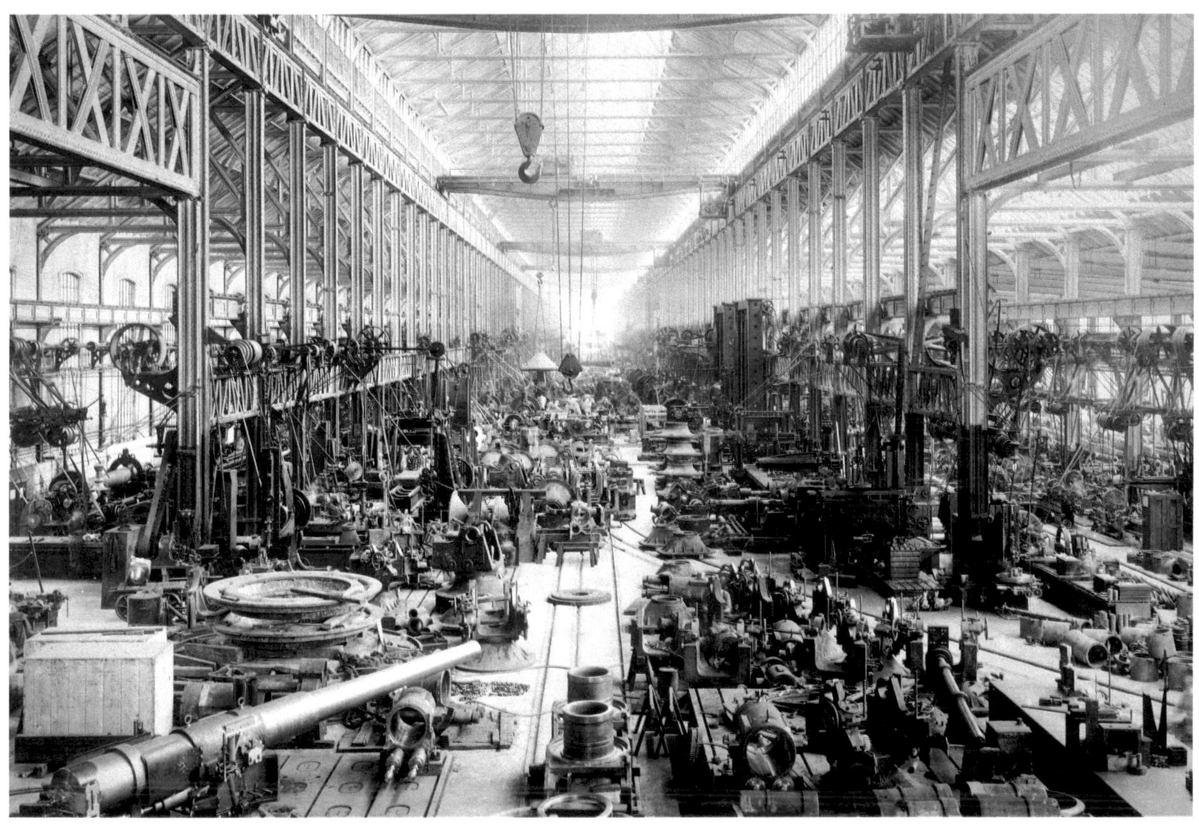

Bethlehem Steel's 6-inch gun shop in 1918
Courtesy National Archives

In the late 1880s when the US began upgrading its domestic capability for manufacturing armor plate and forging heavy guns, there were three principal British suppliers of cannon: Vickers, Armstrong, and Whitworth. In 1896 Vickers bought Nordenfelt-Maxim and began a redesign of the Maxim gun to make it stronger and lighter. The resulting Vickers gun became the standard machine gun of the British army throughout WWI. William Armstrong, based at Newcastle upon Tyne, had pioneered rifled, breech-loading cannon in the 1850s and by the late 19th century was the largest cannon maker in Britain.<41> In 1882 Armstrong merged with Charles Mitchell & Co. shipbuilders, creating a unified provider of warships. Joseph Whitworth and Company of Manchester, though much smaller than Armstrong's company, was his closest cannon-making competitor in Britain and had developed a world-class reputation for high quality ordnance steel and precision forging. As previously described, Joseph Whitworth was selected by US military officers as the model for Bethlehem Steel to follow in the forging of large-caliber gun barrels.

Armstrong and Whitworth were bitter rivals making any merger impossible while both founders were alive. However, in 1897, a decade after Whitworth's death and with Armstrong's armaments hegemony suddenly threatened by Vickers, a merger produced the Armstrong-Whitworth Company. The fact that the US Gun Foundry Board had found Schneider's armor plate to be the best did not mean that Vickers was left out of the armor plate business—quite the contrary. Vickers made everything from field guns to battleships.

The appointment of Jacky Fisher as First Naval Lord was a cause for rejoicing at both Armstrong and Vickers. In 1898 Vickers built its first battleship, the Japanese *Mikasa* that distinguished itself in the 1904–1905 Russo-Japanese War.<42> In that war Vickers sold arms to both sides.<43> The *Mikasa* was Admiral Togo's flagship at Tsushima Straight where the Russian Navy was devastated by the highly trained Japanese Navy.<44> The lopsided outcome of this battle convinced Jacky Fisher to push for the dreadnought concept. By 1897 Vickers had the largest capitalization of any British arms maker;<45> both it and Armstrong-Whitworth would be declaring ordinary dividends of 10-20 percent over the following 10 years.<46> Ultimately, during the economic stress of the post-war years, the great rivals of Vickers and Armstrong-Whitworth would merge in 1927.<47>

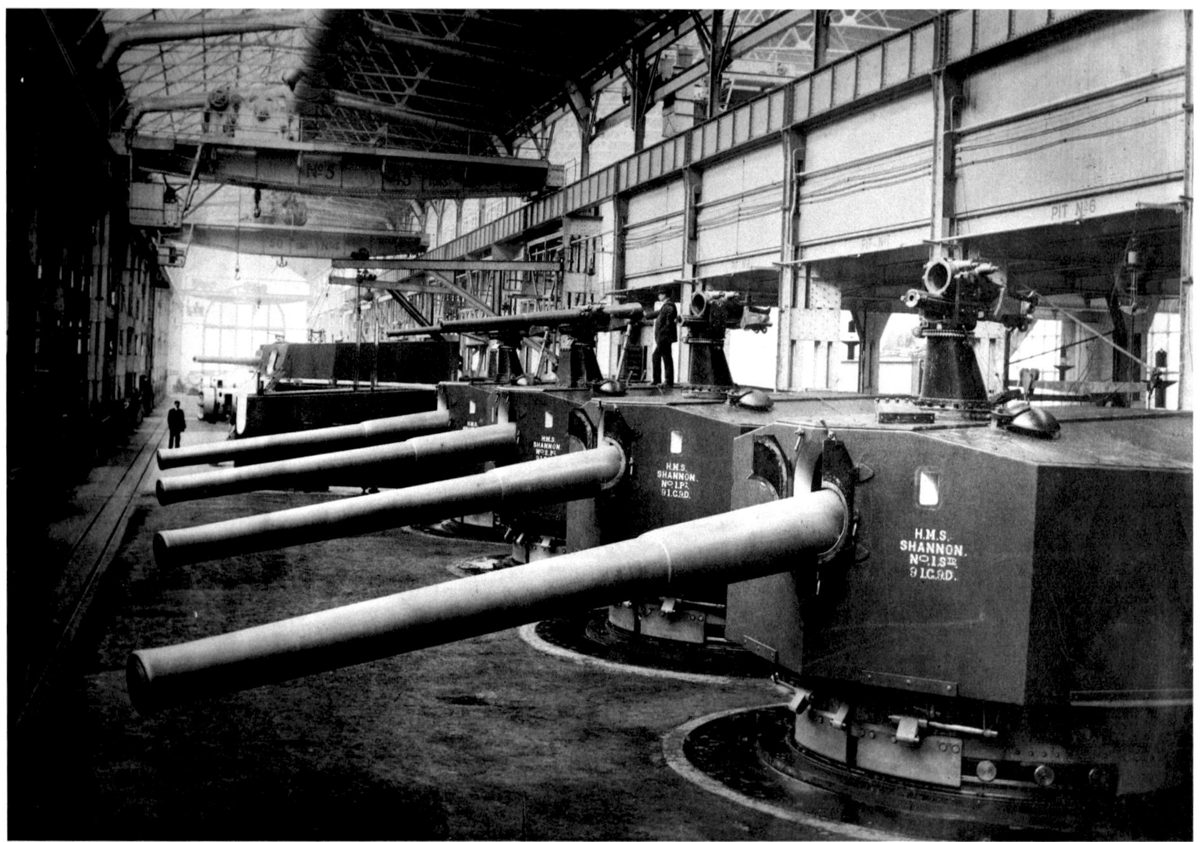

Assembly of the big guns at the Vickers Works for the Armored Cruiser HMS Shannon launched in 1906.
Courtesy Library of Congress, ID 19618

The Schneider steelworks at Le Creusot in Burgundy was not the only arms maker in France but it was the largest and a model of progressive labor management, providing housing, medical care, and pensions to its workers. Le Creusot began its industrial life in coal mining, and in 1782 a Royal Foundry was established under Louis XVI. His queen, Marie Antoinette, founded a crystal works there as well. Cannon were produced in the Royal Foundry until the end of the Napoleonic Wars in 1815, whereupon the foundry fell into disuse. In 1836 the brothers Joseph Eugene and Adolfe Schneider bought it and within two years they had built the first steam locomotive in France.<48> Until its ignominious defeat at the hands of Prussia in 1871, France made all of its ordnance in government arsenals under a veil of secrecy. The war changed all that and private enterprise was encouraged to reinvigorate the military with its innovations. Schneider quickly gained a reputation for high quality gun steel and forging technology. At that time heavy forging was done by steam-driven drop hammers, and Schneider built the largest in the world in 1876 at 80 tons. Later increased to 100 tons, it was dropped from heights of up to 16 feet. Schneider's hammer could reduce an ingot's thickness from 1.4 meters to 0.55 meters in a dozen blows over a ten-day period.<49> The noise and ground shaking it must have caused are hard to fathom. In its technology transfer operation in the 1880s, Bethlehem Steel built an even larger steam hammer based on the Schneider monster. The Bethlehem beast posted a new world record at 125-tons, but it created so much vibration that it threw other machinery out of calibration and was obviously unpopular with the surrounding community. They replaced it after only a few years with a hydraulic press.

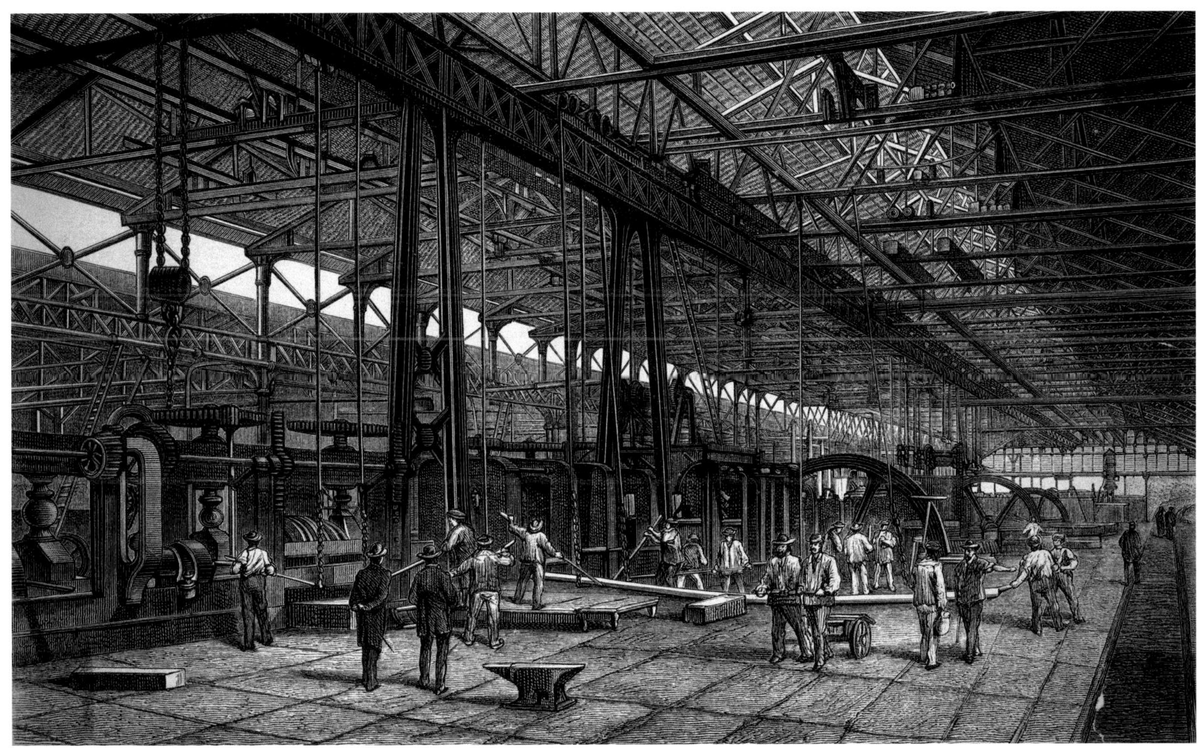

Schneider Rolling Mill at Le Creusot, France c1880
By Permission bpk Bildagentur / Art Resource, NY

Schneider's international reputation continued to grow from these and other innovations and benefited from Russia's rearmament after its disastrous 1904–05 naval war with Japan. Also, in 1906 Schneider guns outperformed those of Armstrong, Vickers, and the German firm of Krupp in a head-to-head shootout for Argentina's business, according to the presiding ordnance committee. Unfortunately, Krupp proved more cunning politically and wrested the contract away. Schneider was similarly outmaneuvered by Krupp in both Brazil and Chile.<50> Krupp was proving to be a fierce competitor both in quality of armaments and in the art of underhanded business practices.

Ironically, the archetype of the Merchant of Death, the Krupp steel and armaments dynasty, was born in the dying embers of failure. At his father's death in 1826, fourteen-year-old Alfred Krupp found himself in charge of a small crucible-steel works with seven idled employees. He had little idea how to make steel, but he was possessed of boundless energy and formidable intelligence in the service of Teutonic grit. The company's "secret recipe" for steel had died with his father, and Alfred had to painstakingly rediscover the process. In four years he succeeded in producing a crucible steel of superior hardness, and in four more years the business achieved solvency for the first time ever. Obsessed with every detail, enduring

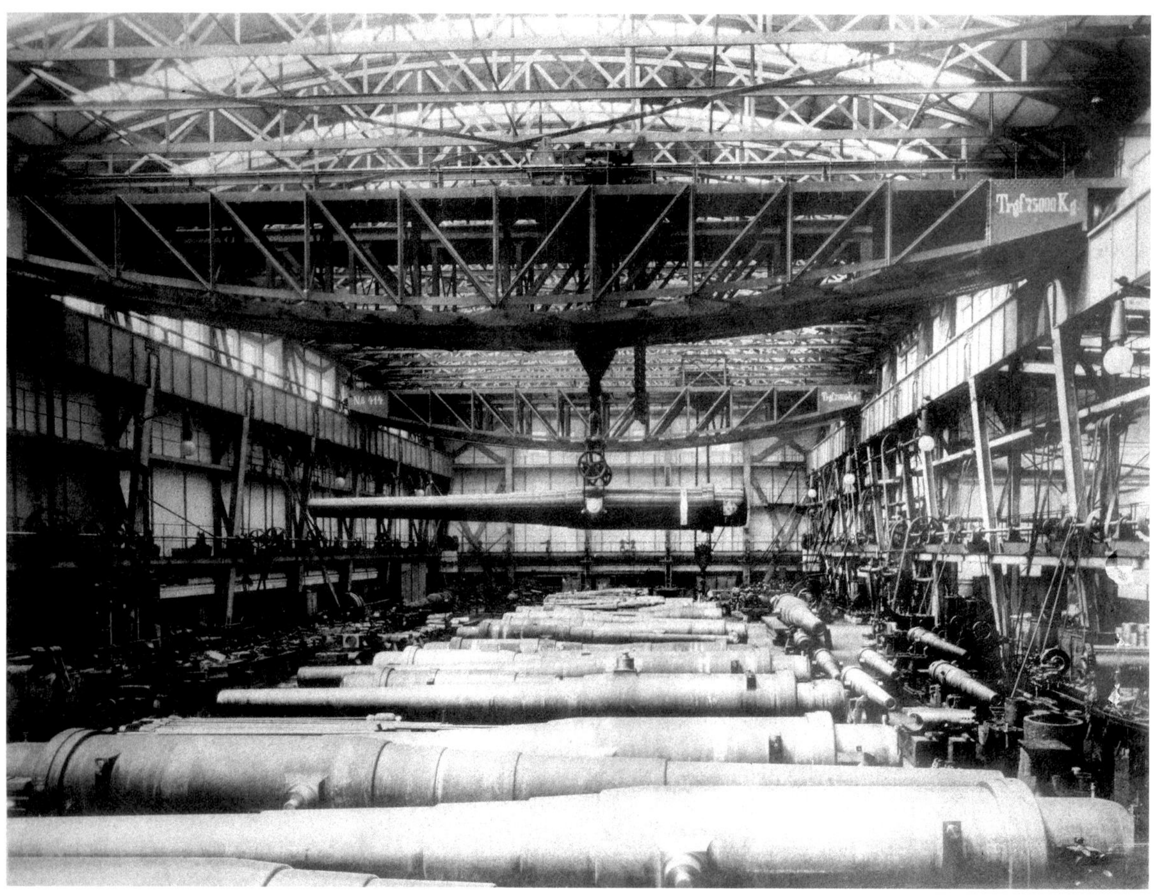

Krupp gun factory in Essen, Germany c1879
By Permission bpk Bildagentur / Ruhr Museum, Essen, Germany / Art Resource, NY

sleepless nights, compulsively writing notes on procedures, equipment, and hazards, Alfred lived and breathed steel. He drove himself relentlessly, the specter of his father's failure always dogging him. As a side project Alfred experimented with forging steel musket barrels, such items customarily being made of iron. From the beginning he had envisioned making cannon of steel if he could sell the notion for muskets. The muskets were rejected as too radical by the Prussian government, but they were willing to consider a steel cannon. September 1847 saw the first delivery of a miniature artillery piece, three years in the making, a three pounder (47mm caliber); this too was rejected by the Prussian military as too costly for too minor a gain over the prevailing bronze barrels. Alfred turned his attention to the profitable railway boom and, with the death of his mother, finding a wife.

Alfred's new wife, Bertha, wooed with the same single-minded intensity that he applied to his business, had no idea what she was in for. Alfred built a wildly fanciful house in the middle of his growing steelworks amid the choking smoke and grime which penetrated the house at every crack. As for spending an evening out at a concert, he exclaimed,

> ***Sorry, it's impossible! I must see that my smokestacks continue to smoke, and when I hear my forge tomorrow, that will be music more exquisite than the playing of all the world's fiddles.***<51>

The biggest steam hammer in the works was in a huge building close to his house; this he named Fritz in honor of his son born nine months later. Unlike his ectomorphic father and his namesake hammer, Fritz grew into a pudgy, irresolute youth and a great disappointment to Alfred. Characteristically, Alfred would try to remake Fritz in his own image, but that would come later. Supplying steel tires for railway wheels, Alfred's own invention, proved highly profitable and would finance his experiments with steel cannon. His breakout in the arms trade came with the ascension to the Prussian throne of Wilhelm I in early 1861. Orders started flowing from Berlin and, with successful showings at several world exhibitions, from Belgium, Holland, Spain, Switzerland, Austria, England, and Russia—especially Russia.<52> In 1863 *Tsar* Alexander II placed an order for so many cannon that Alfred had to build another factory.<53> When word of his new contract

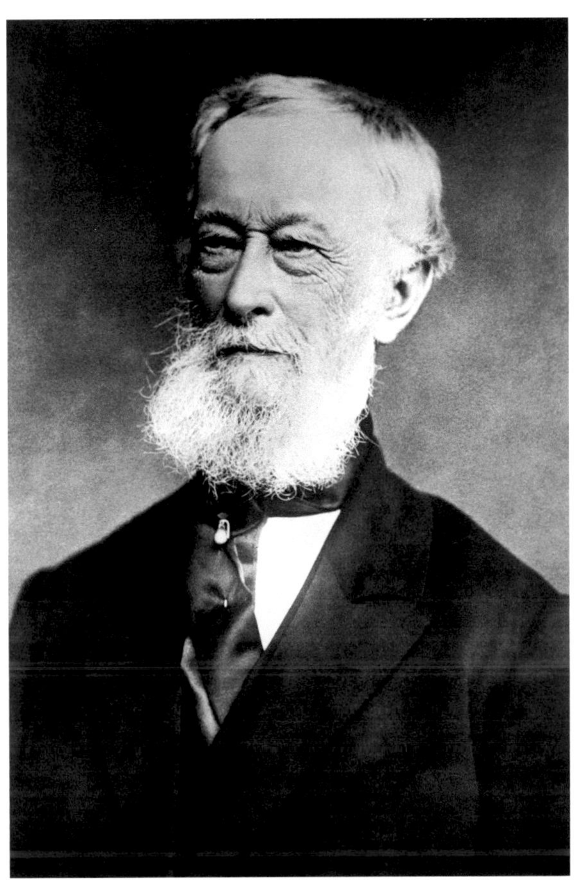

Alfred Krupp (1812–1887)
At age 52 his wife's physician described him as ***a misfit, who attracted attention everywhere because of his unusual height and striking leanness. His features had once been very regular, even attractive, but he had aged rapidly. His face was lifeless, pale, and wrinkled. A thin remnant of gray hair, topped by a toupee, crowned his head. He rarely smiled. Most of the time his face was stony and immobile.***<54>
By Permission bpk Bildagentur / Ruhr Museum, Essen, Germany / Art Resource, NY

with Russia became known, he was dubbed the Cannon King by a Berlin newspaper, and the moniker stuck. With the approaching wars against Austria and France, business mushroomed nearly uncontrollably and Alfred added coal and iron ore mines to his burgeoning empire. Seemingly unaware of the conflict of interest, Krupp was also selling cannon to Austria in the lead-up to the war. Thus began a pattern of business behavior that led finally to Engelbrecht and Hanighen's "Merchants of Death" epithet, and the unique ethical dilemma of private arms manufacturers that lasts to this day.

Business "greed" is not wholly to blame for it. Before 1870, arms were made nearly exclusively in government arsenals. With the rise of mass conscript armies beginning with Napoleon, the cost of operating these arsenals, capable of satisfying sudden massive surges in equipment in war, became prohibitively expensive during times of peace. Governments began to encourage private enterprise to provide excess capacity in time of war. The price of this strategy was to allow them to sell their wares internationally during times of peace. Competition in that market drove innovation that the government arsenals were not funded or fitted to enjoin. Since one could never be quite sure who would be the next enemy, or when, the boundaries of "selling to the enemy" were often blurred—blurred but not erased. Alfred was asked by the Prussian government on the eve of the Austro-Prussian War to hold back orders to Austria but responded that his business obligations had to be met. To express its displeasure, the government frostily denied Alfred's request for a loan to expand his works.<55> Alfred learned nothing, however, and was back trying to sell his cannon to Napoleon III on the eve of the Franco-Prussian War, although without success. The war itself, however, redeemed the Krupp

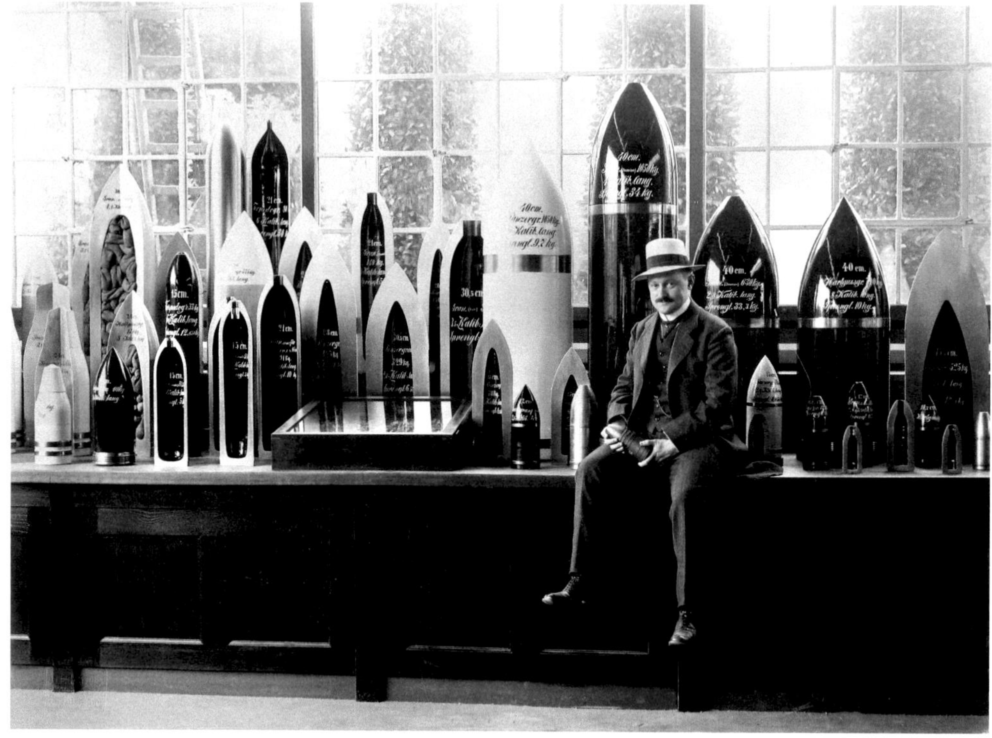

Chief engineer Hans Spaeth in 1912 with a selection of Krupp wares in the Essen factory museum.
By Permission bpk Bildagentur / Ruhr Museum, Essen, Germany / Art Resource, NY

brand. With their superior range and rate of fire, Alfred's steel, breech-loading, rifled cannon routed the surprised French with their muzzle-loading, smooth-bore, bronze cannon. Even the captured and humiliated Louis Napoleon complemented Wilhelm on his artillery.<56> By 1871 the Krupp Works had expanded to a payroll of some 10,000 men and Prussia, now the German Empire, had become Alfred's primary customer for the first time.<57> Both the Krupp dynasty and German nation were now firmly established and would continue their intimate partnership for the next 75 years. And, oh yes, Alfred's single-minded efforts made him rich—very rich. A survey done in 1913 showed that Fritz's daughter Bertha became the wealthiest person in Germany; *Kaiser* Wilhelm II's fortune was less than half that of Bertha's.<58>

With the outbreak of WWI Krupp's fortunes ballooned. From 1914 to 1918 the Krupp workforce nearly doubled. During part of 1916 some nine million artillery shells and three thousand guns were produced—every month.<59> Beyond this stunning performance for the German government, they still managed to produce enough high-grade steel to sell surreptitiously to the French in enormous amounts by way of Switzerland after stops at staging depots where the Krupp trade mark was ground off the goods.<60> Another irony of the prewar international arms trade was that Krupp had sold rights to their artillery-shell fuse to Vickers before the war,

GUNPOWDER AND EXPLOSIVES

War machines would be nothing without the sting of their chemical explosives in shells hurled by chemical propellants. The story of E.I. Du Pont de Nemours & Co. is a fascinating example of the emergence of a giant in the industry. Had it not been for the French Revolution, the company might never have existed. Éleuthère Irénée Du Pont de Nemours arrived in the fledgling United States on January 1, 1800 after fleeing the excesses of a revolution that had run off the rails. At age 28 Du Pont had had a most auspicious background for what he was about to accomplish. He had studied chemistry in Paris under his father's friend, Antoine Lavoisier. Lavoisier had achieved international fame as a chemist and was responsible for the government's production of gunpowder. Du Pont became his assistant and was even being groomed as Lavoisier's replacement. The increasing radicalization of the revolution landed both Irénée and his father, Pierre, in prison for a time.<61> Taking refuge in America, Irénée discovered that the gunpowder in the New World was both expensive and of inferior quality. He was quick to see an opportunity. In 1801 he returned to France to raise capital for his new gunpowder plant. Napoleon Bonaparte, now in power as First Consul and eager to hinder England in any way he could, arranged for French production machines and drawings to assist DuPont's new venture. Moreover, the influence of important persons promoting his business did not stop at the French border. On returning to his new American home, the assurance of orders from the US government for his powder was given by President Thomas Jefferson himself. By 1806 Du Pont's new mill had produced 600,000 pounds of black powder of excellent quality.<62> In the lead-up to the War of 1812 with England, demand for Du Pont's product soared—along with the company's profits, reputation, and future prospects.

and, though Vickers did not pay royalties during the war, British shells of a certain type had the Krupp patent mark KPz 96/04 stamped on every one. German soldiers, seeing duds at their feet, may have wondered what this mark meant. After the war Krupp brazenly sued Vickers in international court for the unpaid royalties and in 1926 won, receiving a settlement of £40,000. If anyone truly deserved the "Merchant of Death" label, it was Krupp.

The great arms makers of Vickers, Armstrong, Schneider, Krupp, and newcomer Bethlehem Steel continued their rivalry through the years of relative peace in the late 19th century, international sales being their business lifeblood during such times. Of course there were many other manufacturers involved in the trade, but these were the most prominent and would continue to supply the new century's two world wars with what they needed. Huge investments were required for these massive works and begged to be recouped. Inevitably, the cycles of innovation and competition ratcheted up the deadliness of the weapons, now viewed as commodities like any other. Marketing intrigues and imperialistic ambitions goaded nations into an upward spiral of war preparations. On the eve of WWI, industrial capacity and mass-manufacturing know-how were at levels inconceivable a hundred or even fifty years prior. No one could have guessed what the combined effects of booming populations, unlimited weapons of unprecedented power, and callow national ambitions would bring, but the world was about to find out.

CHAPTER 4. THE FIRST WORLD WAR

> ***Out of the world of summer, 1914, marched a unique generation. It believed in Progress and Art and in no way doubted the benignity even of technology. The word machine was not yet invariably coupled with the word gun.***
>
> —Paul Fussell, *The Great War and Modern Memory*<1>

MOBILIZATION

Germany's rapidly increasing naval strength had upset the world order, and governments sought safety through strategic alignments and treaties with other countries. Britain forged closer ties to its traditional enemy, France, resulting in the Anglo-French Entente negotiated in 1904. Russia, having had its fleet soundly defeated by Japan in 1905, was intent on rebuilding its military and recovering its position as a world power. France saw Russia as a counterbalance to growing German military strength and helped finance the Russian buildup. This included railway construction in Russian Poland giving Russia increasing mobilization capacity. Likewise, Britain drew closer to Russia, a relationship finally developing in 1907 into the Triple Entente among the three countries.<2> Germany's General Staff under Alfred von Schlieffen, feeling both aggressive and surrounded by growing alliances, was laying war plans. The Schlieffen Plan, begun a dozen years earlier and completed in 1905, posited a swift attack against France through neutral Belgium, thereby avoiding the stiffening defenses in northeast France.<3> Actions taken under this plan did not go unnoticed. As early as 1906 British military attaches in Berlin concluded that Germany was preparing for preemptive war to be launched sometime between 1913 and 1915. In essence Germany decided to strike before her opponents could fully mobilize. It had worked in 1866 and again in 1870. They believed that it would work again. With Helmuth von Molke, nephew of the successful Prussian Chief of General Staff during the wars of German unification, as new Chief of Staff in 1906, the General Staff was increasingly encouraged to focus on military strategy to the exclusion of political considerations. Molke was not the man his uncle was, and with weak civilian oversight this blinkered focus on perceived military necessity propelled Germany towards taking preemptive action.<4> Though Germany allied itself with Austria-Hungary and Italy in the Triple Alliance, Austria-Hungary was no longer strong enough to counter the possibility of an eastern front with Russia. Surprisingly, despite Germany's obsession with planning, little attention was given to the possibility of a prolonged war. Prefiguring Hitler's refusal to properly equip his forces for the Russian winter of 1941–42, even discussing protracted conflict was viewed as defeatist and censured.<5>

The event precipitating the war was the assassination of Austrian Archduke Franz Ferdinand and his wife Sofie on a visit to Sarajevo, Bosnia by Bosnian Serb Gavrilo Princip. Feeling the need to punish Serbia and bring her to heel, Austria-Hungary delivered an ultimatum to Serbia that it could not possibly meet. Events cascaded rapidly after that, as seen in the text box.

WWI OPENING HOSTILITIES TIMELINE		
Sunday	28 June 1914	Austrian Archduke Franz Ferdinand and his wife Sophie assassinated in Sarajevo
Tuesday	28 July 1914	Austria declares war on Serbia
Wednesday	29 July 1914	
Thursday	30 July 1914	Austria declares general mobilization
Friday	31 July 1914	Russia, an ally of Serbia, declares general mobilization to begin next day
Saturday	1 August 1914	Germany declares general mobilization and war on Russia France mobilizes German forces enter Luxembourg
Sunday	2 August 1914	German patrols enter France — first French soldier killed in skirmish
Monday	3 August 1914	Germany declares war on France German soldiers cross border of Russian Poland occupying 3 towns Britain mobilizes army and navy
Tuesday	4 August 1914	Germany invades neutral Belgium Britain declares war on Germany
Wednesday	5 August 1914	
Thursday	6 August 1914	Austria-Hungary declares war on Russia

Mobilization involved moving troops and supplies to the battlefront, anticipated or actual, and the call-up of reserves. A potentially chaotic process, it required intricate planning to work smoothly. On the afternoon of August 1, after signing the mobilization order, Kaiser Wilhelm II received word from his ambassador in London that France might remain neutral in a German-Russian conflict if Germany would agree not to attack France. Wilhelm, thinking that a two-front war might be avoided, ordered von Molke to call off the invasion of France. Molke protested that the mobilization ***had taken a year of detailed planning*** and, once launched, could not be stopped. At 11 PM it was learned that the peace feeler had been a misunderstanding and war-plan orders were reinstated. The extraordinary efficiency and forethought of the German operation can be seen in the fact that the declaration of war against Russia had been delivered to the Russian foreign minister at 7 PM and by midnight German forces had begun staging in Luxembourg and sent patrols into France a full day before even declaring war on France.<6> Only two days later, the invasion of Belgium and Russian Poland had begun. The swiftness of actual troop action on two fronts following mobilization certainly suggests that the full might of Germany's war machine had been poised and merely awaiting some provocation to be unleashed.

Adopting a modified Schlieffen plan of attack, two million men under General Erich von Ludendorff were to sweep through Belgium and race to capture Paris before the French could marshal their forces.<7> Since 1909 Krupp had been perfecting a monster 420-mm (16.5 inch) howitzer for the express purpose of breaching the formidable Belgian forts at Liege.<8> These guns were the most powerful in the world, throwing an 1800-pound high-explosive shell nine

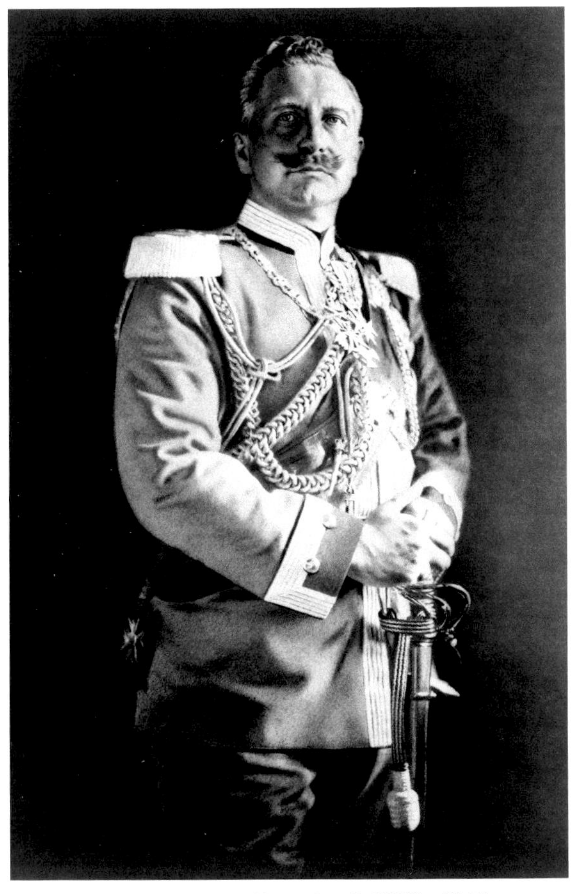

German Kaiser Wilhelm II (1859–1941)
Courtesy Library of Congress, ID 3c06034

miles.<9> Given their chance at last, they crushed the forts utterly, to the abject terror of the Belgians inside, but getting the guns to the fort involved unexpected difficulties. At 43 tons the howitzers, nicknamed Big Bertha for the Cannon King's granddaughter, required a crew of two hundred and transport by multiple rail cars. To the artillerists' dismay the Belgians had blown up a tunnel between the Krupp Works and Liege. Because of the Belgians' unexpectedly stout defense of the forts and the delay in bringing up the Big Berthas, the Germans lost a critical nine days more than expected.<10> In violating Belgium neutrality, German Chancellor Bethmann-Hollweg had breezily declared that ***Necessity knows no law***.<11> This had been too much for the British and had brought them into the war. The Germans were not expecting Britain to aid France but reckoned that, even if they did, they would be too late to make a difference. But the delay over the Berthas gave just enough time for the British Expeditionary Force to reinforce the French troops and form an impenetrable barrier between German forces and Paris. There would be no repeat of the Franco-Prussian War experience. By the end of 1914 the war had bogged down into a continuous line of opposing trenches that ran some 500

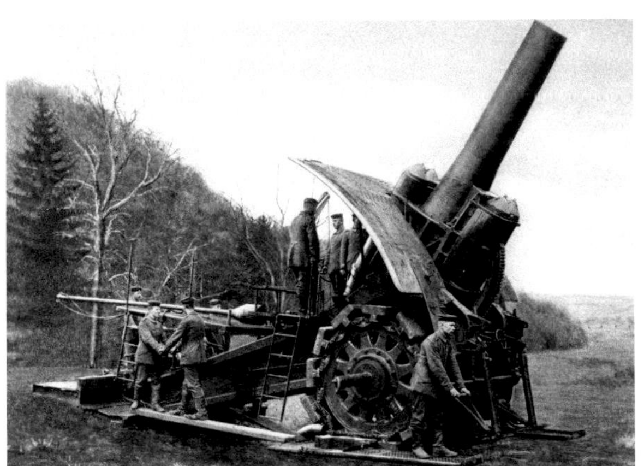

The Krupp-built Big Bertha 42-cm Howitzer near Soissons, France 1918
By Permission bpk Bildagentur Art Resource, NY

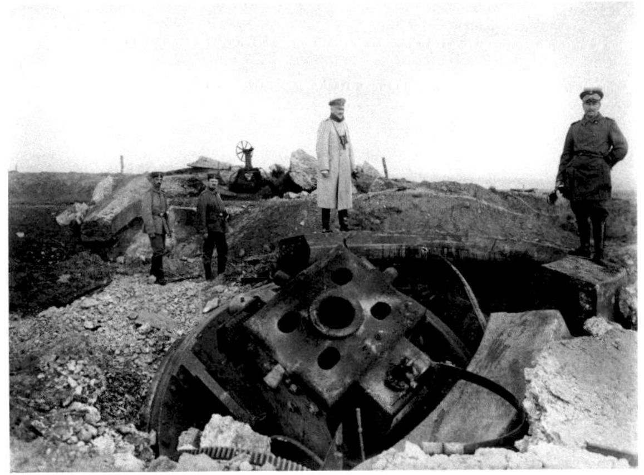

Devastated fortification near Liège, Belgium 1914
By Permission bpk Bildagentur / Kunstbibliothek, Staatliche Museen, Berlin Art Resource, NY

decided to equip the American Army with Enfields chambered for the Springfield .30-06 cartridge. By the end of the war they had made 2 million of these rifles. The few hundred field artillery guns were hopelessly inadequate and arrangements were ultimately made to purchase the latest field and heavy artillery pieces from Schneider, with the understanding that America would supply the steel. Machine guns of the Vickers-Maxim type were both purchased from Britain and made in the US, but neither the American Vickers nor the newly adopted Browning Automatic Rifle (the BAR) became available until September 1918, only two months before the end of the war. Hotchkiss machine guns were procured from France as a stopgap. Small arms ammunition production in the US was made possible only because of the Morgan Export Department's ground work. Artillery ammunition would have to be obtained from British and French sources, again with the proviso that the US would supply the steel. Millions of dollars were spent on nitrate plants (to make the explosive TNT) that failed to produce any product before the end of the war.

Pershing arrived in France in mid-June 1917 and began setting up the American General Headquarters in Chaumont, about 300 miles southeast of Paris. Massive infrastructure had to be built to support a two-million man army in the field. French port facilities had to be expanded; many square miles of storehouses would have to be built; railroads would have to be greatly expanded and refurbished; barracks and training camps would have to be built. All this required enormous quantities of wood and because overseas shipping capacity was critically limited, France reluctantly sacrificed forests to the Americans. This required sending over foresters and the building of sawmills. Communications were critical and Pershing had 200 experienced, French-speaking American telephone operators, "the telephone girls," sent over. Pershing lamented the fact that his troops were having to act as stevedores at the ports when they should have been training for battle; professional stevedores from America were requested and ultimately supplied. Even when the general's requests were granted, it all took a seemingly inordinate amount of time. All the while French and British soldiers were losing men who could not be replaced.

Inadequate shipping capacity was the most critical problem but Pershing endured great frustration with the lack of urgency shown by stateside support. The winter of 1917-18 was to be brutally cold and his repeated requests for winter clothing were at first ignored and then denied by the Quartermaster General for the reason that the clothing was needed at home. Likewise, the Quartermaster General thought that the requested number of trucks was excessive. Pershing wrote to him that he should send every available truck to France and use horse-drawn wagons in the training camps at home. A particular irritation was the fact that it took six months before the training camps in the US were constructed and ready for recruits. Trainees who arrived in France from the States were also inadequately trained. Camps had to be established in France to complete the process. The feeling of crisis in France was acute, at home not so much.

From the start, both the British and the French implored the American commander to use his soldiers as replacements for their armies and to let them be trained by the Allies. This Pershing steadfastly refused to do, insisting that American soldiers be led by American officers. As it turned out to be nearly a year before the first US forces were ready for battle, his policy caused increasing consternation among the Allied high command as the war ground their armies down. With the huge German offensives in the spring of 1918, this issue became critical as Paris itself was then seriously threatened. There was talk that, if Paris fell, France would be knocked out of the war. By the end of April 1918 only a single American division was ready for offensive action.

Some feeling for Pershing's state of mind at this point can be appreciated from the following quote from his memoirs. Though written thirteen years after the events, his frustration and anger are still sharp.

> ***It was depressing to think that ten months had elapsed since our entry into the war and we were just barely ready with one division of 25,000 men. With all our wealth, our manpower and our ability, this was the net result of our efforts up to the moment; all because our people had been deceived by a false and fatuous theory that it was unnecessary in time of peace to make even preliminary preparations for war. Here we were likely to be confronted by the mightiest military offensive that the world had ever known and it looked as though we should be compelled to stand by almost helpless and see the Allies again suffer losses of hundreds of thousands of men in their struggle against defeat.***<21>

Though the Allied armies were at first skeptical of the fighting skills of the American soldier, increasing participation in front-line action won their admiration of the doughboys' tenacity and courage. Considering the unforgiving nature of warfare in the trenches, Pershing's establishment of in-country training camps, unit-commander schools, and schools for every specialty had borne fruit. His near miraculous accomplishment in the face of enormous difficulties was recognized and rewarded in 1919 with the establishment of and his promotion to the new superior rank of General of the Armies, the only person ever to hold this rank on active duty. The only other person to be granted the rank at all was George Washington, promoted posthumously by Congress in 1978. In effect this rank would have the equivalent of six stars, a status none of the great generals of WWII achieved. Something of Pershing's strong character is conveyed in a crisis meeting of the Supreme War Council of Allied leaders on 1 May 1918 during which the old argument for merging American troops into Allied units was pressed yet again. Present were French Prime Minister Clemenceau, British Prime Minister Lloyd George, and Supreme Allied Commander Marshall Foch. This was the time of the great German spring offensive when French desertions were about to turn to mutinies, and there was a very real possibility that France would drop out of the war. Here the meeting is recounted by historian Martin Gilbert:<22>

> ***Lloyd George, while agreeing to a separate American army in principle, told Pershing: 'At the present time, however, we are engaged in what is perhaps the decisive battle of the war. If we lose this battle, we shall need tonnage to take home what there is left of the British and American armies.' This threat had no effect on Pershing, of whom Foch asked angrily: 'You are willing to risk our being driven back to the Loire?' Undaunted by the rhetoric, Pershing replied: 'Yes, I am willing to take the risk. Moreover, the time may come when the American army will have to stand the brunt of this war, and it is not wise to fritter away our resources in this manner.' Foch replied that the war might be over before the American army was ready to enter the battle. The meeting ended with a final altercation: Lloyd George: 'Can't you see that the war will be lost unless we get this support?' Pershing: 'Gentlemen, I have thought this programme over very deliberately and will not be coerced.'***

Pershing's most pressing task was to train his army. At the time infantry training was still centered on rifle marksmanship, but, after the trench-warfare stalemate, the First World War came to be dominated by artillery. In this discipline much had changed.

ARTILLERY

Between the American Civil War and WWI, cannon barrels went from smooth to rifled, and powder went from black to the slower-burning, higher-energy, smokeless variety. Rifling of the barrels imparts a spin to the projectile, stabilizing its flight and giving higher accuracy and range. A slower, more controlled burning of a more energetic powder can push the projectile over a longer distance down the barrel. This means that a longer barrel will impart an even greater energy to the projectile, increasing its range. Artillery range was thereby increased to such an extent that the gunners were no longer able to see their targets, i.e., the big guns went from direct-fire to indirect-fire weapons. Some means of accurately aiming the gun had then to be found. The first requirement was for accurate maps in scales of 1:20,000. At the beginning of the war, maps of northern France proved inaccurate and were only available in the smaller scale of 1:80,000. The first job was to survey the battlefields. This was done by conventional survey techniques behind the battlefront coupled with aerial photographs. The ground survey would then be used to calibrate the aerial photographs and then to create maps. By the end of 1915 some 32 million maps had been prepared at various scales up to 1:10,000 to an accuracy of 15-20 yards.<23> With good maps in hand there needed to be a way of locating and communicating a position on the map. A grid system was developed using markings of letters and numbers to indicate precise locations. The traditional method of referencing locations to landmarks proved unreliable due to the possibility of artillery barrages turning them to dust.

Once gridded maps were available, the science of getting the shot to hit the target came into play. Elaborate firing tables were constructed that allowed for the many variables affecting the fall of shot: air temperature, wind direction, propellant temperature, and the degree of barrel wear, which eventually was periodically measured. Of course, other factors affected accuracy as well such as quality control on the ammunition (which early on was poor), aiming errors, gun-to-gun variations, precision of laying-in (positioning the gun), and so forth. Because the many variables were never under perfect control, getting the shot on target also required seeing where the first rounds fell and correcting the next shots. This was accomplished using forward-observing officers positioned within sight of the target and advising the gun crew by telephone how to adjust their aim. By late 1916 a telephone network had been established allowing any forward observer to connect with any battery in his observational zone for firing corrections. A novel feature of this war was the use of aircraft-borne forward observers. The first aircraft radio transmitters weighed 75 pounds and occupied the observer seat. This put considerable demands on the pilot. In the fall of 1915 the transmitter's weight had been reduced to 20 pounds and had a range of 8-10 miles.<24> The artillery war had become a highly technical one and not everyone was happy about it. One brigadier general complained that,

> ***You damned surveyors with your co-ordinates and angles and all the rest, are taking all the fun out of war; in my day we galloped into action and got the first round off in thirty seconds.***<25>

At the outset of the war Britain's artillery capability was largely confined to mobile field guns with calibers of about 3.3 inches and only shrapnel shells were available.<26> Shrapnel, or metal shards, had been judged suitable for close infantry support, but in the opening battles of the war the heavier German artillery with high-explosive shells took a terrible toll and the British were unable to respond in kind. The Brits had a mere 54 big guns: twenty 60-pounders (5-inch), sixteen 4.7-inch, seventeen 6-inch howitzers, and one experimental 9.2-inch howitzer.<27> The French were in better shape in 1914 with some 5,000 field guns mostly of 75 mm (3inch) caliber. As the front stagnated, the need for heavy-caliber guns increased, and French numbers rose from 300 in 1914 to 5700 in 1918. Insufficient guns and ammunition would remain a problem until the spring of 1917.<28> Lighter, easier-to-manufacture weapons helped fill the gap. British trench mortars, very useful for their short-range, high-arc trajectories, went from twelve being made in the last months of 1914 to more than 2,000 in the last quarter of 1917.

Early in the war ammunition shortages for the big guns were critical. At that time, for every British shell expended, the Germans fired ten. The need at the beginning of the war for high explosives with which to fill shells was also acute; in 1914 there was little British capability for producing amatol, a mixture of TNT and ammonium nitrate.<29> The requirements were staggering. During the 10-month endurance contest over Verdun in 1916, the French alone fired 37 million shells.<30> On the Somme in June of 1916 the preliminary bombardment of German positions saw 1500 guns firing some 1.7 million shells over the space of a week.<31> There was a gun for every 17 yards of front.<32> On July 1, just prior to "going over the top" into the attack, a

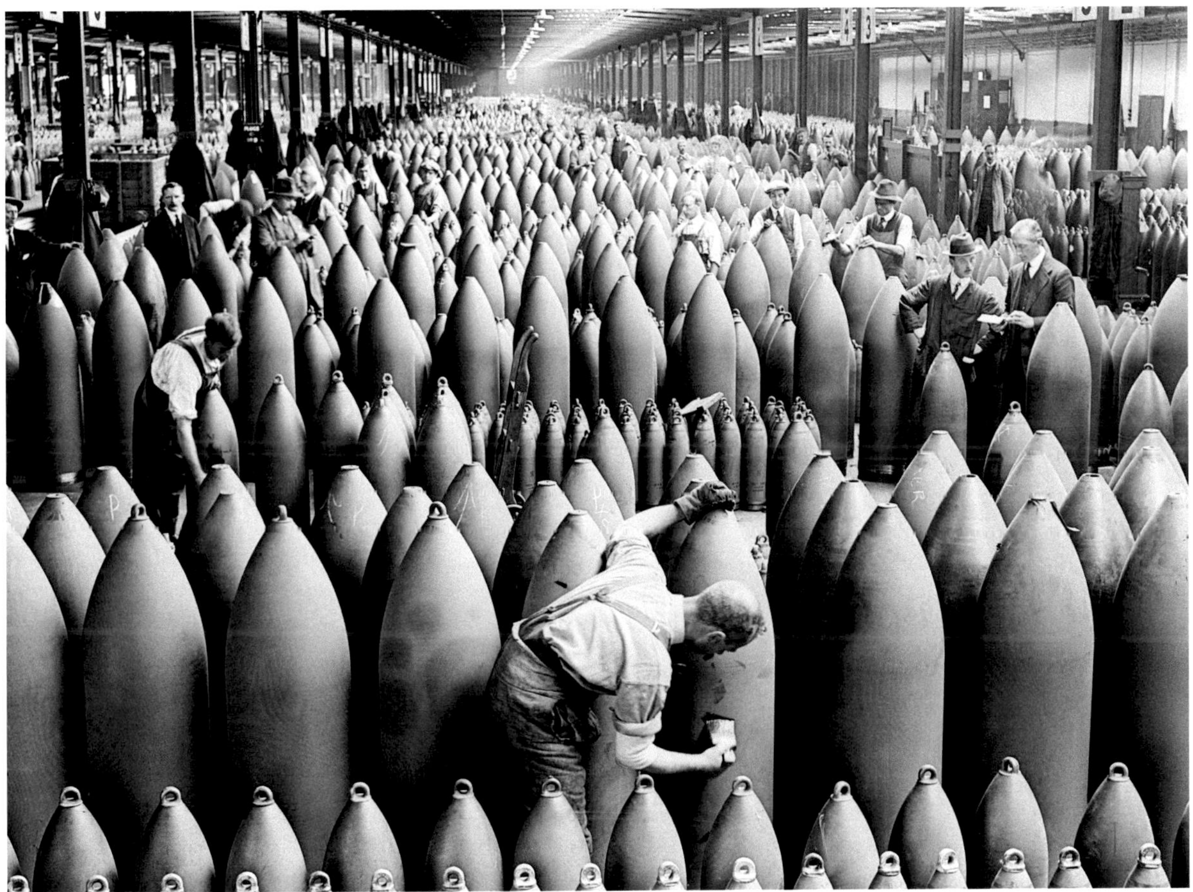

Munitions workers paint shells in the National Shell Filling Factory at Chilwell, Nottinghamshire around 21 August, 1917.
By permission British Imperial War Museum

quarter of a million shells were fired in about an hour creating a continuous roar that could be heard in London more than 150 miles away.<33> At Arras, about twelve miles north of the Somme, in March 1918 a major offensive by the Germans, hoping for a breakthrough before the Americans could make a difference, involved 6600 guns firing more than 3 million rounds on the first day.<34> In the 1918 Meuse-Argonne attack involving 1.25 million Americans, over 4 million rounds were fired from 4000 guns. More than a dozen trainloads of artillery shells per day were required to replenish the guns, and still troops had to be pulled back from the line periodically because of short supply.<35>

The big guns wreaked havoc not only on the enemy. Premature detonations of the shell's high explosive sometimes occurred before the projectile had left the barrel, resulting in the destruction of over 600 French guns in 1915 and the death or injury of their crews. German guns were not exempt from this hazard; that same year 2300 field guns and 900 light howitzers shared the same fate—as many as were lost to enemy hostilities. By German estimates one in two French shells were duds. On the Somme a quarter of British guns failed due to faulty design, materials, or poor quality control and nearly a third of their shells were duds.<36> By 1918 these issues had been largely resolved.

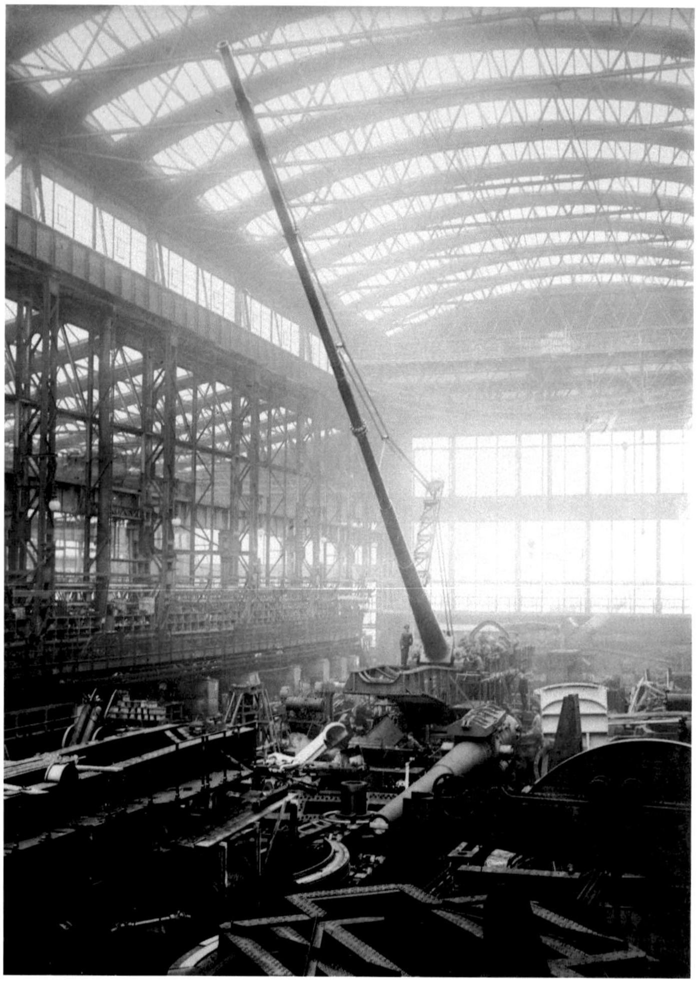

The Paris Gun being assembled in the Krupp works.
By permission British Imperial War Museum

Besides the Big Bertha howitzers first used at Liege, the Krupp factories had another surprise for the Allies —this time for the citizens of Paris. On 23 March 1918 there commenced a bombardment of explosive shells weighing 200-230 pounds hurled from 75 miles away. For the next 20 weeks a shell was launched on average every 20 minutes. The agent of this atrocity was a bizarre 21-cm caliber artillery piece, the Paris Kanone, or Paris Gun. Its barrel was so long that it had to be supported by guy wires. Loading and firing the gun was a major production requiring a crew of 60 naval specialists commanded by a full admiral. The gun had originally been developed for the Imperial Navy and they had retained control of its use on land. The firing of each successive shell eroded the barrel significantly, enlarging the bore. In order to maintain long-rage accuracy, the gun was provided with a complement of shells with increasing diameters to be fired in sequence. After launching 65 shells, the barrel was considered worn out and replaced. Its military value was considered insignificant, but as a terror weapon it delivered. Over a thousand citizens were killed during the course of the siege. Death could rain down without warning at any time — as it did for parishioners of Saint-Gervais-l'Église on Good Friday when 91 were killed and over a hundred injured. To minimize the possibility of the Allies locating the Paris Gun some thirty nearby batteries opened up at the same time to confuse attempts to direction-find by sound. If that failed, the gun would be defended by 40 Fokker fighters on standby.<37> The gun was never discovered and the Versailles Treaty requirement that it be turned over to the Allies was never met.

AIRCRAFT

The military potential of flying machines was recognized early on. France, still smarting from their 1871 defeat at the hands of the Germans, encouraged any military innovation that could even the odds against the more populous and industrialized Germany. Just months before the Wright brothers' famous flight in 1903, French artillery captain Ferdinand Ferber was in communication with them. In 1905 negotiations began to supply France with their airplanes, but the talks foundered on French insistence of exclusivity. Meanwhile flight technology underwent very rapid progress as did its application to military purposes. Wireless telegraphy was adapted to the cockpit and more powerful engines permitted loads of 600 pounds.<38> In April 1912 the first five French Air Force squadrons, each with six planes of different types, were formed.<39> Some controversy attended the question of arming the planes, as some military planners envisioned them only for reconnaissance and artillery spotting. As the war opened, none were armed. Photo reconnaissance required nerves of steel in both the pilot and the photographer. The former had to stall his engine to keep vibrations from blurring the image, the photographer had to stretch over the side with his bulky camera and long lens, loaded with a glass plate in the fierce slipstream, and then the pilot had to restart his engine. By early 1915 glass plates had been superseded by roll film, and planes were fitted with a frame for attaching the camera.<40> Germans knew that recon planes, even unarmed, were a major threat to their security and tactical plans and, not surprisingly, tried to shoot them down. The pilot was constantly busy but the rear seat occupant, when not occupied with duties, had more time to consider the danger he was in. With exploding shells all around his plane, one rear seater used to occupy his mind by repeatedly transmitting ***Sh*t to the Kaiser*** on his wireless to irritate the German radio operators.<41>

The new technology of artillery was matched by that of an altogether new service, aerial forward observing. Airplanes quickly became indispensable for artillery spotting, and it was a harrowing experience for the pilots and forward observers in the rear seat alike. Environmental challenges were bad enough: a ground temperature of 30°F translates to about -40°F at 20,000 feet altitude according to the US Standard Atmosphere,<42> and the wind-chill effects in open cockpits could make conditions unbearable. Enduring the cold and sometimes snow, sleet, or rain in their faces, they had to maintain their positions in the sky within both sight of the target and wireless range with their battery and send their firing corrections in Morse code. All the

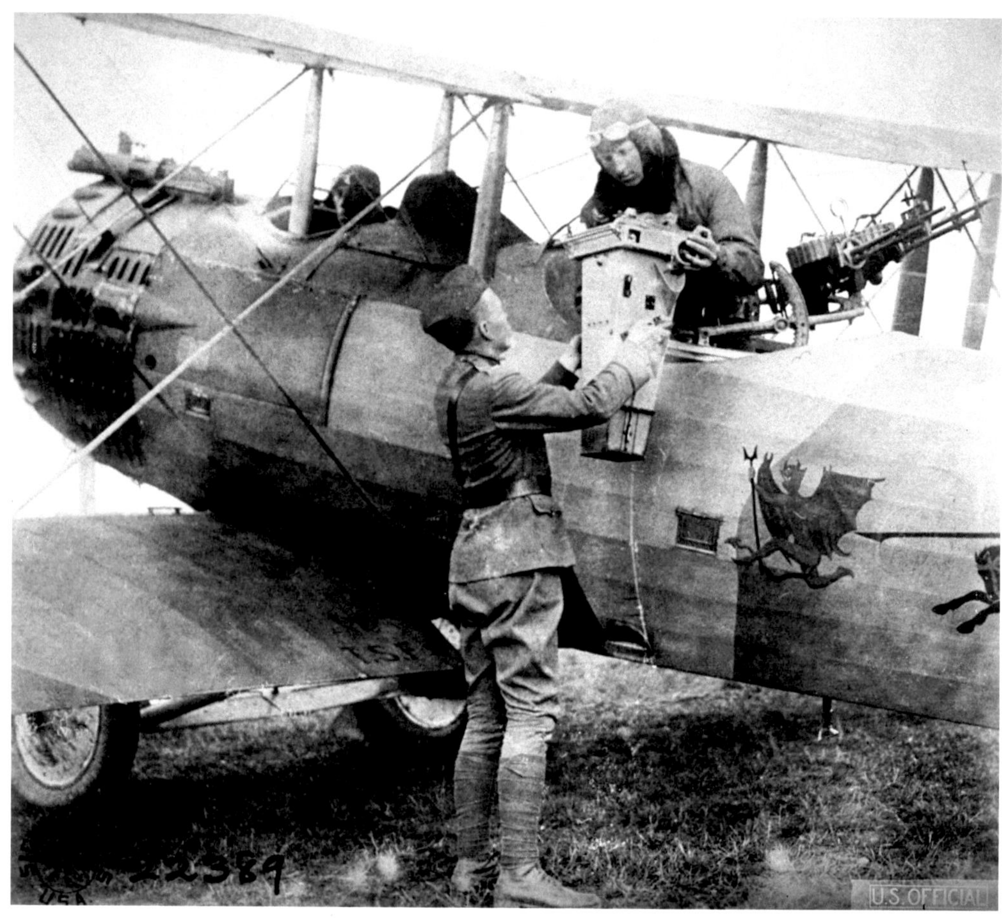

USAAF Observer Lieutenant J. H. Snyder receiving a French 18x24-cm plate aerial camera.
By permission British Imperial War Museum

while they were exposed to the constantly improving anti-aircraft shells timed to explode at altitude. One French publication of the time described their challenges this way,

> ***Weather conditions, guns, machine guns, rifles, planes and shells German and French all conspire to bring them down. The spotters climb into the slowest planes, those least able to defend themselves. Their task requires unfailing courage, utter indifference to death and a particular talent for flying – whatever the conditions, whatever the altitude. When the spotters can't fly, the troops on the ground are paralyzed.***<43>

The famous WWI aerial dogfights evolved quickly. At first there was a gentlemanly esprit between airmen from different sides, even waving or saluting one another on the rare encounters aloft. As the war ground on, pilots were taking potshots at each other using revolvers, carbines and then machine guns. The first confirmed air-to-air kill occurred on 5 October 1914 by French aviator Joseph Franz using a Hotchkiss machine gun mounted on a special tripod and fired by his observer Louis Quenault.<44> Research was undertaken by French aviation pioneer Roland Garros to fire a machine gun through the propeller. His solution was to armor the propeller with steel deflector plates and hope for the best. In fact, he found that only 10 per cent of the rounds hit the plates. Mating a Hotchkiss machine gun in front of the pilot to the fuselage of a fast single seater, the fighter aircraft was born. In April 1915 Garros scored his first kill. By July 1915 German aviation pioneer Anthony Fokker had perfected his synchronizer gear, which timed the firing of a machine gun to clear the propeller blade completely, and the Fokker Eindecker sported the heavier Maxim machine gun. As a result the Germans enjoyed many months of clear superiority in the skies.<45> Not until February 1916 did the French match the German armament with a propeller-synchronized Vickers machine gun.

The first bombing raids consisted of the rear-seat observer dropping a half-dozen 40-mm shells over the side at enemy hangars, but on 14 August 1914 the first serious attempt at heavy bombing by an airplane began with the dropping of a modified 155-mm shell on a Zeppelin hangar at Metz-Frascaty. In a hair-raising episode for the pilot and his bombardier, the plane's engine suddenly cut out amidst a cloud of exploding shells. The pilot, Lieutenant Antoine Cesari, coolly glided his aircraft towards the target as Corporal Roger Prudhommeaux released the big artillery shell at 6500 feet altitude. Just then their engine miraculously restarted and they returned to their base.<46> The bombing of civilians was initiated by the Germans with the dropping of several bombs from Zeppelin airships on Antwerp on August 25, 1914. Six civilians in a single house were killed as they slept; twelve died in all. On January 19, 1915, the Germans conducted a Zeppelin bombing raid on the English coast, killing two at Yarmouth and four at King's Lynn.<47> London was targeted by a single Zeppelin on May 31; 90 incendiary and 30 grenades killed 7 and injured 35. Zeppelin bombing raids continued until nearly the end of the war, but became much riskier for the hydrogen-filled craft after the British introduced incendiary bullets in September 1916. In all 16 German lighter-than-air ships were shot down.<48>

Heavy bombing by the Germans was not limited to Zeppelins. Presaging the London Blitz some two dozen years later, 23 three-seater, two-engine *Gotha* bombers flew from Belgium to bomb London on 25 May 1917. Due to cloudy weather, they did not make it to London and only two dropped their bombs near Dover, but the toll was larger than any previous Zeppelin raid: 95 killed and 193 injured.<49> In October of the same year a fleet of 22 *Gothas* carried out the first fire-bombing raid of London. Fortunately, many of the incendiary bombs were duds; even so, 10 citizens were killed. Nor was Paris left out. In late January 1918 a formation of 31 Gothas dropped 267 bombs over the City of Light resulting in 259 casualties.<50> Representing a quantum jump in bomber design, the German *Riesenflugzeug* (Giant aircraft) had a wingspan more than twice that of the *Gotha*, four engines (on the production Type VI), and a crew of up to nine. It could carry two tons of bombs more than 300 miles and made its appearance in 1917.<51> In the first three months of 1918 the Giants made a number of bombing raids on Britain, but continued malfunctions of incendiary bombs limited the damage they caused. Armed with 3–6 machine guns, the big planes proved formidable to bring down by attacking aircraft. Machine guns, in fact, brought terror to the airman and the soldier on the ground alike.

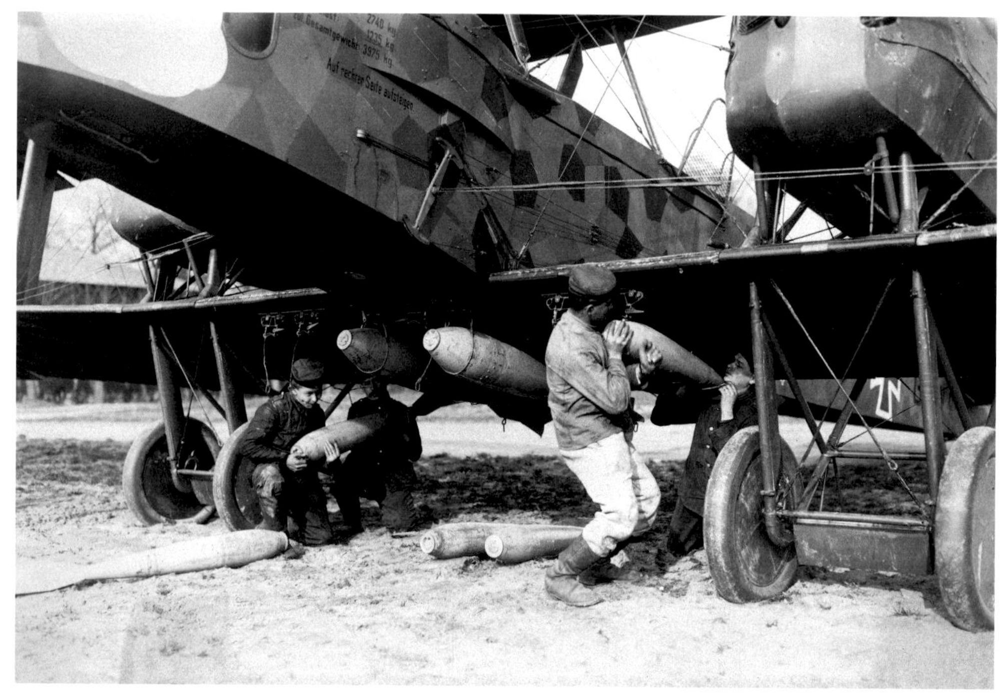

Loading bombs onto a German *Gotha* G V bomber, Western Front 1918
By Permission bpk Bildagentur / Art Resource, NY

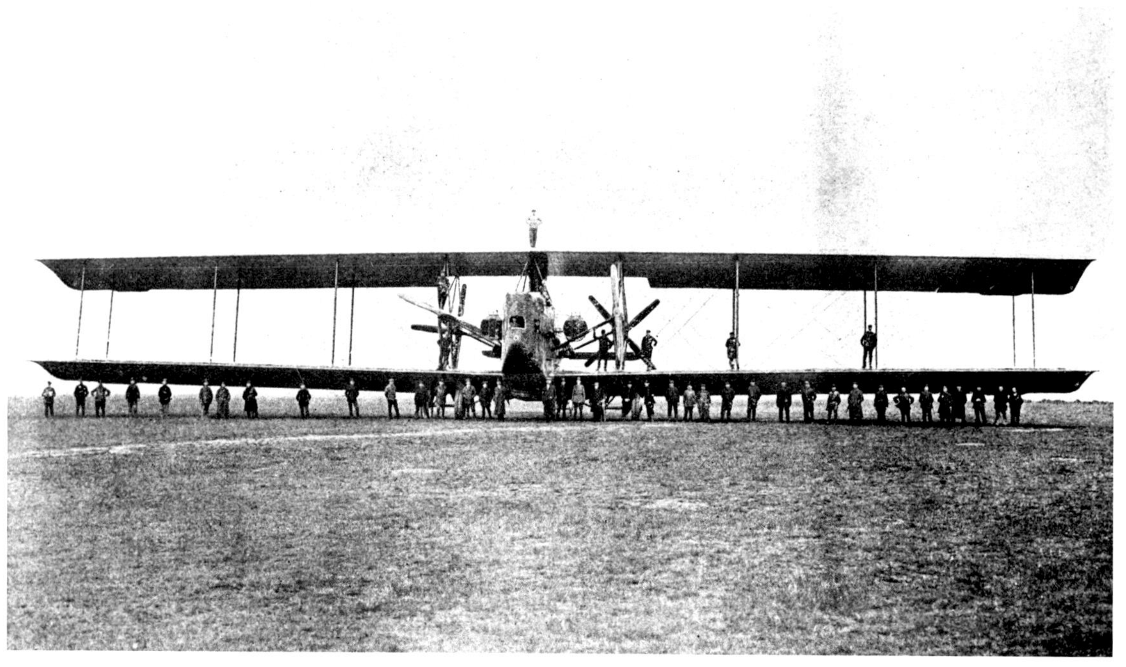

German Riesenflugzeug VIII (Giant) heavy bomber
By Permission of the British Imperial War Museum

MACHINE GUNS

One of the most iconic symbols of the First World War is that of the machine gun, the popularly perceived embodiment of war in the machine age. In actuality, artillery caused two-thirds of combat wounds and was feared by troops the most because of the horrible wounds its blast inflicted and its long reach even behind the lines. Notwithstanding this fact, stories of machine guns wiping out entire units still form a powerful image in our modern collective consciousness. But even if it were not the predominant cause of battlefield death and injury, it was a major cause of the mass casualties that so characterized the war. By the time the war started, Hiram Maxim's invention had stimulated refinements and competitors alike.

In Britain, Germany, and Russia the recoil-operated, water-cooled, belt-fed Maxim gun was adopted with relatively minor changes—as the Vickers in Britain, the MG08 in Germany, and the M1905 in Russia. At the outset of the war, France adopted the M1914 Hotchkiss, a gas-operated, air-cooled design. Austria-Hungary produced the *Schwartzlose*, a home-grown, "blow-back" design in which the rearward forces on the cartridge case act to reload the chamber without fully locking the breach. In the United States the Maxim was finally adopted in 1904, only to be superseded in 1909 by the Benet-Mercie Machine Rifle, a lightweight version of the Hotchkiss and a totally inadequate substitute for the Maxim. This poor choice reflected uncertainty in the tactical role that would be assigned to the machine gun.<52> At war's beginning each of the combatant nations had several thousands of these machine guns. Other guns would emerge during the war: the British Lewis gun, first used by the Belgians and dubbed the "Belgian Rattlesnake" by the Germans for its distinctive chatter,<53> the French Chauchat in 1915,<54> the American Browning Automatic Rifle (BAR) in late 1918.<55> Also in 1918 the German MG18, one of the first submachine guns firing 9mm pistol ammunition would make its appearance.<56>

The opening artillery barrage of the British on 1 July 1916 at the Somme was awesome and terrifying, but ***it was the machine gun which dominated the recollections of those who survived that day, and which has become emblematic of both this battle and, by extension, of the Great War as a whole***.<57> Infantry soldiers, leaving the relative safety of the trenches, entered No-Man's Land weighed down by 66 pounds of equipment, making it a struggle even to get up and over the parapet. They then faced blistering machine-gun fire from the Germans who had survived the bombardment by waiting in reinforced bunkers underground and who emerged to man their MG08s after the barrage ceased. A British observer recalled seeing his comrades ***mown down like meadow grass. I felt sick at the sight of this carnage and remember weeping***.<58> The MG08s had engaged at over 2,000 yards with devastating effect. Some German guns had not had time to be mounted properly and these waited until the British were within 30-50 yards before opening their deadly fire. By the end of the day 19,240 British soldiers lay dead with another 38,230 injured.<59> Machine guns thus established themselves as the premier defensive arm, but their effectiveness also recommended them for use by armored vehicles in the attack.

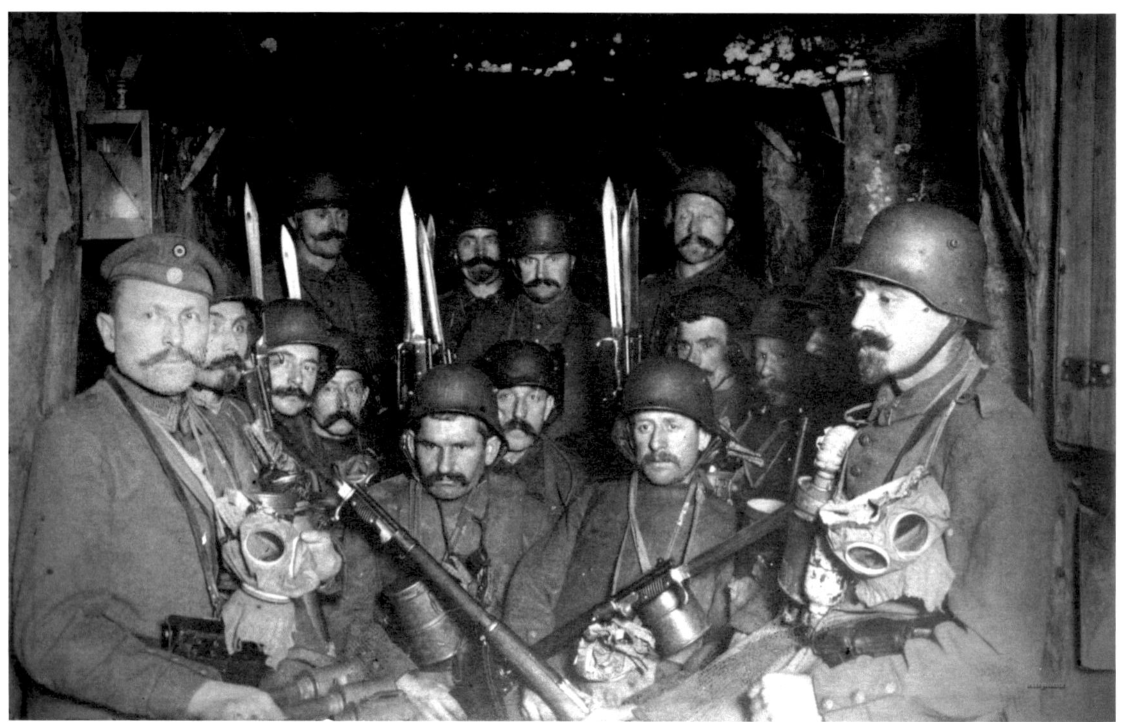

German soldiers in a dugout waiting for the artillery barrage to end before rushing out to engage the attackers.
Original postcard scan courtesy of the Brett Butterworth Collection

TANKS

Armored cars fitted with machine guns were developed by Britain, France, and Austria well before the war and were used early on the Western Front until roads became blocked by trenches. Pershing had even taken American-made armored cars on his hunt for Pancho Villa. But it was the brutal advent of trench warfare that stimulated the invention of "landships," a term used independently by both British and French in their separate development programs. To obfuscate their real purpose, the Brits called them "tanks" and the French called them "tractors." In February 1916, both countries, without knowledge of each other's intentions (despite being allies), committed to producing tanks within days of each other.<60>

Affectionately known as "Mother," the prototype of the first British production-model tank was an ungainly monster. The Mark I was a steel box 8 feet high, nearly 14 feet wide, and more than 26 feet long. It wore armor plate from 6 to 12 mm thick and weighed 28 tons fully armed.<61> Ironically, "Mother" was dubbed a male with two six-pounder guns (57-mm) and four Hotchkiss air-cooled machine guns for armament. The "female" Mark I carried five Vickers water-cooled machine guns and one Hotchkiss.<62> Rather than a turret, the guns protruded from side sponsons, much like those on ships, enabling forward firing. The hurried development left much to be desired ergonomically. Two gearsmen were positioned on either side of the differential to throw one track into neutral and brake in order to make turns. The tank commander, an officer up front with the driver who had control of the throttle, shouted his commands to the gearsmen. Four gunners made a total crew of eight. The gasoline engine was in the center of the hull creating noise and fumes. With poor ventilation and the heat of the engine, the tank interior got unbearably hot, and, in the absence of spring suspension, the bumps and jolts from traveling over the broken ground could cause real injury to a crew surrounded by steel.

Communication between tanks and infantry and among the tanks themselves was accomplished using red, green, and white disks hoisted on a mast above the hull. Codes were established such that different combinations of these disks conveyed the desired information: tank disabled, proceeding to objective, light enemy troops ahead, etc. Long range communication was by means of carrier pigeons. This means proved very satisfactory as long as one sent the birds on their way before sunset; otherwise, they might stop and roost for the night before continuing their journey.<63>

The Mark I was first used to support an infantry assault at the battle of Flers-Courcelette, about fifteen miles northeast of Amiens, on September 15, 1916. Forty-nine tanks were brought up but only 32 actually reached the battle front. Nine tanks became disabled and five got trapped in ditches. The battle did not achieve a lot, a one-mile advance, but given the limited training and no previous tactical experience, the results were encouraging and prompted an order for 1,000 more. Minor improvements were embodied in the Mark II and III, but only about a hundred were made. In March 1917 the Mark IV, with Vickers and Hotchkiss machine guns replaced by Lewis machine guns, made its appearance. A little over a thousand were made.<64> The Mark IV was very similar to the Mark I generally but had better ventilation, escape hatches on sides and top, and a merciful silencer on the engine. The Brits also managed to field a lightweight, “fast” tank, which at 14 tons could travel at 8 miles per hour compared with the Mark IV at just under 4. Dubbed the Whippet, it had a crew of three and was armed with four Hotchkiss machine guns projecting from all four sides of a cupola-like structure. Unlike the Mark IV, this tank’s turning maneuvers were under the direct control of the driver by having a separate engine for each track, though it took considerable finesse to control.<65>

Rivalries within the French military establishment led to two tanks being developed simultaneously, the Saint Chamond and the Schneider. The Schneider was the smaller of the two at about 15 tons. It was armed with a 75-mm howitzer and two Hotchkiss machine guns, one on each side.<66> The Saint Chamond weighed 22 tons and employed a full-length 75-mm gun and four Hotchkiss machine guns on front, rear, and both sides.<67> A third tank was developed by Renault—a two-man, lightweight tank at 7 tons. The Renault FT-17 was armed with a single Hotchkiss 8-mm machine gun in a rotating turret. Its armor, however, was nearly an inch thick (22 mm) and this proved to be adequate protection even against the armor-piercing ammunition the Germans developed to defeat the British tank.<68>

Tanks began to show their potential for ending the trench warfare stalemate in the first significant action near Cambrai on November 20, 1917. Over 400 British tanks were massed in secret and sent into action without prior artillery bombardment. Unfortunately, though advancing some seven miles through the Hindenburg Line, failure to exploit the breakthrough and German counter attacks erased the gains that had been made.<69> Nonetheless, the beginnings of a viable strategy were taking shape. It was not until 8 August 1918 at Amiens that a second mass attack of tanks could be attempted. Here the British threw the whole of their Tank Corps into battle. The new Mark V had been introduced with more power and better maneuverability due to a redesigned transmission that dispensed with the gearsmen. A force of

324 heavy tanks, made up mostly of Mark Vs and 96 Whippets, attacked the German lines. Though by the next day only 145 tanks remained in the fight, the battle was decisive enough for General Ludendorf to describe it as the ***Black Day of the German Army.***<70> The battle marked the beginning of the gradual German withdrawal that would culminate with the Armistice three months later. Although the effectiveness of tanks in ending the war has been questioned, since their entry also more or less coincided with the entry of the American forces, both Britain and France placed orders towards the end of the war for many thousands of additional tanks, an action that speaks for itself.

NAVAL POWER

If artillery, aircraft, machine guns, and tanks came to symbolize the First World War, what happened to the huge naval build-up that so threatened the balance of power in the years leading up to the war? Ironically, the big ships became such important symbols of power that, with the actual onset of war, British and German commanders alike were leery of risking their fleets. Despite the dreadnoughts' big guns and heavy armor, they were vulnerable to being sunk by torpedoes launched by submarines, fast torpedo boats, or destroyers. Admiral Sir John Jellicoe, who was Commander-in-Chief of the British Grand Fleet, bore enormous responsibility for its protection and use. Of him Winston Churchill said that he ***was the only man on either side who could lose the war in an afternoon.***<71> Notwithstanding these enormous risks, there did occur an epic battle that effectively ended further engagements for the rest of the war—the Battle of Jutland.

As Commander-in-Chief of the German Imperial Navy's High Seas Fleet, Admiral Reinhard Scheer hatched a plan to shift the odds in favor of his smaller force. German scouting squadrons had been raiding the British coast and had been chased by Royal Navy battle cruisers with some success. Scheer planned to bait the Grand Fleet with more raiding, but this time, lure them into an ambush by the High Seas Fleet bolstered by submarines. Unknown to Scheer, British Naval Intelligence had the keys to German naval codes in their possession, and Jellicoe knew in advance of Scheer's plan. On the night of 30 May 2016, Jellicoe steamed from Scapa Flow two hours ahead of Scheer to beat him to the ambush zone between Norway and the Jutland Peninsula of Denmark. The next morning, however, British Naval Intelligence wrongly placed the German fleet as still in harbor at Wilhelmshaven on Germany's North Sea coast, so Jellicoe slowed his fleet to conserve fuel. That afternoon the German raiding squadron was sighted and British battle cruisers gave chase. The German squadron turned south to draw the British into the guns of the waiting High Seas Fleet. In the ensuing battle confusion with targets being engaged at distances over 11 miles, three British battle cruisers were sunk along with 11 other ships; 6,784 British casualties were sustained. One German battleship and one battle cruiser were sunk along with 9 other ships; there were 3,058 German casualties. The scale of the battle was unprecedented: 100,000 men in 250 ships engaged for over three days.<72> The relatively fewer losses of the Germans prompted them to declare a great victory. However, while they did have a legitimate claim to victory in the Battle of Jutland, they had lost the war at sea as the High Seas Fleet remained subdued for the remainder of the war. Jellico had not lost the war that "afternoon."

In one important action the dominant British fleet did affect the outcome of the war—its naval blockade of the North Sea. One can view the stalemate on the Western front as a heroic holding action by the Allies while Germany was slowly starved of the food and resources it needed to continue the war. Though Germany claimed to be self-sufficient in agriculture, the reality was that it imported 25 percent of its food.<73> With priority being given to the troops at the front and munitions workers, by 1918 much of Germany's civilian population was starving. Three quarters of a million civilian deaths were ultimately attributed to the blockade, most among the old and weak. The death rate was comparable to the rate among British combatants.<74> Food was not the only critical item in short supply. Important metals like lead, copper, brass and bronze were lacking. By the end of the war lead pipes under the streets were being dug up for recasting as bullets. Lightning rods and roofs were reclaimed for their copper. Churches handed over organ pipes for their brass and bells for their bronze.<75> Effective as it eventually was, maintaining the blockade itself was an enormous challenge for the Royal Navy. The possibility of torpedoes and mines made a close-in blockade of German ports infeasible, but a distant blockade of the whole North Sea would be very difficult to manage. Jellicoe had no real choice but the latter with its logistical nightmare. Constant steaming in the North Sea meant the constant need to refuel and resupply in those vast waters. Destroyers needed to visit their coaling stations at Scapa Flow, a huge natural harbor in the Orkney Islands off the northern tip of Scotland, every three or four days and this meant that at any given time roughly a third of the ships were off patrol.<76> The bad weather of the North Sea and the ever present possibility of submarines made this a grueling duty for the seamen.

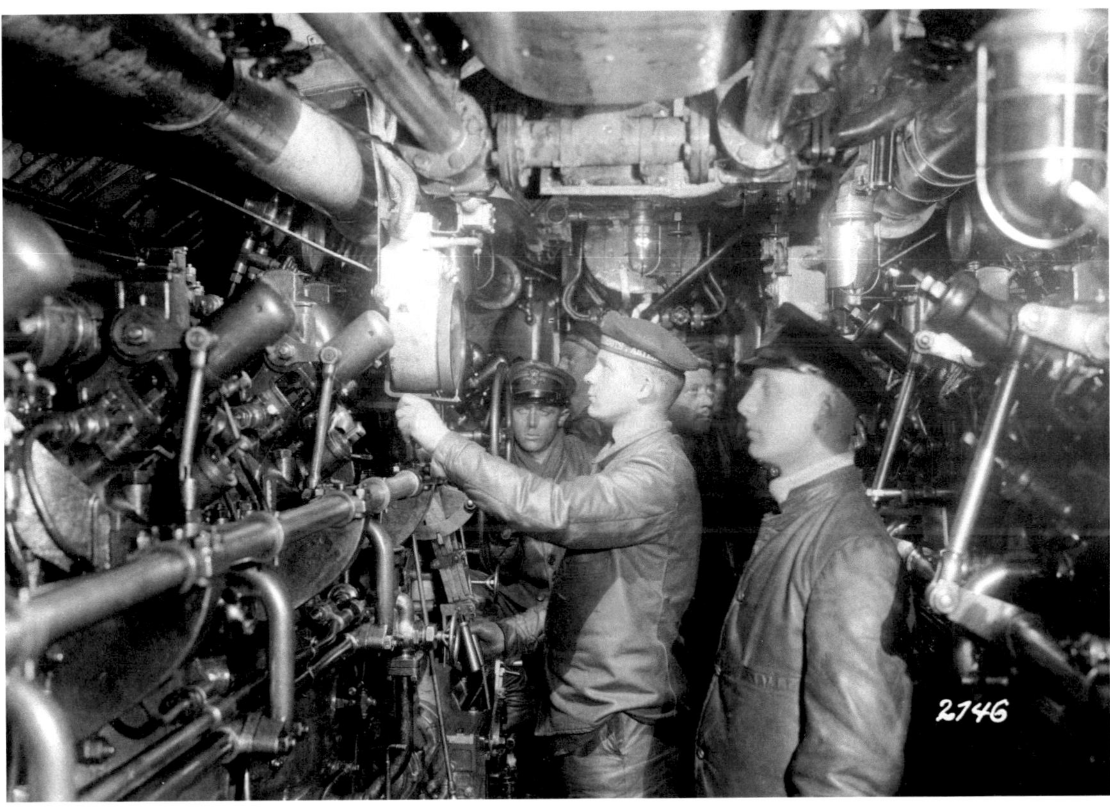

Engine room of a WWI German submarine.
Courtesy National Archives, NA Identifier 17390434

Although the Imperial Fleet saw little further action, the German Navy was not entirely idle. Submarines continued to bedevil the Grand Fleet and commercial shipping. Admiral Tirpitz had not thought highly of submarines, and Germany was therefore the last of the major powers to get them. The upside of this delay was that they possessed the latest technology. The Krupp-built U-1 was commissioned in 1906. By the beginning of the war the Imperial Navy had 24 submarines.<77> By the end of the war they had 176.<78> They provided a continued threat even when the German High Seas Fleet had been largely neutralized. Though not as effective as the British blockade of the North Sea, submarines were responsible for the loss of a quarter million tons of shipping to Britain per month. Overall the lessons of the U-boat experience of the First World War would be interpreted by the future Nazi government as worthy of amplified support in the Second World War, and there they would play a major role.

At war's end the *Kaiser's* beautiful and expensive fleet, the pride of the Second Reich, had counted for little. By the terms of the Armistice, in the closing months of 1918 the Imperial High Seas Fleet was escorted to Scapa Flow where it would be interned during the peace deliberations in Paris. The USS Texas, today the world's only surviving WWI dreadnought-class battleship (See the Gallery Chapter.), participated in this escort duty. En route the Allied escort ships maintained preloaded ammunition elevators and loaded gun tubes as a precaution, but in the end the fight had gone out of the German sailors. They were tired, demoralized, and hungry. At Scapa Flow skeleton crews manned the German ships until on the morning of 21 June 1919 when, at a prearranged signal, the crews opened the seacocks, broke the valves, and abandoned their ships to the watery depths. Word had been received that the German fleet was to be divided among the Allies. Rather than endure such an indignity, the fleet was scuttled. The act did not go uncontested by the British, however. Boarding the battleship Markgraf, British sailors shot the captain in the head as he was ordering the ship scuttled. Nine Imperial Navy personnel were killed and sixteen wounded on that day, but the outcome was unchanged.<79>

THE LEGACY OF THE FIRST WORLD WAR

The First World War became known as The Great War—until a greater one came after—but it set the pattern of industrialized warfare experienced throughout the twentieth century. The German nation so young and vigorous and eager to rival Britain on the world stage had embraced the combined synergies of the new technologies but had harnessed their potential with little regard for human values. They had cruelly trampled their pledge of neutrality with Belgium, wantonly killing 5,521 civilians and destroying 25,000 buildings including cultural treasures such as the 14th-century Louvain Library.<80,81> They violated the Hague Convention of 1907 by placing more than 25,000 mines in the shipping lanes of the North Sea.<82> They initiated the bombing of civilians by airship, by airplane, and by long-range artillery. They disregarded rules of the high seas by sinking ships of neutral nations without warning. They introduced the use of poison gas to the battlefield. They impressed civilians as slave labor under brutal conditions both behind the front lines (62,000 Belgians and French) and in deportations to Germany to the armament works. Between October 1916 and February 1917 some 60,000 Belgians were sent by cattle car to Germany to work.<83> The Germans also bombed Allied hospitals.<84,85> All these actions transgressed the bounds of what had become acceptable conduct in war. In their defense were only the words of German Chancellor Bethmann-Hollweg: ***Necessity knows no law.***<86> It would not be long before preparations were begun for another war with these precedents guiding the way. At war's end not just the seeds of a future conflict had been sown—a new paradigm had been cast. Reliance on technology, the expectation of rolling innovations driven by competition, and attitudes of ruthless total war where civilians were considered as combatants, all had become the sad legacies of the industrial revolution that had both improved lives and enabled destruction and death on unprecedented scales.

When Europe awoke from the nightmare of the Great War, the age of monarchy and imperialism

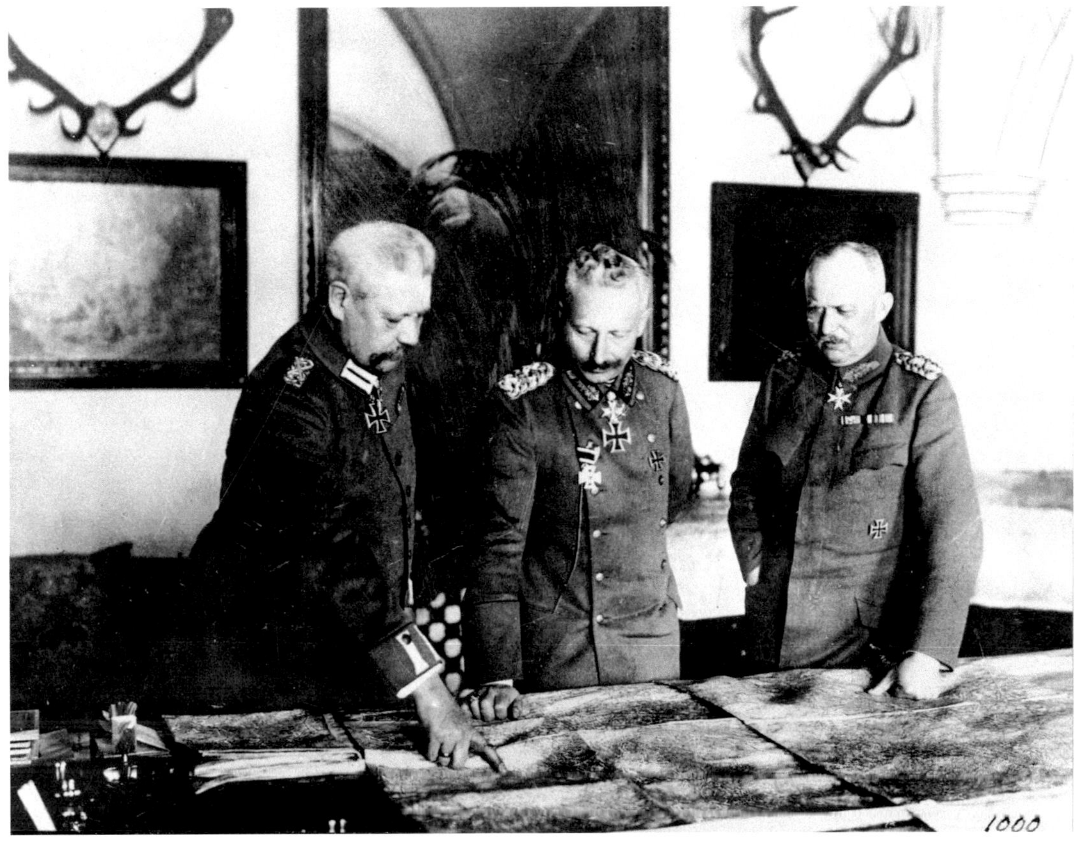

Architects of the new total-war paradigm: Field Marshal Paul von Hindenburg, Kaiser Wilhelm II, and General Erich von Ludendorff in January 1917.
Courtesy National Archives, NA Identifier 17390230

was over. The German Empire, the Austro-Hungarian Empire, the Russian Empire, and the Ottoman Empire had all fallen. Over the course of the century the far-flung colonies would all assert their independence, mostly by force of arms, spawning countless sectarian conflicts in the process with some still playing out today. But the appalling death toll did not put an end to the suffering. ***The wounded men of all nations were to be a legacy of war which ended only with their deaths, or with the deaths of those who had lived with them and guarded their broken bodies or minds, or both.***<87> Some eight million men were left physically invalided in the wake of the war with another estimated million men psychologically crippled.<88> There were other legacies of consequence. The ruined landscape of northern France and Belgium today still harbors an estimated 300 million unexploded shells filled with high explosives, shrapnel, and various poison gases.<89> In France two million acres of deadly ground is still officially sequestered. Germany experienced a debilitating hyperinflation following the war in the early 1920s. At its worst a dollar traded for more than 4 trillion German marks.<90> Workers would be paid several times a day as the value of the currency changed by the minute. This crisis was followed by the worldwide depression of the 1930s. Through all this chaos, a corporal from the German trenches of France, Adolf Hitler, found fertile ground for his political movement to restore Germany's prestige. From the moment of his appointment as Chancellor in January 1933, preparations by Germany for the next war began—in violation of the Versailles Treaty.

Having suffered brutal invasions twice in 44 years by the German nation, France began considering new defensive fortifications as early as 1920. Under Minister of War André Maginot the new defensive line was begun in 1929 and soon was known by the minister's name. Nominally stretching from the North Sea to the Mediterranean, the heaviest fortifications were in northeastern France protecting the border with Germany and in southeastern France protecting the border with Italy, the former being called the Maginot Line Proper and the latter the Little Maginot Line. There were a few fortresses along the Belgian border, known as the Maginot Extension to be bolstered by mobile defenses, but the French did not want to offend their Belgian ally or make them feel that they had been cut off by too heavy a buildup. Cost was also an issue as these constructions occurred during the depression. The Belgians too were building new forts and rearming the old to protect their border with Germany. However, when in 1936 the new Belgian King announced his country's neutrality and refused the presence of French troops until war had broken out, the French were hung out to dry. It was then too late to construct adequate fortifications. Neither the French nor the Belgians had believed that the Ardennes Forest was a possible route for a modern mechanized army to pass and so had not fortified that corridor.<91> As the Ardennes was to be precisely the route of the main attacking German forces in 1940, this assessment turned out to have been a colossal misjudgment with disastrous consequences.

CHAPTER 5. THE SECOND WORLD WAR

I was sleeping so soundly I did not hear the alarm at dawn, but a maid shook me violently by the shoulder and said: 'Wake up! The Germans are coming again!

—American playwright Clare Booth in Brussels on the morning of 10 May 1940<1>

MOBILIZATION

The relative economic strength of the warring powers in WWII, as measured by their GDP and crude steel production between the wars and during the war, is shown in the accompanying charts. They reflect a number of significant historical currents. The contraction of the European economies in the post-WWI period is evident. By 1933 the American economy had suffered a similar fate as a result of the Great Depression. As of 1940 all the economies had begun to recover and Japan had overtaken France. From 1933–1943 the steel industry revived due to renewed economic activity and rearmament attending the onset of the new war. The performance of the US economy and steel industry from 1933 to 1940 relative to the other major powers is especially noteworthy. How could the Axis Powers, especially tiny Japan, have possibly expected to prevail against US economic might? The answer is—only if the US failed to mobilize its economy for total war in a timely way, as had been the case in WWI.

The story of America's mobilization for WWII is extraordinary by any standard and is unforgettably told by Arthur Herman in his book *Freedom's Forge*. This heroic accomplishment,

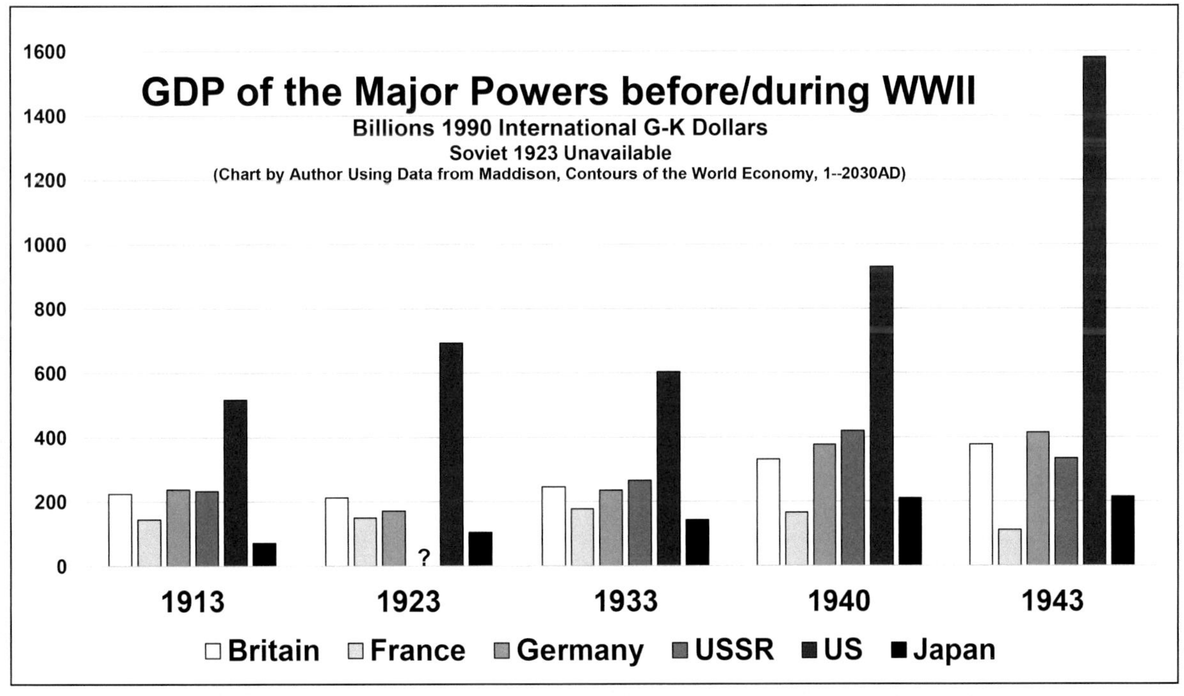

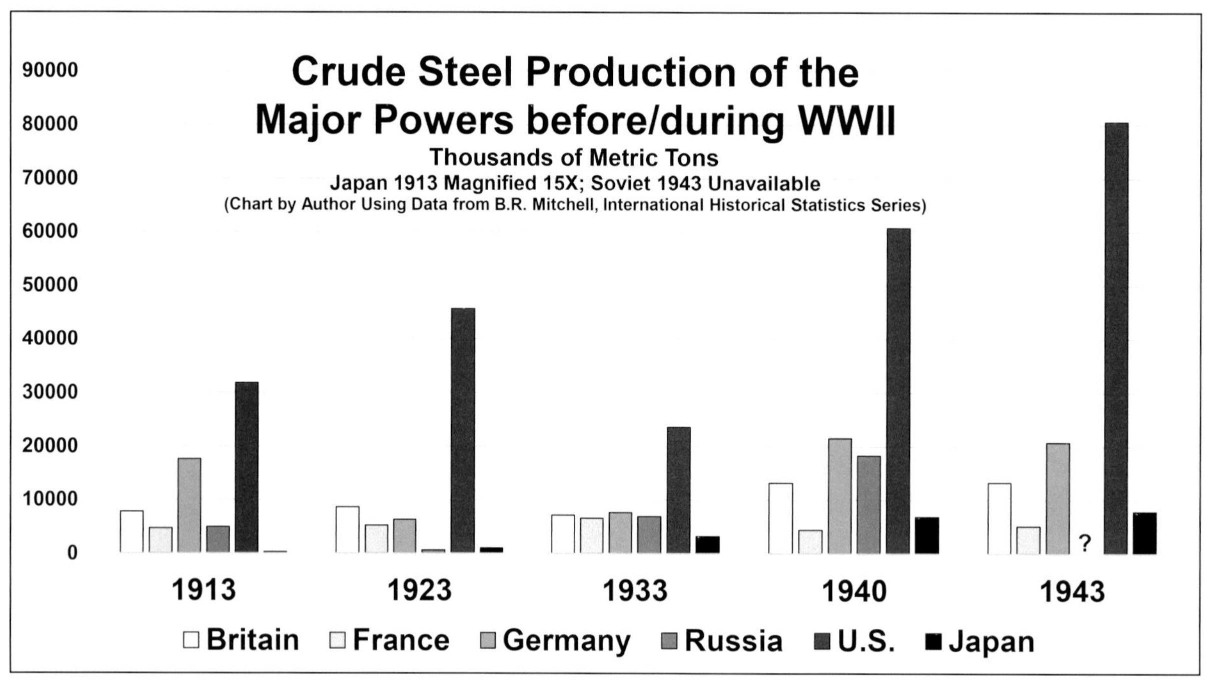

in Herman's words, was achieved by

> ***... the American businessmen, engineers, production managers, and workers both male and female who built the most awesome military machine in history: the arsenal of democracy that armed the Allies and defeated the Axis. Americans produced two-thirds of all Allied military equipment used in World War II. That included 86,000 tanks; 2.5 million trucks and a half million jeeps; 286,000 warplanes; 8,800 naval vessels; 5,600 merchant ships; 434 million tons of steel; 2.6 million machine guns; and 41 billion rounds of ammunition—not to mention the greatest superbomber of the war, the B-29; and the atomic bomb.***<2>

In the dark days of 1940 when Hitler's forces had crashed through the "impenetrable" Ardennes Forest, driven the British army off the continent at Dunkirk, and captured Paris in only nine days, British Prime Minister Winston Churchill telegrammed President Franklin Roosevelt and warned him that Germany might well capture the Royal Navy, the largest in the world. As a former Secretary of the Navy under Woodrow Wilson, Roosevelt knew the dangers that posed to the US. Though anti-war sentiment in the US was strong, he had to act. During WWI, Roosevelt had held a ring-side seat to the mobilization debacle and remembered the crucial role played by Bernard Baruch as Chairman of the War Industries Board. Baruch was summoned and offered his old job. At age 69, Baruch averred that the job was beyond his abilities and that he knew of but one man who could do it—William S. Knudsen.

From a penniless Dutch immigrant in 1900, Bill Knudsen had worked his way up to being president of General Motors, and he had a positive genius for the workings of mass production. As Henry Ford had found to his chagrin, the only limit to mass production was demand: in his case, how many black Model-T's he could continue to sell year after year. Knudsen had worked for Ford but ran afoul of Ford's prickly personality. Having moved to Chevrolet, he introduced the ideas of improving models year by year and offering them in different colors; Chevrolet had zoomed past Ford because of it. Now faced with a war which would create virtually unlimited demand, Knudsen was confident that he knew what to do. After a phone call from the President in July 1940 asking for his help, he resigned his position with GM before even meeting with Roosevelt or knowing what would be expected of him. Such was his feeling of obligation to his adopted country that had given him such unlimited opportunity. As it turned out, Roosevelt wanted Knudsen to serve on the freshly formed National Defense Advisory Commission that included Edward Stettinius, Jr., president of U.S. Steel (and son of the WWI figure at J.P. Morgan's Export Department). The commission had no powers and not even a chairman. From this confused start, Knudsen would work his organizational magic and get the country on the road to full mobilization. However, just as Edward Stettinius, Sr. had found himself toxic to Wilson's Progressives because of his ties to big business, so would Knudsen find disfavor among Roosevelt's New Dealers and especially First Lady Eleanor.<3> One month after Pearl Harbor, Knudsen was effectively fired by virtue of the appointment of a superseding board by the President. Sadly, he learned of it from a news ticker and not from the President himself. Fortunately, Secretary of War Henry Stimson, recognizing Knudsen's unique skills in this time of crisis, lobbied Roosevelt to make Knudsen a Lieutenant General (three stars) in the Army. Roosevelt assented, and immigrant William Knudsen became the only civilian in the history of the country ever to be elevated to this rank. His factory-floor problem solving skills proved critical time and time again in troubleshooting production bottlenecks throughout the war.

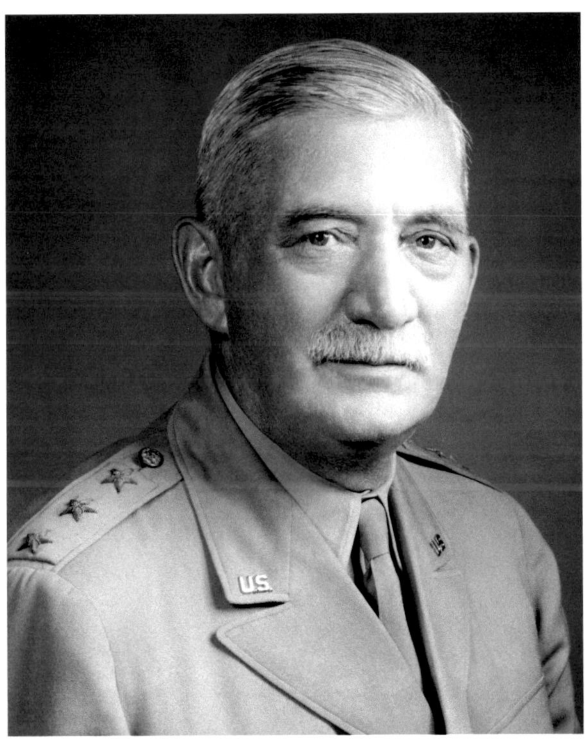

US Lieutenant General William S. Knudsen (1879–1948)
Courtesy National Archives

Probably the most serious shortcoming of the US WWI mobilization had been a lack of cargo ships to transport the enormous burdens of war across the ocean. The tonnage required this time would be far greater as the US would be supplying its allies instead of the other way around. Enter the irrepressible maverick, Henry J. Kaiser. Having gotten his start with road building in California, Kaiser had made his name by assembling a consortium of six heavy construction companies and building Hoover Dam across the Colorado River, completed in 1935. Next came the Bonneville and Grand Coulee Dams across the Columbia River in 1938 and 1942, respectively. The Grand Coulee was the largest concrete structure in the world. The dams would provide critical electricity needs of the war, including the plutonium reactors at Hanford, Washington, that produced the nuclear explosive for the Nagasaki atomic bomb. But

with no shipbuilding experience Kaiser would find his entry into this industry a tough sell. His consortium had, however, built shipyards, and in their desperation the British turned to Kaiser to build cargo ships. Existing American shipyards were overwhelmed with orders for US Navy ships and could not manage more business. Meanwhile German submarines were exacting a terrible toll on British ships. In December of 1940 the British and Kaiser signed the largest ever US peacetime contract.<4> The shipyard they built in Richmond, California, would also be the first to build the famous Liberty Ship, the first mass-produced ship. These ships were destined to ferry US equipment and supplies all over the world during the war. Kaiser went on to build other shipyards, and his crews competed to turn out the ships in the shortest time. Their record was launching the Liberty Ship Robert E. Peary in little more than four and a half days, though this blistering pace could not be sustained.<5> By the end of the war 2,708 Liberty Ships had been built primarily in fourteen shipyards.<6>

There was another war machine with a craftsman-like construction tradition that would yield to mass production—the warplane. In the opening weeks of the war, Germany had insulated its homeland from land attack by occupying northern France, and it would be years before the

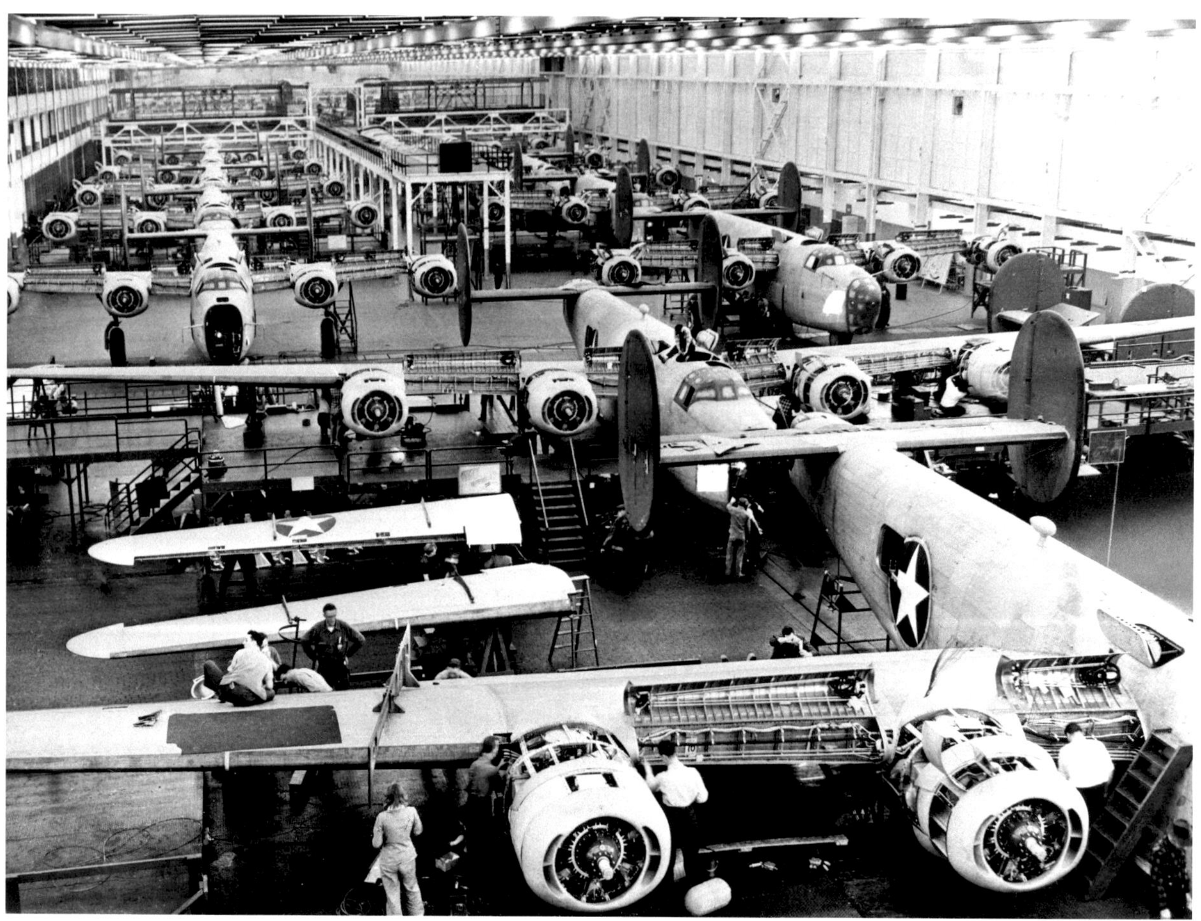

B-24s on the assembly line at Willow Run.
Courtesy National Archives

Allies would be ready to mount a ground invasion. This meant that, in the interim, heavy bombers would be needed and lots of them. When Charlie Sorensen, a self-made Danish immigrant like Knudsen and Henry Ford's production man, proposed making the four-engine B-24 Liberator bomber using automotive-style mass production, even Knudsen was skeptical. Knudsen knew that an automobile had some 15,000 parts, but had learned that the bomber had thirty times as many. Sorensen was known as Cast-Iron Charlie, a nickname that suited his personality as well as his metal working skills, and he insisted that he would build the entire plane, or nothing at all. He may well have later regretted his inflexible brashness. It was then October 1940; by the following February Ford had a contract to build the plane.<7> Ford built a huge new plant west of Detroit at Willow Run, ultimately occupying nearly 5 million square feet.<8> However, the design of the plane turned out to be a moving target—575 significant changes in the first year—requiring new dies and even new production jigs.<9> This was not the way Cast-Iron Charlie had built cars and decidedly not what he had had in mind when he had raised his hand to volunteer, but combat-experience feedback necessitated the changes, and a way would have to be found to accommodate them. It was now July of 1942 and the public was beginning to wonder where the bombers were. Some newspapers were rebranding the plant ***Will It Run?***<10> Newly minted Lieutenant General Bill Knudsen figured out how to do it. Ford could freeze the design, build the plane, and the improvements would be effected in separate plants as field modifications.<11> Ten modification centers were created with commercial airlines supplying their maintenance personnel to make the changes. Sorensen's bold gamble was vindicated, and the first Ford-built Liberator became a reality in September 1942.<12> By the beginning of 1944 five hundred B-24s a month were rolling off the assembly line at Willow Run.<13> Stunningly, mass production had decreased the labor required from 200,000 to just 18,000 man-hours, a more than 90 percent reduction.

America's economic mobilization in the Second World War seems to defy rational explanation. As early as January 1942 some 25,000 contractors were hard at work supplying the nation's need for war materiel. But they in turn were dependent on 120,000 subcontractors.<14> General Motors alone had hired 18,000 subcontractors.<15> Boeing relied on 1400 subcontractors to build its B-29 bomber.<16> Henry Kaiser used about 600 subcontractors for his shipyard. Most of the men building his ships had never even been on a ship before the war. More than 40 percent of world armaments in 1944 was generated by the United States—half again as much as the combined output of either its allies or its foes.<17> Of course, one can see the potential for imbalance reflected in relative GDP (above) even before the war, but the same was true before the First World War. The point is that effective mobilization is not a foregone conclusion. Historian Arthur Herman puts it down to the efforts of production-genius Bill Knudsen and his crew, expressing it this way:

> ***They had created, in effect, an almost self-perpetuating mechanism that fed upon its own individual dynamic elements. Theorists of the science of complexity would call it emergence. Economists have another term: 'spontaneous order.' It was the most powerful and flexible system of wartime production ever devised, because in the end no one devised it. It grew out of the underlying productivity of the American economy, dampened by a decade of depression but ready to spring to life. Out of what seemed like chaos and disorder to Washington would come an explosion of innovation, adaptation, and creativity—not to mention hard work—across the country.***<18>

TANKS REDUX

With the further development of tanks and aircraft after the Great War, the terrors of static trench warfare faded into history—only to be replaced by other terrors. Germany lost WWI in part because it did not realize the potential for tanks to break the stalemate. At the end of that war Germany had only 20 tanks of its own making, the A7V, and 30 captured British Mark IVs. Contrast this with the thousand British Mark IVs, hundreds of Schneiders, and several thousand Renault FT light tanks built by the Allies during the war.<19> Because of this enormous imbalance, young German officers of that war, destined to be leaders in the next, knew viscerally the terror and panic that these big machines could instill in their troops. One of these officers, Heinz Guderian, was a lieutenant in charge of a wireless signal unit early in the First World War and a logistics, intelligence, and General Staff officer later in the war.<20> In the postwar period Guderian, an embodiment of the Prussian military ideal—passionate but disciplined, visionary but practical, developed both theory and training regimens for fast-moving, mechanized forces.

Though the Versailles Treaty had expressly forbidden Germany to possess tanks, in 1927 development proceeded in secret<21> following the relaxation of oversight by the Inter-Allied Military Control Commission.<22> In 1929 a joint Soviet-German proving ground and training center was set up in Russia where they could be tested in secret as well.<23> By 1934 the first production model *Panzer Kampfwagen* (*Pzkpfw*) I, a 6-ton light tank armed only with twin machine guns, began field tests in the Soviet Union. Hitler officially renounced the Versailles Treaty in 1935, and proving grounds were then established in Germany. In 1936 the 7.9-ton *Pzkpfw* II with 20-mm automatic cannon entered production.<24> Next came the *Pzkpfw* III and the Krupp-built *Pzkpfw* IV in 1937. The *Pzkpfw* III was armed with a high-velocity 37-mm gun and the IV with a low-velocity 75-mm gun.

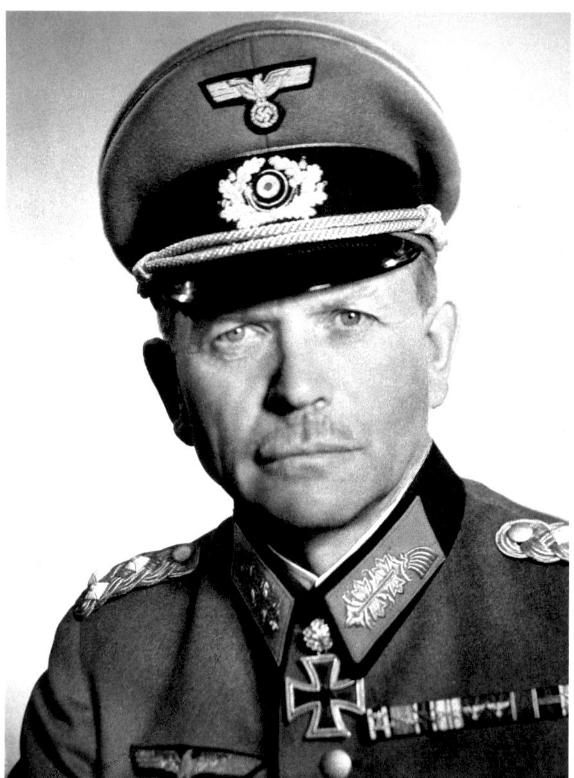

German General Heinz Guderian (1888–1954) in 1944
By Permission Bundesarchiv, Koblenz, Germany

When Czechoslovakia was annexed in 1939, the Skoda Works came into German hands, providing them with nearly three hundred Model LT-35 tanks with 37-mm high-velocity guns. By the outbreak of war in September 1939, Germany had 1,493 Is, 1,223 IIs, 98 IIIs, and 211 IVs—about 3300 tanks in all.<25> They were put to the test in Hitler's invasion of Poland on 1 September 1939, the reopening of world war as it turned out. The combined might of the tanks, motorized infantry, and close air support demolished the Polish forces. Though they fought bravely with what they had, their 225 largely outmoded tanks, 360 aircraft,<26> tenacious artillery, and skilled horse cavalry were no match for the armored Blitzkrieg tactics—Guderian's tactics. Comprised of many who had experienced the humiliation of Versailles and some (like Guderian) who had lost their ancestral Prussian homes on land that the treaty had given to Poland, the new German Army was out for blood and showed no mercy.<27>

When the decision was made to attack France by an armored thrust through Belgium's dense Ardennes Forest, Guderian was selected to put his theories into action against a more able foe. He was to command three panzer divisions and a motorized infantry regiment to spearhead the attack.<28> Of the ten panzer divisions in all (about 2,800 tanks) participating in the attack on 10 May 1940, seven divisions (about 1,260 tanks) crashed through the Ardennes. In the force as a whole some 70 percent of the tanks were the light Pzkpfw Is and IIs. Numbering about 3650, French tanks were more numerous overall but, in the main, not organized into massed armored units as were the German panzers.<29> Furthermore, they were dispersed over the length of the border from the English Channel to Switzerland.<30> The French Somua 35S tank not only had thicker armor than any of the German tanks but its 47-mm guns had armor-penetration performance superior to that of any of the German tanks including the short-barreled 75-mm gun of the Pzkpfw IV.<31> But the Somua was handicapped by the poor visibility afforded by the one-man turret and poor communications with other tanks. The German tanks were connected by radio nets and each tank commander had an innovative throat microphone that made speech intelligible above the engine and battle noise. Local numerical superiority and aggressive tactics overwhelmed the French and British forces and the speed of the invaders gave no time for regrouping or relief by distant reserves. The consequence was an encirclement of British and French forces by Guderian's panzer divisions turning west and other German forces squeezing from the north, resulting in the near complete defeat of the Allied armies at Dunkirk. Given the relatively few panzers and their unimpressive armor and firepower, this was a major triumph of Guderian's Blitzkrieg strategies—but perhaps the last strategic success, depending as it did on surprise and the relatively confined spaces of Western Europe.<32>

The glow of success could not last; innovation always spurs competition. Then began the ceaseless cycles to create bigger and better tanks that would last throughout the war. The *Pzkpfw* IV would give way to the *Pzkpfw* V Panther, the *Pzkpfw* VI Tiger I, and the *Pzkpfw* VI Tiger II. The Allies, on the other hand, tended to overcome performance shortfalls with increased production numbers. In the period 1939–1945 Germany produced 24,242 tanks, which compares with the 76,186 built by the Soviet Union and the 80,140 made in America.<33> German engineering genius lay in building superlative prototypes, which later turned out to be difficult to manufacture and maintain—products of what one might call boutique engineering. On the other hand, American and Soviet engineering genius lay in the grittier business of designing (or rather, redesigning prototypes) for optimal mass production and maintenance—an adaptation of the American automotive miracle. It was significant that "when Hitler's massive invasion force stood poised on the Soviet frontier in Jun 1941, it deployed 3350 tanks and 650,000 horses."<34> The horses performed the prosaic but vital function of logistics. Allied logistical needs, on the other hand, were served by American trucks.

ARTILLERY INNOVATIONS

Although Hitler did not formally repudiate the Versailles Treaty until 1935, already in 1932 any pretense of compliance was growing thin. Indeed, for some in Germany, it had been a matter of covert resistance from the start. A 1932 Krupp internal report extolled the company's role in Germany's rearmament, listing as examples a ***3.7-cm gun for armored cars; 5-cm gun for armored cars; 7.5-cm heavy anti-tank gun, tank turrets ZW38, heavy field howitzer 18; heavy 10-cm gun 18; gun carriage and limbers ...*** and boasting that these early projects were only possible because ***the firm, acting on its own initiative and believing in a revival, has, since 1918, retained at its own expense its employees, practical knowledge, and workshops for the manufacture of war material.***<35> Hitler, with his war of words against labor unions and communists, was just the man that Krupp and other industrialists were looking for and they backed him in the elections of 1933 with millions of reichsmarks. In spite of their support, Hitler's coalition won only a slim majority in the *Reichstag*, far short of the two-thirds majority required to pass the law enabling his dictatorship. However, his new rich industrialist friends gave him the wherewithal to bribe his way to its passage after all.<36> Krupp's idle production capacity, so unselfishly preserved for just this moment, was about to be unleashed yet again.

In 1936 Hitler paid Gustav Krupp a visit at his factory works in Essen. Gustav was the husband of Bertha Krupp, the granddaughter of Alfred Krupp and then current owner of the firm. Gustav had been hand-picked by *Kaiser* Wilhelm II as the most suitable mate for Bertha; indeed, Wilhelm had permitted Gustav to adopt the surname Krupp to keep the dynasty going. Hitler was already brooding over how to smash through the formidable Maginot Line as Krupp's Big Bertha had done to the Liege forts and posed the problem to Gustav. Gustav's chief designer, Erich Mueller, responded with a "railway" gun of caliber 800-mm (31.5 inches). It was a railway gun in a very peculiar sense. Weighing 1,329 tons and half a football field long, it had to be broken down into pieces that rode on trains of special railway trucks and constructed on site by gantry cranes also dismantled and carried by train.<37> Among the two dozen ancillary trains required were ammunition cars, crew coaches, two anti-aircraft batteries, and security detachments.<38> At the site, parallel sets of custom tracks (eight rails) had to be laid in a curve to enable lateral aiming by moving along the track. It took a crew of 1,720 highly trained men, under the command of a general officer, three weeks to prepare the gun to fire.<39> Hitler loved it—and ordered three. Sources differ,<40> but it appears that only two were ever built, Heavy Gustav and Heavy Dora, named for Mueller's wife. The gun was deployed to the Crimea in summer 1942 and engaged the fortress at Sevastopol, firing 48 shots over two weeks. In one case a concrete-penetrating round blew up a magazine, but the Soviets claimed that no primary targets were disabled. Nevertheless the fortress fell. The barrel was worn out and sent back to Essen for relining, but the big guns were never used again. Only pieces of the guns were found after the war and sent to Aberdeen Proving Ground, Maryland. (See author's photograph of an 800-mm shell in Chapter 8.) The cost of each gun was the equivalent of 25-28 Tiger tanks,<41,42> which the Wehrmacht could have put to far better use. The Nazis had set the record for caliber with Heavy Gustav; they also would make a try for a range record with their next bizarre weapon.

Hitler had one more "Vengeance" weapon beyond his V-1 cruise missile and his V-2 ballistic missile (both discussed in my second volume, ***Weapons of Mass Destruction***)—the lesser known V-3. Consisting of two batteries of 25 guns each, the V-3 was intended to hurl 300-

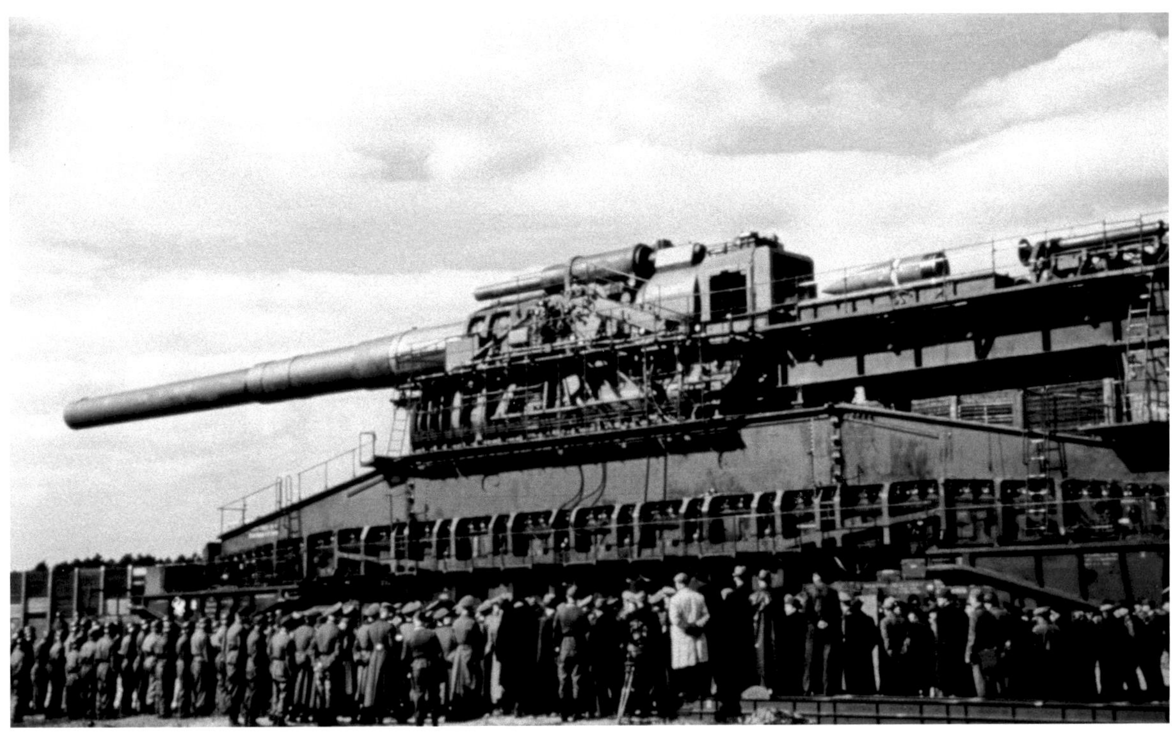

Heavy Gustav 80 cm (31.5 inches) rail-mounted gun being demonstrated for Hitler.
By Permission bpk Bildagentur / Bayerische Staatsbibliothek, Munich / photographer Hans Hoffman / Art Resource, NY

pound, 6-inch projectiles on London at the rate of 200 every hour from a fortress at Marquise-Mimoyecques near Calais, one every 18 seconds. The fifty guns with barrels nearly 500-ft. long were to be ensconced in a subterranean concrete fortress and permanently aimed at London some 90 miles away. Work began toward the end of summer 1943 on the fortress that would comprise a warren of access tunnels, rooms for storing ammunition, and troop support facilities. Calculations indicated that a muzzle velocity of about 5,000 feet/sec would be necessary, and it was to be achieved by arranging side chambers along each barrel in which propellant charges would ignite as the projectile passed so as to continually accelerate the projectile along its path in the barrel. Test barrels were set up at a proving ground near Magdeburg and near the Baltic coastal town of Misdroy. Timing of the charge ignitions along the barrel turned out to be a challenge as did regular barrel-bursting incidents. Meanwhile massive construction work was proceeding at the Mimoyecques fortress and an artillery battalion was being trained to operate the guns. The Allies had started aerial reconnaissance looking for possible V-1 launch sites and had spotted the activity at Mimoyecques in September 1943. In November of 1943 the US Ninth Air Force bombed the site, damaging the western battery beyond repair. The wily Germans left the superficial damage untouched but continued working below ground until the following July when the RAF bombed the site with five 12,000-lb. bunker-busting bombs that caused irreparable damage. By August Allied ground forces had penetrated to the site. Though the V-3 site was overrun, the proving ground work continued until the end of the year. British attempts to complete the demolition of the site after their occupation did not entirely succeed, and one can visit the Mimoyecques Fortress today to see remnants of the tunnel system.<43>

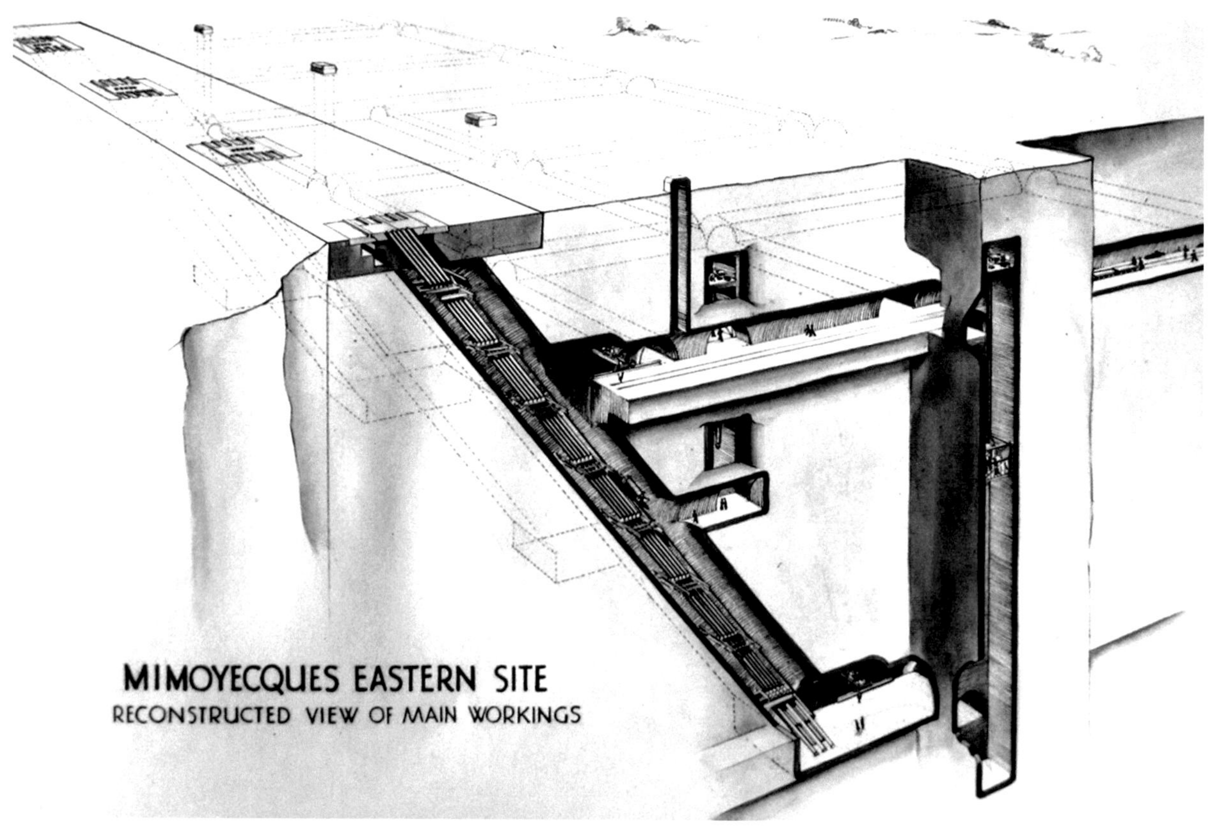

British wartime drawing of the Mimoyecques Fortress plan.
Courtesy British National Archives

Not all artillery innovations in WWII were as flashy and impractical as the Heavy Gustav and the V-3. Besides improvements in reliability and speed of operation, one of the most important changes was in their mobility. The new combined-arms tactics, first of Germany and then of the Allies, demonstrated the necessity of motorizing the artillery to keep up with and support tanks and motorized infantry. Tank chassis proved useful for mounting guns and commonizing both parts and maintenance resources. When the decision was made to consolidate all tanks into panzer units for shock effect, the German assault gun or *Sturmgeschütz* was developed specifically to support infantry.<44> With no turret, the *Sturmgeschütz*, or *StuG*, had a lower profile and was easier and cheaper to produce. They proved effective both in their infantry support role and as able tank killers. With upgrades in their guns as the war wore on, combined with so many different tank chassis versions, the taxonomy of German *Sturmgeschütze* grew complex indeed, as will be seen in Chapter 8. In keeping with the American philosophy of mass production of limited types of weapons to reduce the logistical burden, the US fielded only two main self-propelled artillery pieces, the M7 105-mm and the M12 155-mm self-propelled guns. The M7 was dubbed "The Priest" because of its pulpit-like machine gun turret. Its chassis was based first on that of the M3 Lee tank and later on that of the M4 Sherman. Shown here is the M3-chassis version.

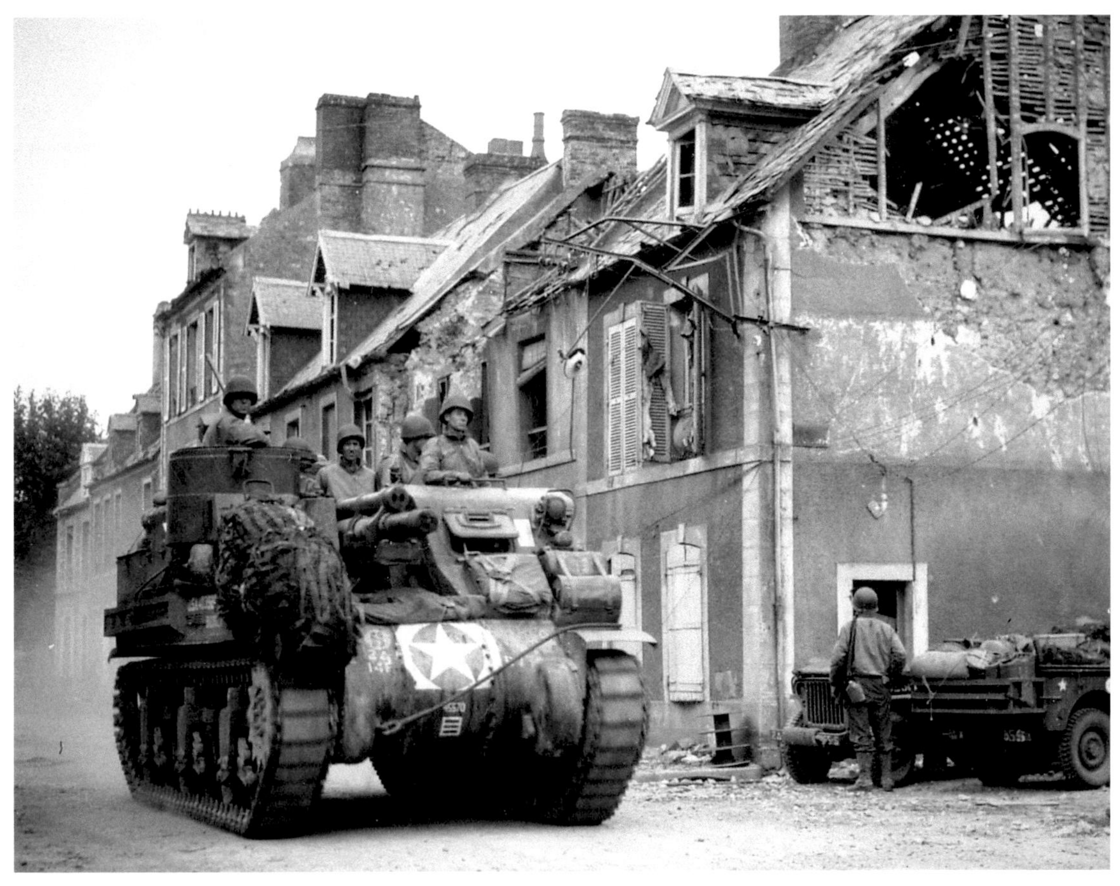

US self-propelled 105-mm howitzer, M7 "Priest"
Courtesy National Archives

Unquestionably, the most important practical innovation in artillery during the Second World War was the proximity fuse. In a US attempt to even the odds of ships against attack aircraft, Vannevar Bush's National Defense Research Committee (NDRC) instituted Section T in August 1940 with Merle Tuve as its head to investigate the possibility of developing an artillery-shell fuze that would explode when close to an enemy aircraft.<45> Timed fuzes were too difficult to program in a fast moving tactical encounter and contact fuzes placed too stringent a requirement on gun accuracy. Tuve, a physicist with the Department of Terrestrial Magnetism at the Carnegie Institution in Washington, D.C., proved to be a brilliant and tenacious project manager who took the program from concept to production in two years' time. The fuze itself was a marvel of miniaturization, incorporating vacuum-tube radio transmitter and receiver, essentially a tiny radar set tucked into the nose of an artillery shell. As work on the fuze grew more promising, laboratories moved from the Carnegie Institution to the Bureau of Standards and then in the spring of 1942 to its own building in Washington where it became the Applied Physics Laboratory administered by the Johns Hopkins University.<46>

The story of the proximity fuze development is another example of the American mobilization miracle. Production of the little wonder increased from 500 per day in September of 1942 to

70,000 per day in the last months of the war. More than 2,000 subcontractors were involved. By 1945 the Sylvania Company alone was producing over 400,000 vacuum tubes per day;<47> compare this with the US 1940 yearly output for vacuum tubes of all types of only 600,000.<48> During the war some 22 million proximity fuzes were manufactured. This was no mean task as the requirements for safety and reliability of each fuze were strict. Each fuze employed safety switches to insure that it did not arm until 0.3–0.5 seconds after leaving the gun barrel and that it would self-destruct if it missed the target, lest friendly forces accidently feel its sting.<49> Also, secrecy was so intense that subterfuge had to be employed in the procurement of fuze components. The plastic nose pieces, for instance, were procured through the Johns Hopkins Hospital as "rectal spreaders."<50>

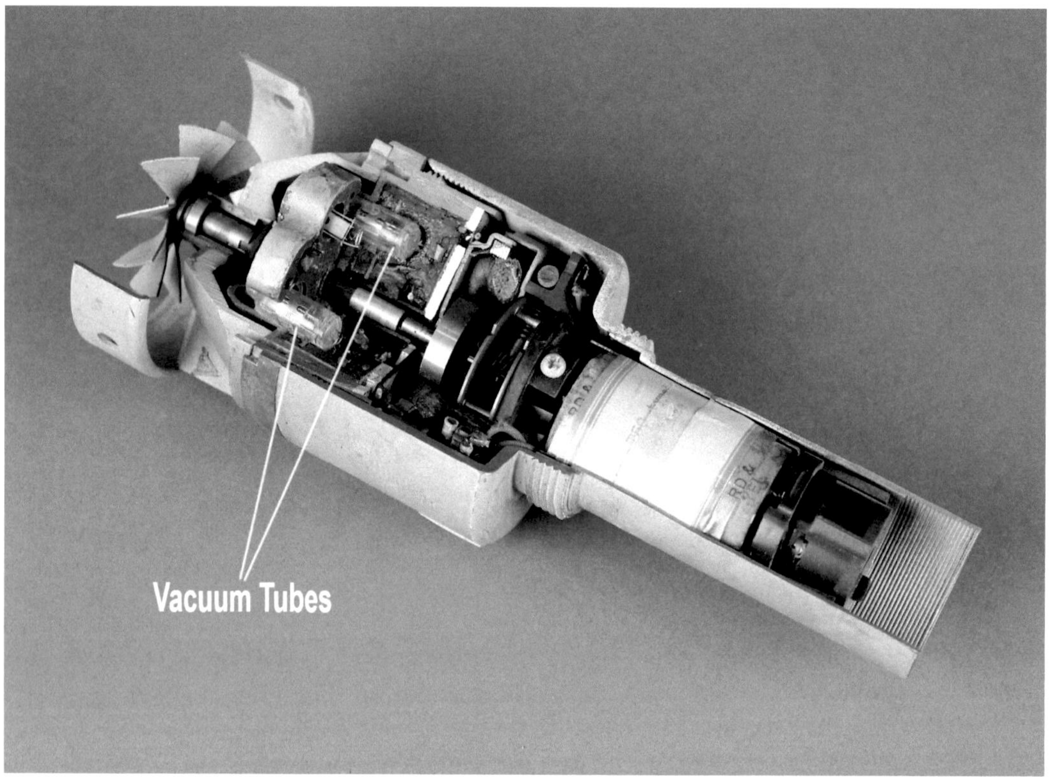

Type T-89 proximity fuze, one of many developed during WWII. This one was used for several types of bombs. At upper left are turbine blades that rotate the shaft of the generator at lower right to provide electricity for the radar circuitry. The vacuum tubes are clearly visible at center left.
Courtesy National Institute of Standards and Technology Digital Collections, Gaithersburg, MD 20899

Secrecy also influenced its deployment, being confined to shipboard use at first where duds were not likely to be recovered and analyzed. Eventually they were employed against the German V-1 cruise missile and brought down 85% of the V-1s making it across the English Channel.<51> The fuzes were also adapted to the Army's Field Artillery to burst at tree-top level and used in the fog in the Battle of the Bulge. The proximity fuse was a harbinger of the smart munitions that would come to dominate the battlefield toward the end of the century.

The proximity-fuze development program contributed to winning the war in a less direct way as well. Commander William "Deak" Parsons was assigned the duty of liaison with both the Navy and the NDRC for the fuze project. Vannevar Bush was so impressed with his leadership and technical skills that he recommended him to General Leslie Groves to serve as Ordnance Chief and Associate Director of the atomic bomb project at Los Alamos. Parsons thereafter played a lead role in developing the uranium bomb, "Little Boy," that was dropped on Hiroshima. In fact, he was the weapons officer aboard the Enola Gay and armed the bomb from a cramped and precarious perch in the bomb bay after takeoff from Tinian Island.

THE CHANGING FACE OF NAVAL POWER

In the First World War, sea power was all about battleships and this mentality carried over, to some extent, into the Second World War. Four members of America's last battleship class, the Iowa class, were built during the war and two more had had their keels laid when the war ended. The two unfinished ships were subsequently scrapped. There was also a larger, Montana class ordered by the Navy, but aircraft carriers edged them out in priority and they were never built. The Japanese attack on Pearl Harbor had changed everything. Projecting such devastating power over some 4,000 miles—in secret—was something new. The air-power lesson of 7 December 1941 was driven home in the following few days. On the 8th, the US Army air forces in the Philippines were largely destroyed on the ground by Japanese Nell and Betty bombers escorted by Zero fighters that had flown from Taiwan, a round trip of more than a thousand

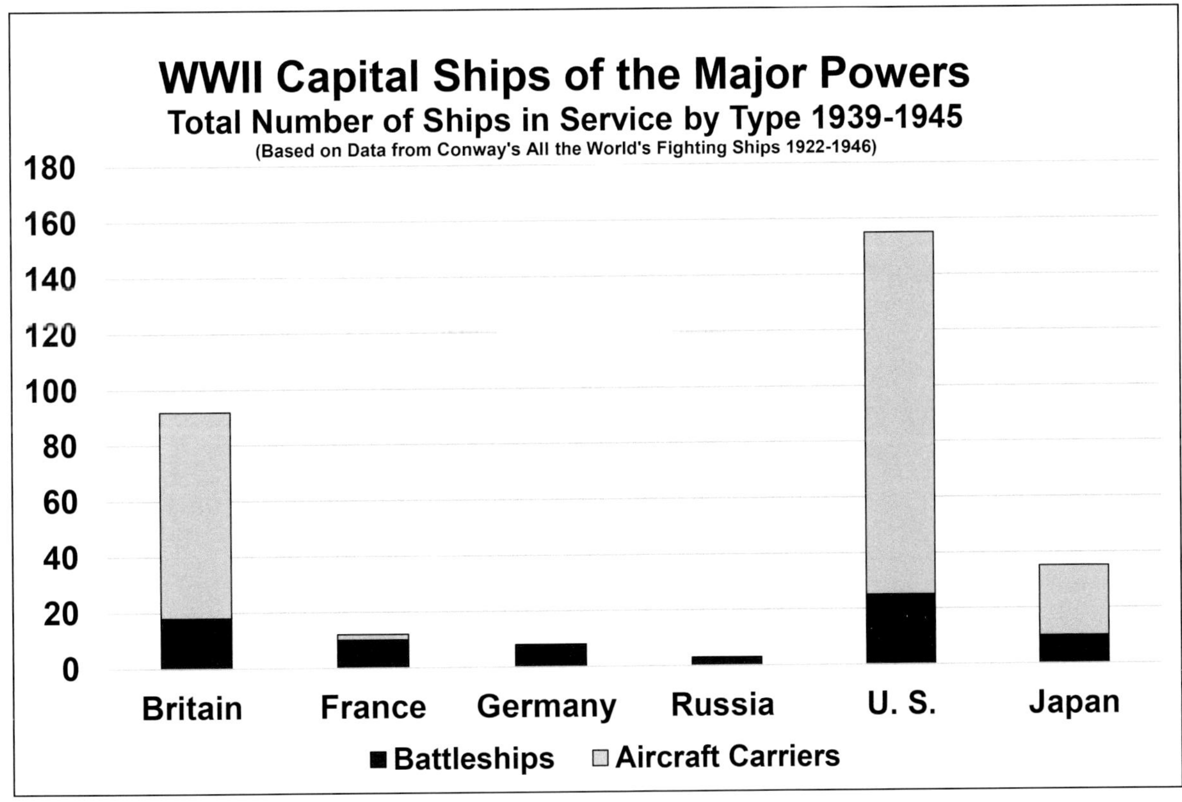

miles. Two days later one of Britain's latest and most powerful battleships, the HMS *Prince of Wales*, and the battlecruiser HMS *Repulse* were attacked and sunk by bombs and torpedoes in the Gulf of Siam by nearly a hundred Nell and Betty bombers that were more than four hundred miles from their base. Until then, safety at more than two hundred miles from a Japanese base had been the rule of thumb.<52> As Guderian had changed the face of warfare on land, so did the Japanese for naval warfare. The accompanying chart tells the tale of the nearly complete displacement of battleships by aircraft carriers during the war. Engaging an enemy 10 miles away with 16-inch guns was dicey, but F4F Wildcats could engage the enemy with close-range accuracy after flying from a carrier hundreds of miles away. With their ability to launch either bombs or torpedoes or both, carrier aircraft proved superior to naval guns in versatility as well as range and effectiveness.

AIR POWER

The story of air power in WWII is an amalgam of rapid technical developments, their countermeasures, and moral ambiguity. During the war the term "strategic bombing" gradually came to be a code word for the bombing of civilian populations. For the most part, fighter aircraft, while undergoing considerable development during the war, remained committed to tactical military targets. Though there was some bombing of cities in World War I, the limitations of the aircraft of the time served as a constraining factor. Foreseeing the inevitable direction of aviation technology after that war, Italian Air Force commander Giulio Douhet published *The Command of the Air* in 1921. His book proposed that, under the conditions of total war, the only viable strategy would be to maximize one's bombing capability to deprive the enemy of his ability to mount counter-bombing offensives. Spending on anti-aircraft defense would be wasted due to the three-dimensionality of the airspace and the high speeds of the aircraft. Such investments should instead be channeled into the bomber force.

Though his death in 1930 denied him the grim satisfaction of seeing the fruition of his ideas, his book became influential among European air commands. Douhet's prediction of the course of future wars was eerily prescient, but he had failed to appreciate how technology would even the scales for defense against bombers. Radar helped the British to marshal their thin fighter resources to defeat the Nazi bombers and, until the technology of long-range fighters was developed, similar German defenses likewise took a terrible toll on British and American bombers.

A key premise to Douhet's dark vision was the presence of total war, and the First World War demonstrated how quickly conflicts in the industrial age could devolve into absolute war. The bombing of the small Spanish town of Guernica by Hitler's Condor Legion in 1936 was the opening act on the modern bombing of civilian populations for the rest of the century. The taboo having already been transgressed in the First World War, the quantum jump in the destructiveness of the new air power did little to inhibit its continued use. The threshold for further atrocities fell lower with each event. Nanking followed, then Warsaw, Rotterdam, London, Coventry, Hamburg, Berlin, Dresden, Tokyo, Hiroshima, and Nagasaki.<53> Both the British Royal Air Force and US Army Air Force had begun the war with a policy of confining bombing to strictly military targets, at first only targets at sea,<54> but the devil could not be restrained for long. Poor communication, poor navigational skills, and poor weather produced the inevitable mistakes and those mistakes lead to cries for retribution. In Rotterdam the Dutch defenders were given an ultimatum on 13 May 1940 to surrender or suffer "complete destruction." Bombing ensued even though the Dutch had surrendered; their capitulation message had failed to reach the Luftwaffe in time.

Even before plans were finalized for Operation Sea Lion (the invasion of Britain), the Luftwaffe had begun making exploratory raids, training runs, on eastern and southern England during July and August 1940. Small-scale bombing runs tested small numbers of large bombs vs. large numbers of small bombs vs. incendiary bombs. Analogous raids by the British on the nights of 25–26 and 29–30 August 1940 caused the death of only ten Berliners but prompted Hitler to vow revenge against British cities. Whether in genuine pique or as a ruse to initiate previous attack plans in a more palatable way for the German public, the heavy bombing of Britain began in earnest on September 7 with over 300 German bombers and 600 fighters pounding the city of London.<55> They had just jumped onto the slippery slope of total war where combatants and citizens alike were fair game. [See my book, ***Weapons of Mass Destruction***, for more on the air war of WWII.]

RADAR

James Clerk Maxwell's 1864 mathematical abstraction describing the propagation of electromagnetic (EM) waves was a marvel. EM waves are different than water or sound waves, which are basically propagating disturbances of the water or air itself. EM waves can propagate in a vacuum and they are characterized by their wavelength, or alternatively, by their frequency. The character and properties of these waves vary enormously with wavelength from the very short (gamma rays), at less than a millionth the breadth of a human hair; to X-rays, at 100-1000 times longer; to visible light, another thousand times longer; to microwaves and radio waves, with wavelengths ranging from millimeters to millions of meters. In 1887 German physicist Heinrich Hertz, guided by Maxwell's brilliant formulation, devised a way to generate radio waves with wavelengths from several tenths of a meter to several meters. Significantly he also showed that these waves could be reflected from solid objects.<56> Building on Hertz's discovery, Marconi changed human society forever when he demonstrated his transatlantic wireless telegraph in 1901 and, by the 1920s, commercial radio ended the isolation of farm and hamlet. Activity to exploit the new technology was brisk in many countries and the development of devices for direction finding and ranging developed simultaneously in different countries though following somewhat different paths. Early recognition of the military implications of radio waves led to secrecy; but, surprisingly, no evidence ever surfaced of espionage playing a critical role in the spread of the technology prior to the war.

Development of **radar**, **r**adio **d**etection (or direction finding) **a**nd **r**anging, was paced by the growth of the electronics industry. The electron, which had secretly played such a central role in Hertz's experiments, had not yet even been discovered at the time of his experiments. Only in 1897 did British physicist J.J. Thomson demonstrate its existence, and it solved the riddle of many unexplained phenomena. For example, Thomas Edison in 1880 had found that a current would pass from a hot filament in an evacuated bulb to a second electrode in the bulb when a positive voltage was applied to this electrode relative to the filament. Known as the Edison Effect, or thermionic emission, (negatively charged) electrons are literally boiled off the hot filament and, thus freed, are attracted to the positive electrode. In 1906 Lee de Forest put a third electrode bent into a serpentine shape in the bulb between the filament and the Edison's second electrode, which became known as the plate or anode. It was found that, by applying a suitable voltage to this serpentine electrode, dubbed the grid, the current flowing through the grid between filament and plate could be controlled or regulated. This invention, called the triode vacuum tube, later turned out to be the enabler of amplifying circuits necessary to radio and, by extension, to radar.<57> More refinements followed: a fourth electrode, creating a tetrode vacuum tube in 1919, was followed by a fifth electrode creating a pentode, each addressing some limitation caused by the previous generation device. For the next thirty years amplifier circuits relied on the pentode for wide frequency-band operation. According to physicist/writer Louis Brown, ***Radar would have been impossible without the pentode.***<58>

Between the wars radar was being developed simultaneously and independently in the United States, Britain, France, Germany, the Soviet Union, and Japan.<59> But it was in Britain that the first networked air-defense system was fielded. Scottish physicist Robert Watson Watt, a descendant of steam-engine developer James Watt, had devised a technique of locating thunderstorms by the detection of lightning electrical activity using direction-finding loop antennas and triangulating the data of two or more stations. It was possible to determine the direction of the electrical disturbance by manually rotating a loop antenna to maximize the signal, but Watt ingeniously employed two stationary loop antennas at fixed at right angles and used the relative signal strength from each to ascertain the direction of the discharge. This invention formed the basis of his radar receiver, which he coupled with a pulsed transmitter, allowing a determination of range as well as direction.<60> He picked a wavelength of 50 meters thinking that it would resonate roughly with a bomber's wingspan, but the wavelength turned out not to be critical because of the multiple reflecting surfaces of an aircraft. To avoid interference by other communication signals, he found wavelengths in the 7.5-15 meter range more practical. By the fall of 1935 Watt had demonstrated the ability to track a bomber from 40 miles away. With his success Britain authorized the construction of five stations at the mouth of the Thames River which took the name Chain Home. Transmitters were linked to horizontal dipole antennas, which were just wires stretched between 100-meter high steel towers at several heights, to determine a bomber's altitude as well as its direction and range. Receiving antennas were placed on 75-meter wooden towers.

In 1938 the Chain Home system began operating with RAF personnel, and other stations were added along the British coastline. By the summer of 1940 with German air raids commencing, the system was already obsolete but still served to save Britain from disaster. They enabled the efficient scrambling of scarce fighter plane resources against the incoming bomber groups. Equipment operating at ever shorter wavelengths and ever greater power would be developed during the war. Already in 1939 Japan had devised a system operating at 10-cm wavelength and had done research on wavelengths down to 1.5 cm.<61> The Brits fielded their first 10-cm set in late 1940 with some help from the Americans.<62> In the United States enormous resources were poured into radar. Prominent among the efforts was the Radiation Laboratory (Rad Lab) at the Massachusetts Institute of Technology where at peak operations during the war nearly 4,000 scientists, engineers, and technicians toiled. After the war the accumulated researches of the Rad Lab were published as a 28-volume set of books that provided a treasure trove of electronics expertise to the economy after they were declassified.<63> As a measure of importance the US government attached to radar, expenditures for field sets totaled $2.8 billion, nearly half of which represented equipment designed by the Rad Lab. By comparison $3 billion was spent on the B-29 bomber, the most expensive project of the war, and $2 billion on the Manhattan Project to develop the atomic bomb.<64> Warfare without radar had become unthinkable.

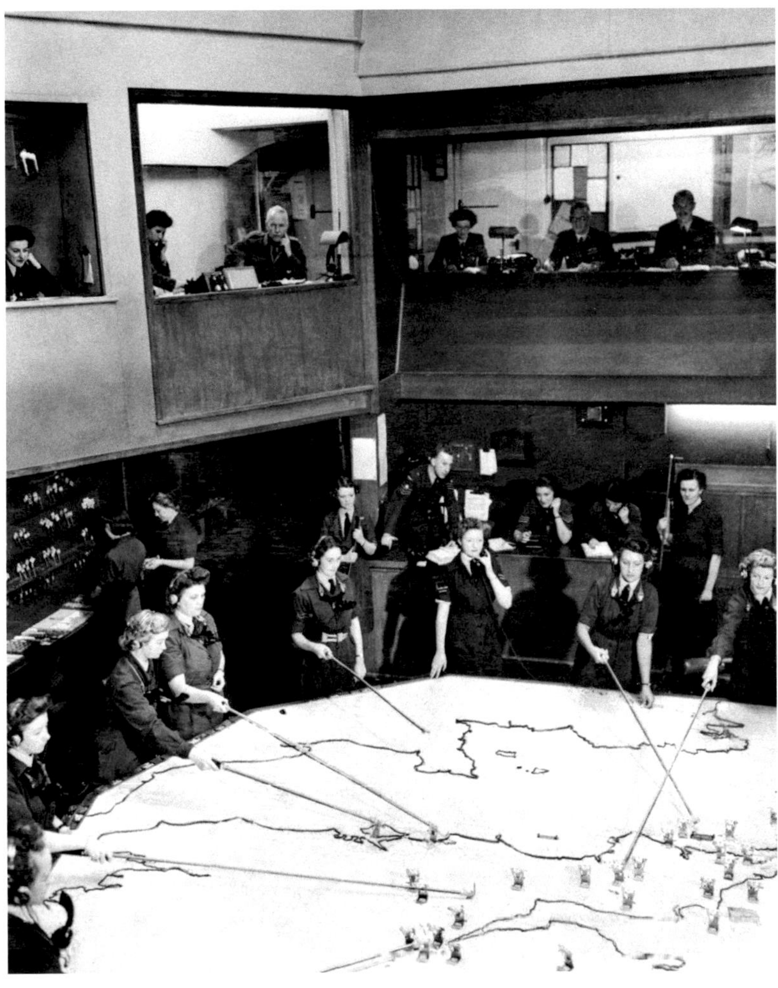

The Operations Room at RAF Fighter Command's No. 10 Group Headquarters, Rudloe Manor (RAF Box), Wiltshire, showing WAAF plotters and duty officers at work, 1943.
By Permission British Imperial War Museum

ATOMIC BOMB

The most portentous innovation of the Second World War was, without doubt, the atomic bomb. The course of its US development and deployment in WWII and after were the subjects of my first two books, ***The Neutron's Long Shadow*** and ***Weapons of Mass Destruction***, and will not be repeated here. The Nazi atomic bomb program, however, received little attention in those volumes. While the fear of Nazi Germany's possible lead in such development spurred the US program to life and sustained it during the early years of frustration, by the spring of 1945 it had become fairly clear that there was no possibility of a German atomic bomb deployment during the war. Likewise, Japan, though engaged in some nuclear research that might have led to a bomb, was even further away from building one. General Leslie R. Groves, who commanded the Manhattan Project, was always of the opinion that Japan lacked the muscular industrial infrastructure required to produce the exotic fissile nuclear explosives. About Germany, he was not so sure. In January 1939 German scientists Otto Hahn and Fritz Strassman had published their discovery of nuclear fission in uranium subjected to neutron bombardment. Germany was preeminent in physics, chemistry, and engineering—all of which were necessary to developing a bomb—and they had a robust industrial base. In fact, they did have a secret nuclear weapon project headed by Werner Heisenberg, one of the most prominent atomic physicists in the world and a Nobel laureate. So why, amid all their successful wartime technical innovations, did they fail in the one weapon that might have won the war for them?

After the war a number of German scientists, including Heisenberg, claimed that they had fully understood the bomb-making requirements but had deliberately held the atomic project back so that Hitler would not get it—a kind of passive resistance, not an outright refusal to work on it. At the time, despite the improbability that they would have risked such behavior under a brutal totalitarian regime, there seemed to be no way to refute their claims. In 1992, however, the US government declassified the records of internment of leading German scientists, including Heisenberg and Otto Hahn, at a British country manor called Farm Hall. At the end of the war in Europe these ten scientists had been transferred from their various detention centers to Farm Hall where they spent the next six months together in gentile confinement. Gen Groves had ordered this internment and transcripts were secretly made of their conversations. Before dinner on 6 August 1945, Hahn was informed by the American officer-in-charge that an atomic bomb had been dropped on Hiroshima. Hahn then informed his compatriots at dinner, and the transcripts of the discussion that followed provide a Rosetta Stone for understanding the true thoughts and feelings of those present. One has to have considerable knowledge of bomb physics to be able to interpret the proceedings, but, fortunately, physicist/writer Jeremy Bernstein in his book *Hitler's Uranium Club: The Secret Recordings at Farm Hall* has provided just such a critical commentary.<65> Though at the time of the Manhattan Project, Bernstein had not yet begun his professional career, he later was involved in the US nuclear enterprise and had personal relationships with many of the Project's participants. Thus he was well positioned to write this remarkable book. The book reproduces the relevant transcripts and essentially provides an exegesis in the margin of what is being said.

There is plenty of subtlety in this issue, but it seems to come down to several considerations. The Germans used heavy water for the moderator in their test reactor and their main supply from Norway was cut off by Allied actions. Americans used a graphite moderator, but they had understood that it had to be cleansed of even small amounts of neutron-absorbing impurities such as boron and cadmium. Heisenberg explained their rejection of graphite based on measurements that had used impure material, but at least one German scientist had understood this limitation and made measurements on suitably pure material. Why Heisenberg did not know or make use of this latter information is a mystery. Second, Bernstein is convinced that Heisenberg's Farm Hall reactions to the Hiroshima bombing showed that he had only a superficial understanding of the true bomb physics and had grossly overestimated the amount of fissile material required. Heisenberg was an extraordinarily gifted theorist but was not good with numbers—not a good trait in a director of an engineering project. Third, the resources needed for undertaking such a complicated and uncertain development were extremely high and required commensurate levels of personal boldness. In a wartime, totalitarian environment such boldness might entail considerable personal as well as professional risk in the event of failure. In America the pressures on the principal scientists were enormous but involved no risk of physical punishment. This factor may have inhibited a full commitment on the part of the German scientists to such an uncertain task. Finally, Hitler was no scientist but believed in the brilliance of his intuition in all matters, and to him physics was a Jewish science that had no real place in the New Germany. This meant that development funds for this quintessential physics project might have been limited in any event. Certainly any such all-out program would have been handicapped by the loss of the many Jewish scientists he had disenfranchised or driven away.

The ten scientist-detainees were all good men but conflicted between their professional and patriotic ideals and obligations. So it is quite understandable that their memories, recent though they were, of their wartime activities and motivations were colored even in their own minds by the unexpected defeat of their country. Bernstein expressed it this way:

> ***... a common theme among the so-called 'good Germans' ... [was that they] wanted Germany to win the war, but Hitler to lose it. I do not think any other formulation describes the mixed motivation of these people.***<66>

CHAPTER 6. IN THE SHADOW OF THE GREAT WARS

American policy sought to stop Russian or communist expansion, an attitude that allowed American leaders to assume the role of world policemen. But no matter how far overseas America extended her power, the technological revolution had overcome distance and, despite anything America could do, national security was constantly in jeopardy.

—Stephen E. Ambrose, *Rise to Globalism*<1>

After WWII there was rapid demobilization of American conventional forces and an initially slow build-up of nuclear forces. The Soviet Union was slow to demobilize and the United States used its nuclear monopoly to counterbalance an increasingly aggressive Soviet foreign policy. Though the world war was over, threats remained, leading to a constant state of semi-mobilization for the rest of the century. A good measure of the extent of economic mobilization for war is the fraction of a country's GDP that is committed to the military. The persistent postwar state of mobilization in the United States is reflected in the accompanying chart.

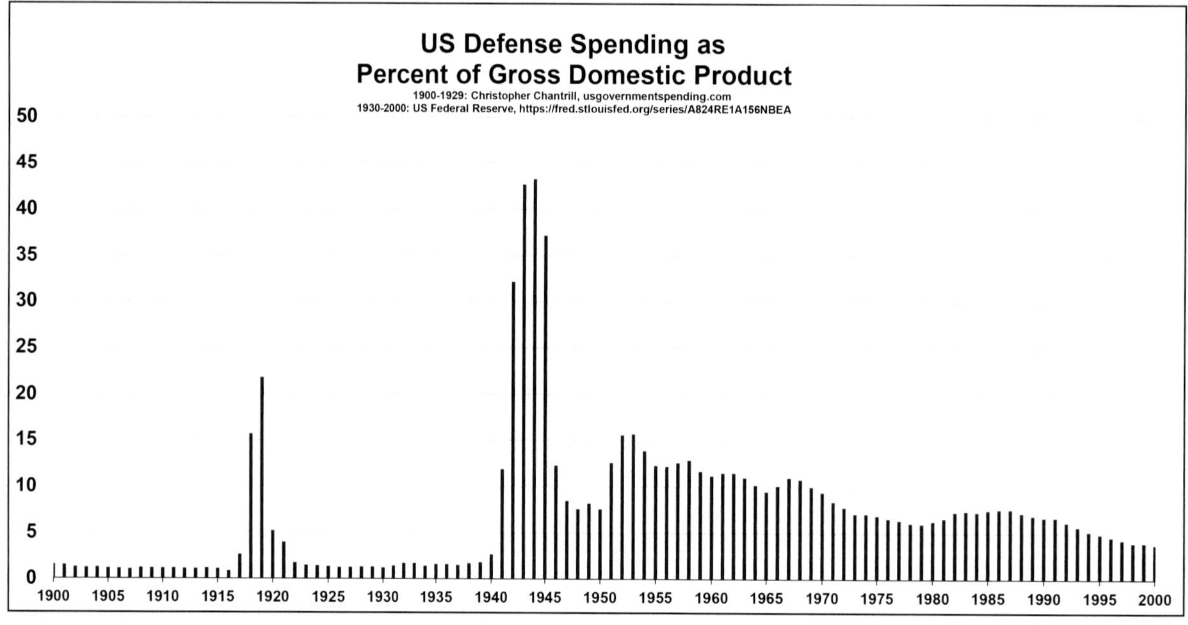

Though WWII was the last world-wide all-out war of the twentieth century, there were still many regional wars. The Correlates of War Project (COW), started in 1963 at the University of Michigan by political scientist David Singer, seeks to develop a comprehensive critical body of knowledge of war to serve as a solid foundation for analysis. Their enterprise requires considerable judgment and scholarship not least because much of their work concerns third-world countries where records may be scant and combatant affiliations may be ambiguous. The COW online database lists an astonishing 35 distinct wars with some 3.5 million military deaths from 1945 to 2000.<2> Of these the United States was directly involved in six: Korea, Vietnam,

Laos, Cambodia, the Persian Gulf, and Kosovo. Of course, the US was indirectly involved in a number of others by supplying arms and advice. Some of these wars were a continuation of long standing grievances between competing ethnic groups; some were the result of independence movements growing out of the age of imperialism; some were the result of tensions created by superpower agreements in the aftermath of the two world wars. Of course, from the perspective of the combatants and civilians involved, a regional war looks very much like a world war.

COLD WAR

Achieving the total defeat of Germany and Japan had exhausted the Allies as well as the Axis Powers, but there was little respite in its aftermath. The uneasy alliance between the Soviet Union and the West that had been so critical to victory quickly unraveled. The war had left Europe and Japan in ruins with starving, dehoused, and demoralized populations. Stalin viewed this condition as fertile ground for the spread of communism, and conflict with the West emerged everywhere. The spectacular use of the atomic bomb against Japan left Stalin at a disadvantage and he ordered his war-torn country to develop one with the highest priority. The detonation of the first Soviet atomic bomb in August 1949 alarmed the United States, which had demobilized its own wartime forces and had come to rely on the monopoly of atomic weapons to even the odds against overwhelming Soviet conventional forces. Fearing the loss of the US technological edge, President Truman pushed for the all-out development of the hydrogen bomb, which was successfully tested in November 1952. Less than a year later, the Soviet Union claimed to have their own thermonuclear bomb. By the end of the decade both sides had nuclear tipped intercontinental ballistic missiles. The resulting arms race continued unabated until the collapse of the Soviet Union on Christmas Day 1991. For nearly half a century the world had been menaced by the possibility of a nuclear Armageddon that may well have erased the gains of modernity. [Please see my previous book, ***Weapons of Mass Destruction*** for more on the Cold War.] Significant conflicts with "conventional" weapons nonetheless also marred this period.

KOREAN WAR

The wisdom of relying on the US monopoly of atomic weapons as a cheap substitute for conventional forces was tested on 25 June 1950 when North Korean forces poured across the 38th parallel into South Korea. Less than a year previously, the Soviets had tested their own atomic bomb, prompting the US to shift its strategy from nuclear monopoly to superior numbers of bombs. In addition, the hydrogen bomb was being developed. Now Truman was faced with the failure of his exotic bombs to deter the aggression of a third-world country. During its drive south, North Korea had very nearly expelled US forces from the Korean Peninsula altogether. Truman rattled his nuclear saber by sending nuclear-capable B-29 bombers to Guam with atomic bombs, though secretly not accompanied by their nuclear cores. This had no effect on the communist forces who continued to consolidate their gains. What stopped them was Gen. Douglas MacArthur's daring landing at Inchon little more than two months later. This bold maneuver subsequently pushed the North Koreans back to the Chinese border. There followed a hard-fought three year conventional war that ended with the original

38th parallel as the border. The US fought with those leftover weapons of WWII that had not yet been scrapped. Although new tanks were being developed after WWII, they were not available in time to be used in Korea. B-29s lay in abundance in the Arizona desert, but they had not been maintained because no one had imagined any future need for them. So severe was the shortage of operable planes and parts that missions would consist of a dozen planes instead of the hundreds that had been massed against Japan.<3> Gen. Curtis LeMay, who had so successfully brought the bitter fruits of total war to the Japanese, was now commander of the Strategic Air Command and not directly involved in the new war, but he advised a swift, intense action against the North Koreans. His advice went unheeded. After the war, LeMay expressed his dismay at the result of slowly increasing the violence inflicted on the enemy:

> ***So we go on and don't do it and let the war go on. Over a period of three-and-a-half or four years we did burn down every town in North Korea and every town in South Korea ... And what? Killed off 20 percent of the Korean population ... What I'm trying to say is if***

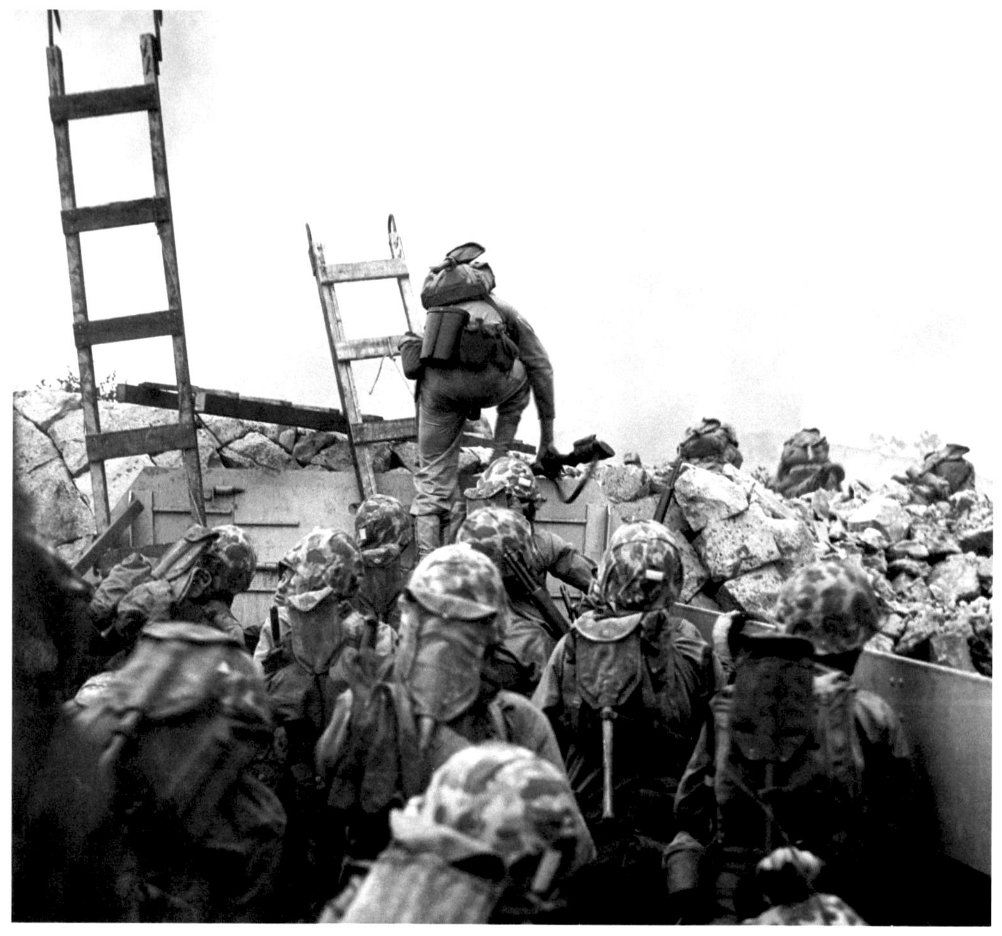

In a scene eerily reminiscent of "going over the top" in WWI, Marines storm ashore at Inchon using scaling ladders in Korea during McArthur's bold 1950 initiative.
Courtesy National Archives, NA Identifier 5891320

once you make a decision to use military force to solve your problem, then you ought to use it and use an overwhelming military force. Use too much and deliberately use too much so that you don't make an error on the other side and not quite have enough. And you roll over everything to start with and you close it down just like that. And you save resources, you save lives—not only your own but the enemy's too, and the recovery is quicker and everybody's back to peaceful existence hopefully in a shorter period of time.<4>

But the awful memory of total war was still fresh, and it had seemed worth trying something short of it. The debate on the wisdom and effectiveness of managing a middle way short of total war has continued ever since. The United States had learned the limits of its power, both nuclear and conventional. More than two million people died in Korea and hopes for peace had been dashed. As historian T. R. Fehrenbach expressed it,

The American public, and that of Europe, learned that the postwar world was not the pleasant place they hoped it would be, that it could not be neatly policed by bombers and carrier aircraft and nuclear warheads, and that the Communist menace could be disregarded only at extreme peril.<5>

These lessons and more would be hammered home in America's next war.

VIETNAM WAR

Communism was viewed as an increasing threat to the western democracies in the late 1940s. In 1949 Chinese Communist forces under Mao Tse-tung succeeded in banishing the Nationalists led by Chiang Kai-shek to the island of Formosa (now Taiwan), despite his receiving aid from the US. Both the Soviet Union and Communist China next helped North Korea in its aggressive bid to unify the Korean Peninsula under its Communist regime. Now the Soviets and the Chinese began giving arms to Vietnam's Communist leader Ho Chi Minh to oust the French colonialists from Vietnam. With North Korea's invasion of the South, Communist encroachments in Eastern Asia and even the Pacific became more than a theoretical possibility. It was feared that Japan, the Philippines, and the Pacific island protectorates could fall successively in what began to be called the Domino Theory of growing Communist hegemony. To counter this perceived threat, President Truman began in 1950 to send financial assistance to the French to combat Ho Chi Minh. Since the Korean and Vietnam problems were coupled through the agency of the Chinese, it was thought that bolstering support for the French would diminish China's ability to arm North Korea.

US aid to the French climbed rapidly from $10 million in 1950 to over $1 billion in 1954.<6> After a two-month siege in 1954, Ho Chi Minh's growing army managed to capture the fortress at Dien Bien Phu from the French. There followed a peace conference in Geneva where both sides agreed to a provisional partitioning of Vietnam at the 17th parallel with the promise of elections in 1956. However, infiltration of South Vietnam by North Vietnamese forces continued to feed the insurgency in the South. In 1955 the Americans assumed the task of training the South Vietnamese Army from the French, and from that moment the inexorable escalation of American involvement began. By November 1961 there were some 900 American soldiers in Vietnam; one year later there were more than 11,000, largely in an advisory role.<7> In his January 1962 State-of-the-Union speech, President John Kennedy, referring to Vietnam, proclaimed:

> ***The systematic aggression now bleeding that country is not a 'war of liberation' — for Vietnam is already free. It is a war of attempted subjugation and it will be resisted.***<8>

Kennedy did have a point. One of the provisions of the Geneva Accords was a one-year resettlement period when the populace could choose whether to live in North or South Vietnam. The North's iron thumb caused a million citizens to choose the South compared to less than 100,000 for the North. And this huge disparity was in spite of an autocratic and corrupt South Vietnamese regime led by Ngo Dinh Diem.<9>

In August 1964, as a result of a reported attack by North Vietnamese torpedo boats against the US destroyer Maddox in the Gulf of Tonkin, a resolution was passed by both Senate and House of Representatives authorizing then President, Lyndon Johnson, ***to take all necessary measures to repel any armed attack against the forces of the United States and to prevent further aggression.***<10> The gloves were off. Before committing US combat troops, President Johnson decided to test Ho Chi Minh's resolve by bombing the North. The campaign, begun in March 1965 and code-named ROLLING THUNDER, would continue until November 1968. By the end of 1967 some 864,000 tons of bombs had been loosed over North Vietnam, compared with totals of 635,000 tons in Korea and 503,000 tons in the Pacific phase of WWII. Though causing considerable damage and hardship, losses in food and materiel were more than made up for by the Soviets and Chinese. Another bombing campaign, LINEBACKER, was ordered by President Richard Nixon in the spring of 1972 with the goal of interdicting supplies from China. This time smart bombs, guided by television and laser, were used for the first time and had a devastating effect on North Vietnamese infrastructure. Targets like railway bridges and tunnels that were previously restricted because of their proximity to politically sensitive areas were now hit. Radar defenses of Hanoi, now among the most sophisticated in the world, were neutralized by new countermeasures. LINEBACKER was halted toward the end of October when it appeared that progress was being made in peace talks but resumed in December for 12 days when the talks faltered. At the end of January 1973 a cease-fire agreement was signed in Paris, but by then the Watergate scandal was unfolding, seriously hampering the administration's ability to enforce the cease-fire. By August 1974 Nixon had resigned and in April 1975 the South Vietnamese Capital of Saigon fell to the North Vietnamese Army. The war was over, and a tiny third-world country had bested the United States, albeit with crucial help from the major Communist powers.

Historian Guenter Lewy has estimated that some 1,313,000 people, 28 percent of whom were civilians, lost their lives in Vietnam as a result of the war years 1965-1974, though considerable educated guesswork was required in arriving at these figures.<11> The Vietnam War proved to be the most politically divisive, second only to the Civil War, that the United States had ever experienced. As American troop strength in Vietnam grew from 180,000 at the end of 1965 to over half a million in 1968 so did the casualty rate, which caused major political unrest at home.<12> Though American involvement began with genuine national security concerns, growing distrust of government motives and competence pushed public opinion into two irreconcilable camps. The "doves" viewed the war as one of national reunification and even an extension of the struggle against imperialism, and the "hawks" viewed the war as a noble defense of

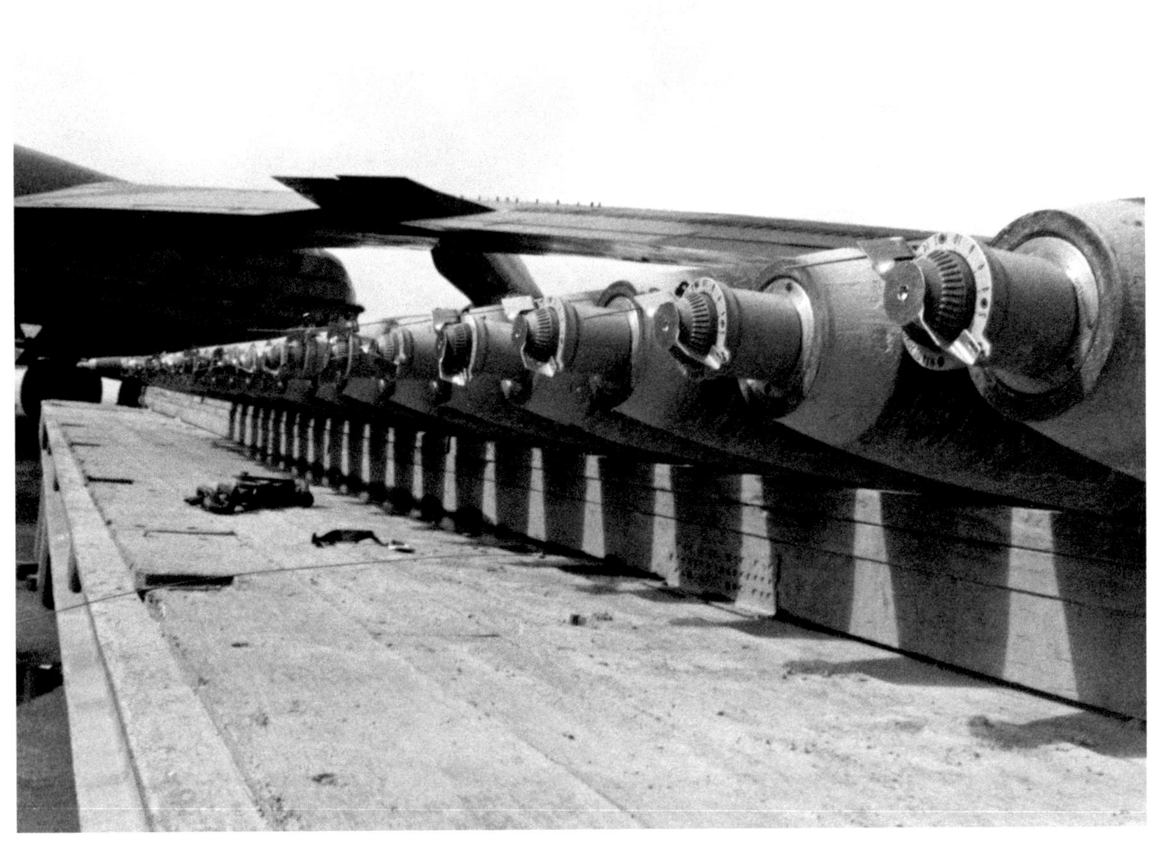

Some of the 175,000 tons of bombs used against North Vietnam in Operation LINEBACKER. Over the course of the war, the US dropped a million tons of bombs in Vietnam.
Courtesy National Archives, ID DF-SN-84-11607

democracy against brutal totalitarianism on the march. Extreme polarization of the two sides became an exercise in confirmation bias, something not unknown today. Lewy put it this way:

> ***The reality of the Vietnam War was composed of myriad events; the intricacy and variety of the scene was such that visiting 'hawks' and 'doves' could each observe, investigate and leave, assured of the wisdom of the view each held upon arrival. Like pieces in a kaleidoscope, the 'facts' of the Vietnam war could, and still can, be put together in a multitude of configurations which in turn lead to different political and moral judgments and conclusions.***<13>

The Americans brought iconic new weapons to this war. While helicopters found some use in the Korean War in scouting, rescue, and medical evacuation operations, they came into their own as combat weapons in the Vietnam War. The Bell UH-1 "Huey" was a mainstay of airborne assault missions, and the Bell AH-1 "Cobra" attack helicopter was used for both escort and close-air-support roles. Although tanks, including the M48 Patton and the M60 Patton, were used in a limited way in Vietnam, they represented only incremental improvements over WWII tanks. The weapon that perhaps generated the most controversy was the M-16 rifle, the first US battle rifle to abandon the .30 caliber bullet and dispense with a wooden stock in favor of a plastic one.

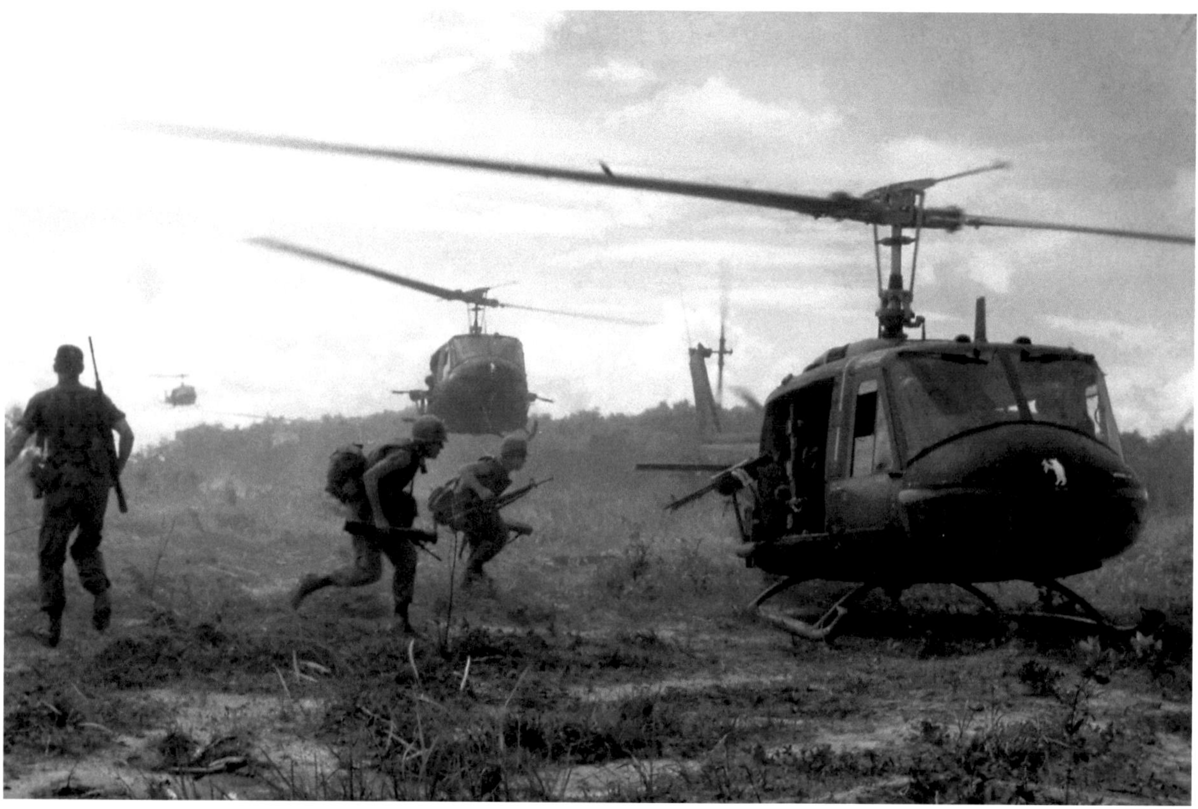

Use of helicopters like the "Huey" enabled routine assault and extraction missions behind enemy lines.
Courtesy Pixabay Images

In 1957 the Army had adopted an update to the WWII semi-automatic M1 rifle, the selectively automatic M14 that was chambered for the new 7.62 mm NATO round (still about .30 inch in caliber).<14> The Air Force had only separated from the Army in 1947 and still used Army supply for its small arms. For their security operations the M14 was considered too bulky. They preferred the small-caliber, high-velocity concept incorporated into the new Armalite AR-15 rifle, licensed to Colt and chambered for a 5.56 mm round (about .22 inch caliber). The Defense Department's Advanced Research Projects Agency (DARPA), the same organization that later developed the Internet, purchased 1,000 rifles from Colt in December 1962 for testing by US Army advisors imbedded in the South Vietnamese Army. The resulting report found them much better suited to the small stature of the Vietnamese soldier and the best "all around" shoulder weapon in Vietnam. There followed a number of anecdotal reports of its devastating wounding effects on the human body.<15>

In May 1962 the Air Force officially adopted the AR-15 as the M16 and the US Navy followed suit for its SEAL forces.<16> The rifle had lower weight, with lower recoil providing better control, and allowed the soldier to carry more rounds for the same combat load. DARPA field results also seemed to suggest higher lethality of the round. Though the M14 rifle was now being produced at the rate of 300,000 per year<17> by the Springfield Armory and three civilian contractors, a report extolling the virtues of the AR-15 landed on the desk of the new Secretary of Defense, Robert McNamara, who ordered head-to-head tests of the AR-15 with the M14 with a deadline only three months away. After two months the tests broke down in rancorous stalemate between the advocates of each rifle, and the Inspector General was called in to adjudicate. The IG found that the AR-15 was superior to the M14 in many respects but lacked the latter's reliability of function, ammunition, and dim-light sighting. They recommended continued development to overcome these deficiencies. Secretary of the Army Cyrus Vance nonetheless recommended that 50,000–100,000 AR-15s be procured for Army airborne and Special Forces. Because of its black plastic stock and black metal parts, it quickly became known as the "Black Rifle."

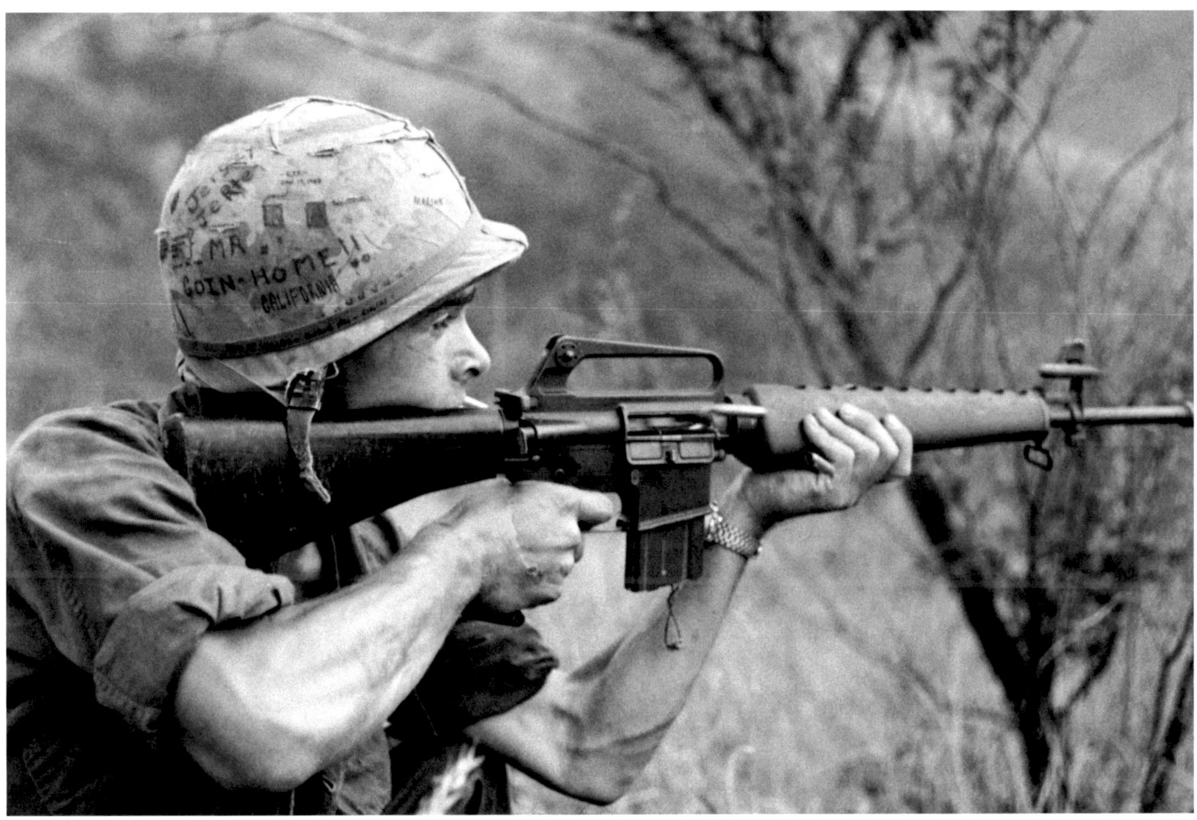

US Private Michael J. Mendoza firing an M16 rifle during the 1967 Operation Cook in Vietnam.
Courtesy National Archives, NA Identifier 17331452

The Army Ordnance Department had long-standing procedures for thoroughly testing arms and ammunition under wide-ranging battlefield conditions and correcting any defects before placing the weapon in the hands of a soldier in harm's way. McNamara, fresh from his success as Chairman of the Ford Motor Company, was brimming with confidence that he could cut through what he considered to be an ossified Army bureaucracy to expedite the fielding of new weapons. For the procurement of new weapons systems he instituted the position of Product Manager (PM), to be held by a bright young military officer, who would bear responsibility for each system and who would be answerable directly to the Office of Secretary of Defense (OSD). In effect, the PM would have the responsibility, but OSD the authority.<18> The net result was that decisions were made by OSD to override concerns with reliability, bypassing what McNamara considered to be ponderous and unnecessary testing, and field the weapon anyway. Proper operation of the M16 turned out to be sensitive to the type of gun powder, or propellant, used in its ammunition. The most egregious action authorized by the PM, but almost certainly dictated by OSD, was to certify ammunition tested with one type of propellant but send to Vietnam ammunition with a different propellant.<19> By the summer of 1966 reports were coming back from Vietnam that the new rifles were experiencing stoppages and malfunctions.<20> By May of 1967 a congressional committee was investigating the M16 program. A retired Army officer was sent to Vietnam to gather facts, and he reported that about 50 percent of the soldiers queried had had serious malfunctions with their M16s.<21> In the officer's opinion, the rifle required redesign and remanufacture as a result of hurried development. The result was that the PM, who on paper was responsible for the fiasco, was reassigned.<22> Subsequently fixes were found and adopted to mitigate the malfunctions, and the M16 rifle ultimately proved its worth. The Black Rifle debacle, however, stands as a cautionary tale against short-circuiting traditional development wisdom and experience.

PERSIAN GULF WAR

Operation LINEBACKER, in the last phase of the Vietnam War, started a seismic shift in the way wars are fought. Smart weapon technology continued to advance until its next major use in the Persian Gulf War in 1990–1991. By then the Global Positioning System (GPS) was coming online. Though the system was not fully functional until the last of 24 satellites was launched in 1995, 16 were in place and their positions were optimized for the Persian Gulf. In the featureless desert of southern Iraq and Kuwait, GPS proved crucial in both ground and air operations. The system enabled precision bombing, unit positioning, and artillery fire control 24 hours a day, even in the frequent sand storms.<23> Besides GPS, stealth technology in concert with laser-guided smart bombs played a big role, especially in the opening battles when F-117 Nighthawk fighter-bombers were charged with destroying the formidable Iraqi air-defense system—Bagdad had sixty surface-to-air (SAM) batteries and 3,000 anti-aircraft artillery (AAA) guns. The Nighthawks' stealth technology enabled them to pass virtually undetected to accomplish their mission, but their pilots' nerves were nonetheless tested because the frustrated Iraqis fired random barrages of missiles and AAA rounds into the air in hopes of snagging one.<24> Several hundred Tomahawk cruise missiles were employed, some fifty fired from WWII battleships Wisconsin and Missouri. The battleships also fired nearly 1,100 rounds from their 16-inch guns at Iraqi positions in Kuwait before falling silent forever.<25> The cruise missiles were not yet guided by GPS but used a combination of terrain-contour matching, which compared radar altitude readings to topographical maps at distances greater than 8 miles from the target, and digital scene matching, a technique that compared real-time television images to those stored in the missile's memory, at distances closer than 8 miles.<26>

Despite these high-technology wonders, some 93 percent of the bombs dropped by the US during the war were conventional "dumb" bombs that hit their targets only 25 percent of the time. Clouds, haze, smoke, and enemy fire also took their toll on the accuracy of smart munition delivery.<27> Despite these circumstances, pilots made a concerted effort to spare civilian targets, and all these factors combined to make a significant break from past bombing campaigns. According to journalist/historian Rick Atkinson:

> ***During the Normandy invasion in 1944, one civilian died for approximately every four tons of bombs dropped, in Vietnam in 1972, one died for every fifteen tons. In the gulf war, if the Iraqi civilian tally is accurate, there was one noncombatant death for roughly every thirty-eight tons.***<28>

The march of technology that had enabled massive bombing earlier in the century was now beginning to mitigate its indiscriminate consequences.

CHAPTER 7. EPILOGUE: TRIUMPH AND TRAGEDY

> **[Heisenberg's uncertainty principle]** ***fixed once for all the realization that all knowledge is limited. It is an irony of history that at the very time when this was being worked out there should rise, under Hitler in Germany and other tyrants elsewhere, a counter-conception: a principle of monstrous certainty.***
>
> —Jacob Bronowski, *The Ascent of Man*<1>

The twentieth century was characterized by the rise and fall of the paradigm of total war. It waxed as a result of humankind's newfound technological and industrial capabilities. These remarkable new powers, in turn, were made possible by a profound change in how human beings relate to their world, by a shift of authority from tradition to observation, and by the unshackling of their creative genius. War has always been part of the fabric of human existence so it was not surprising that these new tools would find their way into this realm as well. Humankind's natural competitiveness and hubris then drove the cycles of innovation until, with the development of the hydrogen bomb, civilization found itself at risk of self-immolation. Only the major powers had the resources to develop such weapons in the beginning though proliferation among many nation states demonstrates the growing ease of obtaining them. So far, the major powers have acted rationally in their own self-interest and drawn back from the brink. The total war paradigm has thus seemingly passed. As technology spreads, non-state groups may eventually get access to weapons of mass destruction, possibly through intentional transfer from third-world nations like North Korea or Iran, and the world may be in for some nasty surprises. However, the major powers will need to be involved for there to be another world war. With luck that prospect will remain improbable and the nuclear genie will stay in its bottle. Regional conflicts are not likely to end, but it is probable that further advances in smart weapons technology, while not preventing them, will continue to lower the civilian-to-combatant casualty ratio. If that happens, the twentieth century will retain its crown of infamy among the ages.

If nothing else, the study of history reaffirms the importance of the individual. On the face of it, it seems improbable for single individuals to make a difference to the affairs of the some 108 billion human beings that have lived on the earth.<2> Nikolaus Copernicus, Isaac Newton, James Clerk Maxwell, Otto von Bismarck, Albert Einstein, *Kaiser* Wilhelm II, Alfred Krupp, John Pershing, Adolf Hitler, Josef Stalin, William Knudsen. Winston Churchill, and many others all played important roles in the story unfolded here—individuals both brilliant and tragically flawed. Had each not lived, it is quite possible that the course of history might have been different. In all probability, if any of the scientists had not lived, their discovery would eventually have become someone else's discovery. Even granting that, timing could have been everything. If Bismarck's unique statesmanship had been absent, perhaps Germany would not have been unified when it was. If a less flawed and feckless individual than Wilhelm II had been Kaiser, perhaps both world wars could have been avoided altogether. Suppose Albert Einstein's stunning discovery of the mass/energy equivalence, crucial to realizing the implications of nuclear fission, had not been published in 1905. Perhaps, the atomic bomb would not have been ready in August 1945, and the war may have ground on into 1946 (as was expected by Pentagon planners) with untold more deaths and suffering. If Bill Knudsen had not so skillfully set in motion the American war mobilization, perhaps the war would have lasted another year or two. Obviously, if Adolf Hitler had not lived, history would have been different. The relatively recent field of Complexity Theory has found many instances of physical phenomena where insignificant changes to one part of a system can alter the course of the entire system. The inordinate influence of individuals on history suggest parallels with this theory. At the November 1943 Tehran Conference with Stalin and Roosevelt, Winston Churchill, despite being convinced of his own historical destiny, glimpsed the twin realities of human significance and insignificance when, late into the night to his physician Lord Moran he mused,

> ***Stupendous issues are unfolding before our eyes, and we are only specks of dust, that have settled in the night on the map of the world.***<3>

A speck of dust he may have been, but his inspired leadership may well have saved his country from a brutal Nazi occupation and assured a safe staging ground for the Normandy invasion. Churchill understood this reality as well when he said, ***We are all worms, but I do believe I am a glow worm.***<4>

The twentieth century was both the brightest period in the evolutionary journey of humankind and the darkest. Heir to three centuries of discovery of what the mind can do, humanity used that growing awareness to virtually end the timeless cycles of famine and epidemics of disease; to engineer a walk on the moon; to discover that solid matter is overwhelmingly empty space; to discover that our enormous life-giving sun is but one star out of billions in our galaxy, a galaxy that itself is only one of billions of other galaxies each having billions of stars. The wondrous new powers were used to chart the history of the universe from the tiniest fraction of a second when it was an unfathomable cauldron of light, electrons, and neutrinos, to its distant future of cold, lifeless darkness; and to infer that all the visible matter in the universe with which we are familiar is dwarfed by twenty-five times as much virtually undetectable dark matter and dark energy.<5> These are stunningly magnificent accomplishments. And yet, these same miraculous gifts were placed in the service of chauvinistic ambitions that twice plunged the world into paroxysms of death and destruction on unprecedented scales. This is the great mystery of human existence: its agony and its ecstasy.

Remarkable as human capacities are, still there are limits to the human ability to grasp and manage complexity. As noted in Chapter 2, Isaac Newton, though founding his revolutionary theory of gravitation, lamented that he could not reconcile the success of his mathematics with an intuitive understanding of the physics. More than two centuries later, Niels Bohr, in accepting the 1922 Nobel Prize in Physics for his work on the structure of the atom, echoed the same sentiment when he said that we must ***content ourselves with concepts which are formal in the sense that they do not provide a visual picture of the sort one is accustomed to require of the explanations with which natural philosophy deals.***<6> Richard Feynman, 1965 Nobel laureate in physics for his work on quantum electrodynamics, confessed to students that ***I have no picture of this electromagnetic field that is in any sense accurate.***<7> And Feynman was a physicist who attributed much of his success to his tenacious struggles to visualize the phenomena he studied. The scale of the horrors in the First World War, the first war of mass production, and in its successor was a consequence of the peculiar dichotomy of the human mind which allows it to follow a logical progression of small-scale steps, on the one hand, but miss completely the large-scale consequences on the other.

Harold Nicolson was a member of the British delegation to the 1919 peace talks at Versailles and lifelong diplomat. As junior member of the Foreign Office, it had fallen to him late on the night of 3 August 1914 to deliver Britain's declaration of war to the German Ambassador Prince Karl Max

Lichnowsky.<8> At the end of the war, in his duties at Versailles, Nicolson saw firsthand the enormous complexities, conflicts and contradictions facing Lloyd George, Clemenceau, and Wilson—the most influential members of the Council of Five. He was only too aware of their personal shortcomings. Recognizing the ominous portents created by the Versailles Treaty, after the war he wrote:

> ***All that I hope to suggest is that human error is a permanent and not a periodic factor in history, and that future negotiators will be exposed, however noble their intentions, to futilities of intention and omission as grave as any which characterized the Council of Five. They were convinced that they would never commit the blunders and iniquities of the Congress of Vienna. Future generations will be equally convinced that they will be immune from the defects which assailed the negotiators of Paris. Yet they in their turn will be exposed to similar microbes of infection, to the eternal inadequacy of human intelligence.***<9>

The twentieth century taught us that our future will be determined both by the as-yet-unfound pearls of rationality and by the ***eternal inadequacy of human intelligence***. Between those poles lie the triumph and tragedy of the human experience. The lingering vestiges of the turbulent century's war machines, both conventional (in the next chapter) and nuclear (in my previous books), serve as reminder and metaphor of our brilliance and folly.

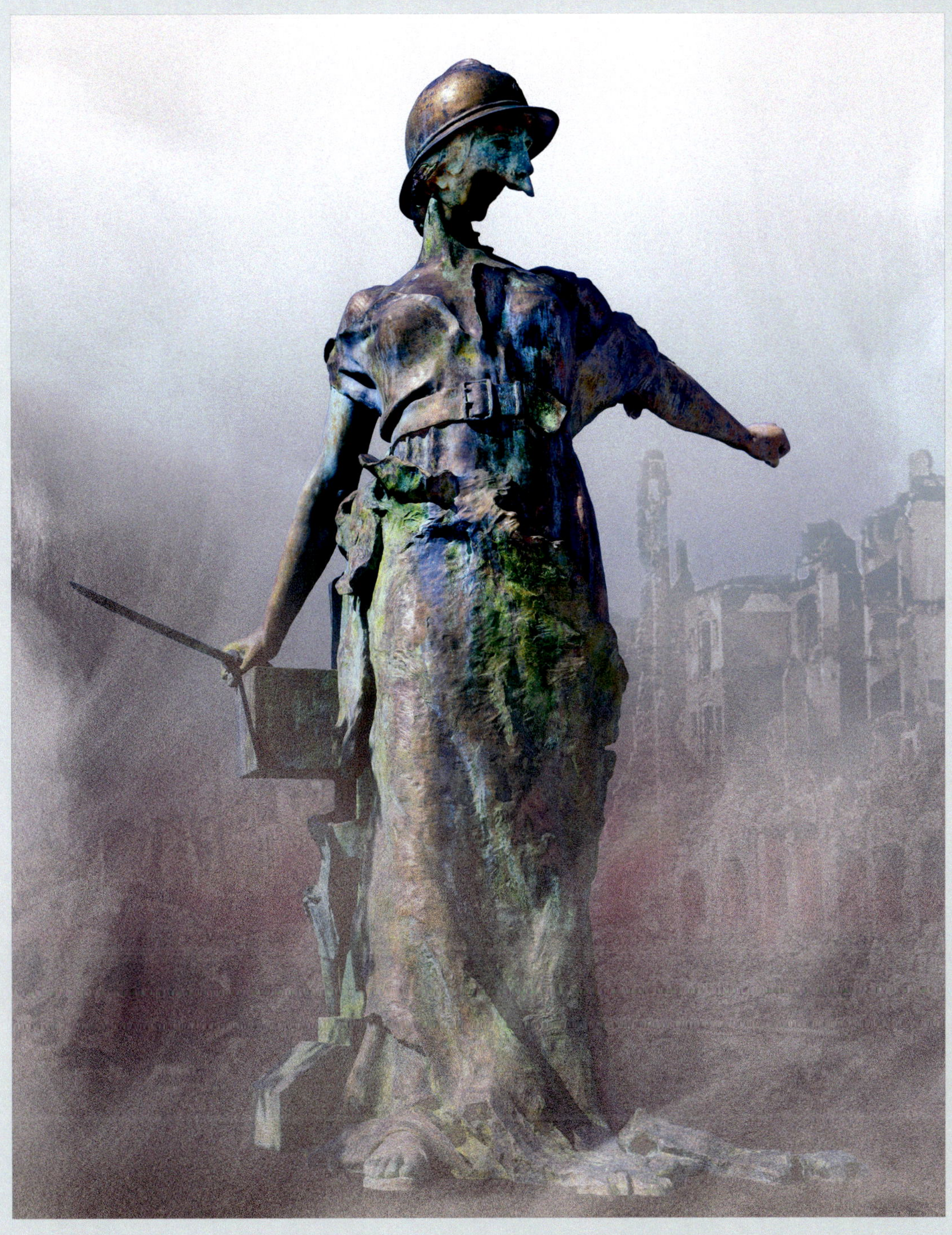

I Am War

The above image uses a photograph of the Dresden destruction of February 1945 as a backdrop to my photograph of a replica of Le Monument aux Morts by Edmond de Laheudrie in Trevieres, France about 7 km from Omaha Beach. The statue was originally dedicated in 1920 to the memory of 43 soldiers from the village that were killed in WWI. Two days after D-Day 1944 its face was damaged by an Allied artillery shell. There could be no more poignant symbol of the recurrent horrors of war. The replica statue stands at the WWII Memorial in Bedford, VA, USA.

CHAPTER 8. GALLERY: PHOTOGRAPHS BY THE AUTHOR

> ***... rather than giving us verifiable access to the real, memory, even and especially in its belatedness, is itself based on representation. The past is not simply there in memory, but it must be articulated to become memory.***
>
> —Andreas Huyssen, *Twilight Memories: Marking Time in a Culture of Amnesia*<1>

In the second chapter of this book the world was shown to have been transformed by a relatively few highly intelligent, creative, and hard-driving individuals. In Chapter 3 the new technological advances were seen to have been harnessed in the service of national ambitions by other, reckless individuals pursuing grandiose, imperialistic agendas that led, ultimately, to two world wars. These wars were different from prior wars, both qualitatively and quantitatively. They were marked by advanced technology multiplied by mass production, and they placed onerous new demands upon those who were compelled to fight them. Previous wars required courage and endurance; the new wars required, in addition, much greater analytical thinking under pressure by operators of the new war machines. These new machines were complex and dangerously unforgiving of operational error, bad design, or faulty ordnance. Yet, despite the unprecedented degree of injury and death they caused, soldiers, sailors, and airmen rose to their respective challenges—a testament to the extraordinary adaptability of human beings. In this last chapter, portraits of the war machines of the twentieth century, in all their detail and brutality, bear witness to the enormous complexity of these machines and also speak volumes about their now absent human partners. As Huyssen suggests in the epigraph above, powerful memories require powerful representations. It is my hope that the preceding narrative combined with the following images will shape a deeper appreciation of how modernity's astonishing rise, coupled with humanity's improvident nature, produced the barbarous realities of twentieth-century warfare.

Note: Dates at the end of the caption titles indicate the dates of introduction into the inventory or final testing if not adopted.

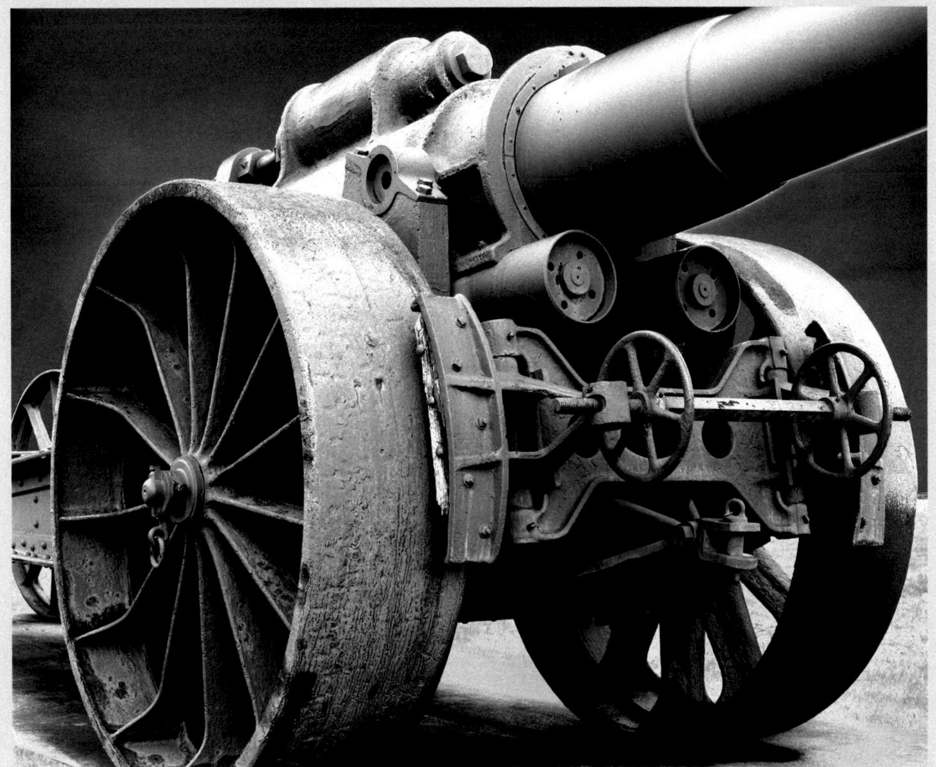

US 6-inch towed cannon 1905 Prior to WWI the US Army had not envisioned a large role for heavy artillery. As static trench warfare settled in and heavy field artillery proved its worth, they were forced to find readymade expedients. This gun was originally designed as a seacoast defense weapon and was tested at the new (October 1917) Aberdeen Proving Ground with various wheeled carriages such as the one shown here.<2> *Former Collection of the US Army Ordnance Museum, Aberdeen Proving Ground, MD*

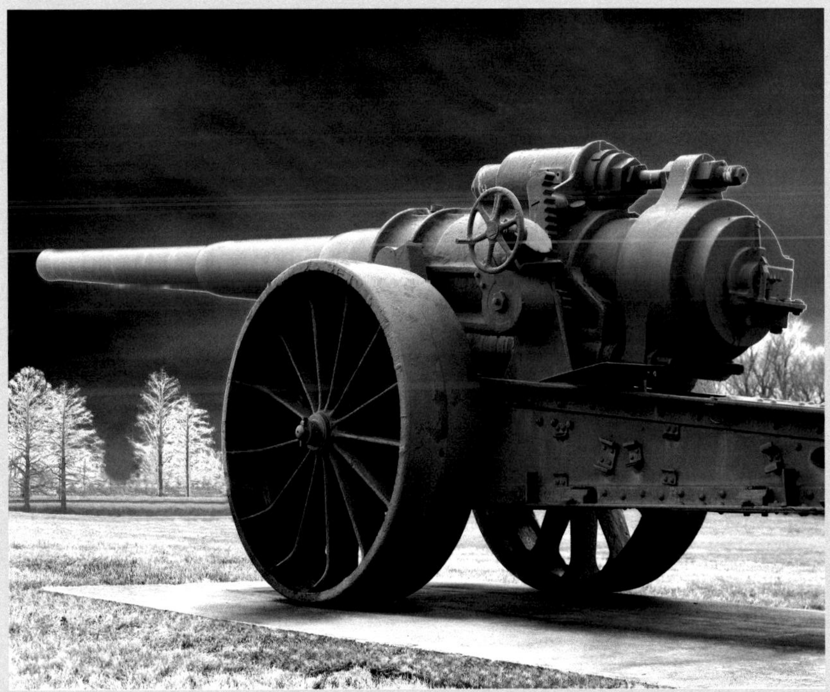

US 6-inch towed cannon 1905 The 1905 carriage for the 6-inch seacoast defense gun was tested in October of 1918 and found to be impracticably heavy.<3> These tests occurred barely a month before the November Armistice, illustrating how late the US was in achieving mobilization. Though this gun failed to be fielded, the Navy did successfully adapt five 14-inch battleship guns to railway carriages in early 1918. The railway guns pounded German positions in the closing weeks of the war.<4> *Former Collection of the US Army Ordnance Museum, Aberdeen Proving Ground, MD*

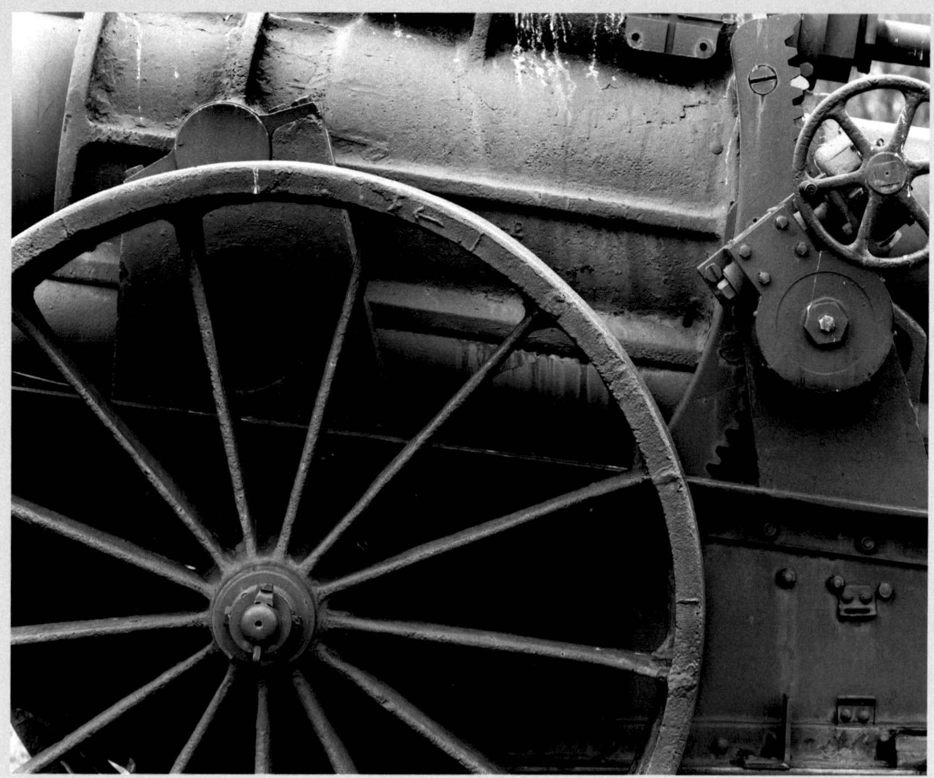

Detail, US 6-inch towed cannon 1905 As a result of the rejection of this gun/carriage combination, the Army adopted the **French 155-mm cannon 1917**, seen later.<5> Because of the US late entry into the war and its tardy mobilization efforts, the American Army mainly used British and French artillery and ammunition while supplying those countries with the steel required for their fabrication. *Former Collection of the US Army Ordnance Museum, Aberdeen Proving Ground, MD*

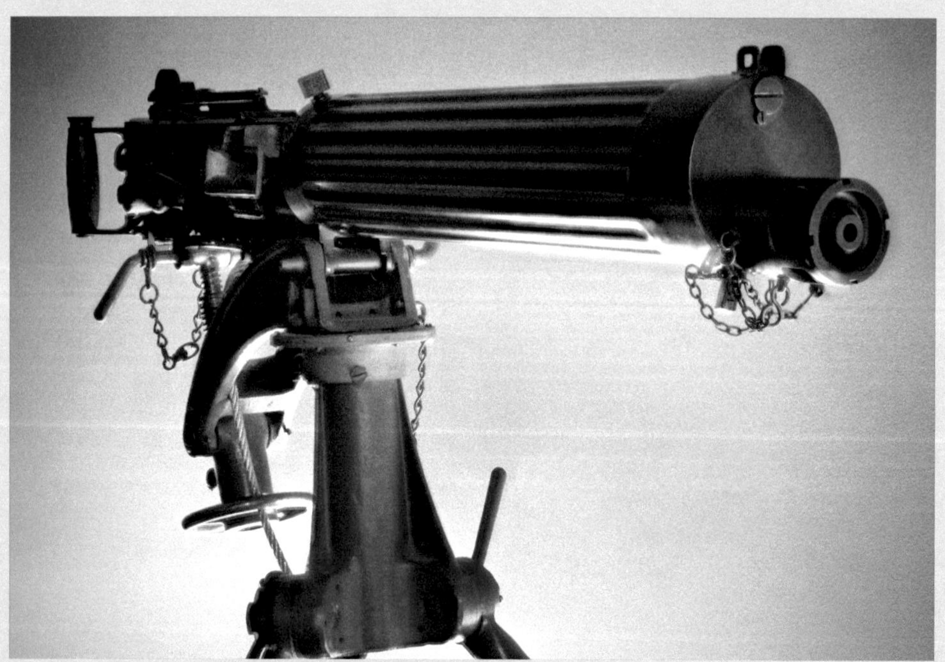

Vickers (improved Maxim) machine gun 1912 The Maxim machine gun or variants were used by all sides during WWI. Vickers redesigned the receiver to be lighter and more compact and further lightened the gun by replacing the heavy brass water cooling jacket surrounding the barrel by a corrugated, thin steel version. The gun proved so successful that it remained in use by the British Army until 1968.<6> Though the US adopted the Maxim in 1904 and later the Vickers, less than 300 had been acquired before it was superseded by the lightweight Benét-Mercié Machine Rifle that proved more mobile but less battle worthy.<7> *Collection of the British Imperial War Museum*

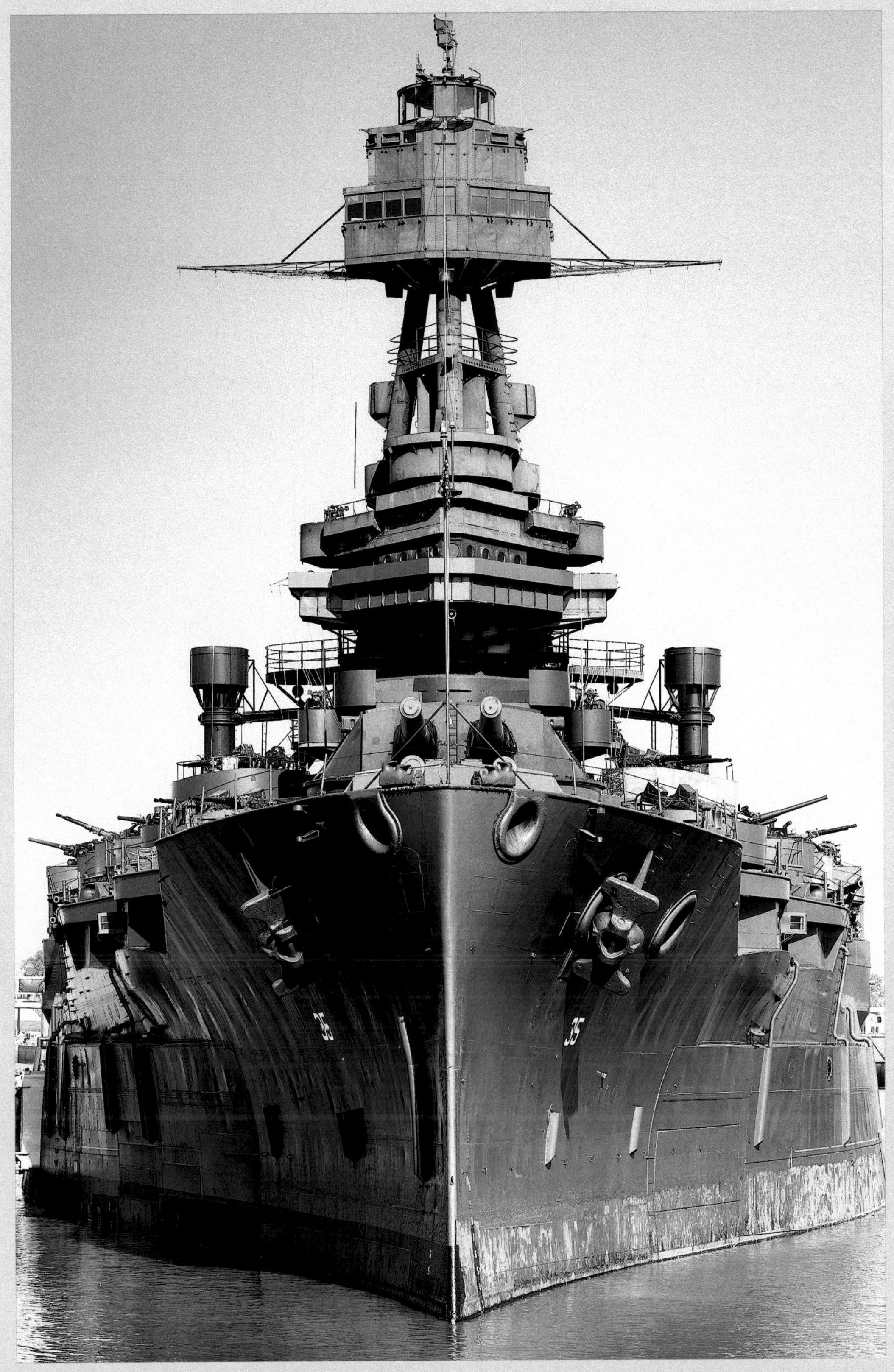

USS Texas battleship BB-35 1914 The sister ship of the USS *Texas*, the USS *New York*, was the first of this class of two ships. The *New York* class was the fifth class of dreadnought-type battleships built by the US. The first, the USS *South Carolina*, was launched only two years after HMS *Dreadnought* so the US Navy was far more up-to-date than the US Army when the war was enjoined in April 1917.<8> When commissioned on 12 March 1914, the *Texas* was the largest battleship in the world, displacing 28,367 tons fully loaded.<9> It is the last dreadnought-era battleship in the world. *At Battleship State Historic Site, Houston, TX*

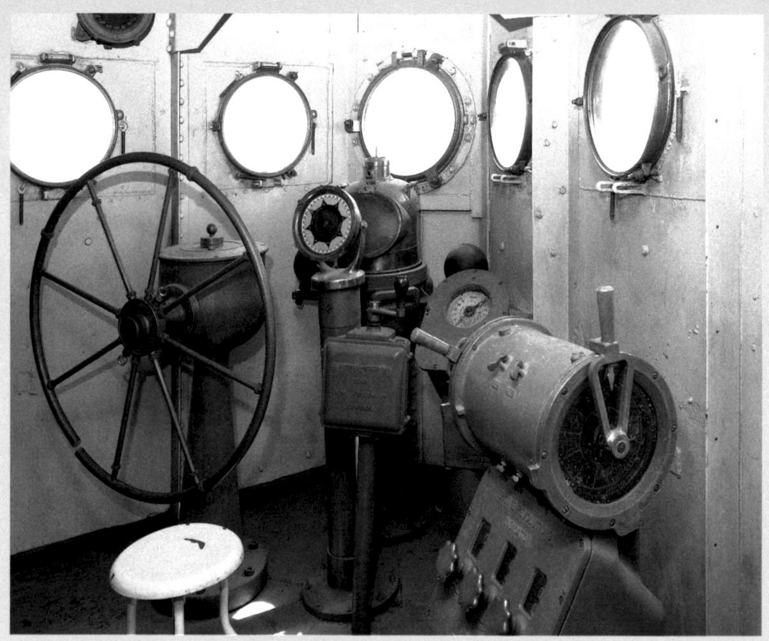

Pilot House, First Bridge, USS Texas 1914 While supporting the Normandy invasion with bombardment of German shore positions off Cherbourg, the Texas wheelhouse was hit by two German shells. One, fortunately, was a dud; at 240mm caliber it would have caused quite a mess. As it was, one sailor was killed and ten injured. Two more sailors sustained concussion injuries from the *Texas's* own guns. The guns of the Texas were never silent during this crisis.<10> *At Battleship State Historic Site, Houston, TX*

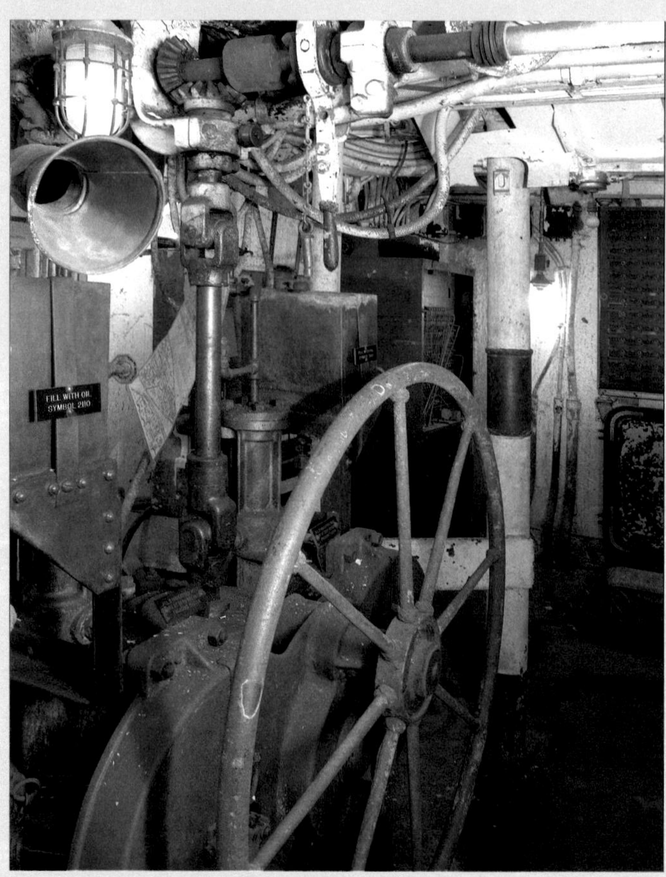

Auxiliary helm, Central Station, USS Texas 1914 Central Station was the most protected room on the ship. It was designed to provide emergency command and control of the ship even if the ship were damaged. *At Battleship State Historic Site, Houston, TX*

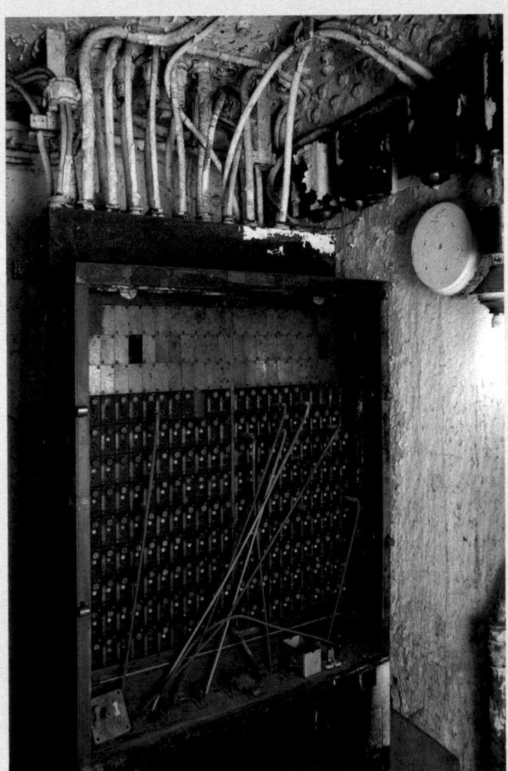

PBX Telephone Switchboard, Central Station, USS *Texas* 1914 Communication was vital in combat conditions aboard ship to coordinate damage control, offensive actions, and maneuver. Amazingly, sound-powered telephone headsets were employed for maximum reliability and this switchboard permitted voice contact from Central Station to any part of the ship — only 35 years after the telephone was invented. *At Battleship State Historic Site, Houston, TX*

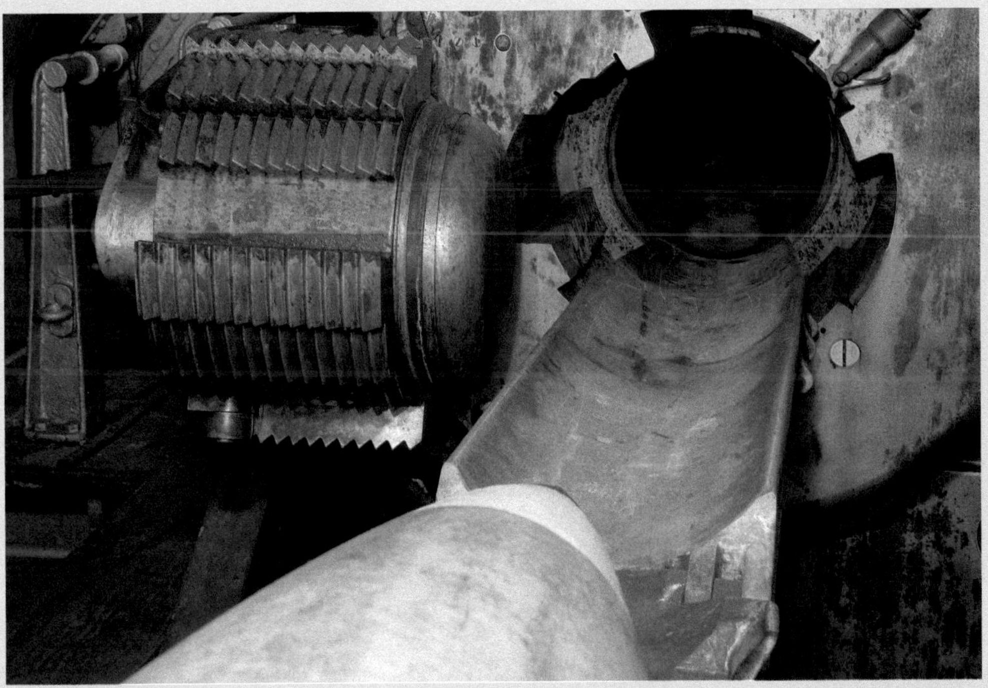

14-inch projectile loading, USS Texas 1914 The USS *Texas* carried ten 14-inch guns in five turrets, three forward and two aft. Previous classes had only 12-inch guns. When commissioned, its ten magazines could hold 1,000 armor-piercing projectiles.<11> Each projectile weighed 1,500 pounds. The threaded breach block at left alone weighs several tons. Behind each projectile rammed into the breech would follow four silk bags of cordite (smokeless) powder weighing a total of 420 pounds.<12> All this barely 25 years after the invention of smokeless powder necessitated complete redesigns of gun barrels. *At Battleship State Historic Site, Houston, TX*

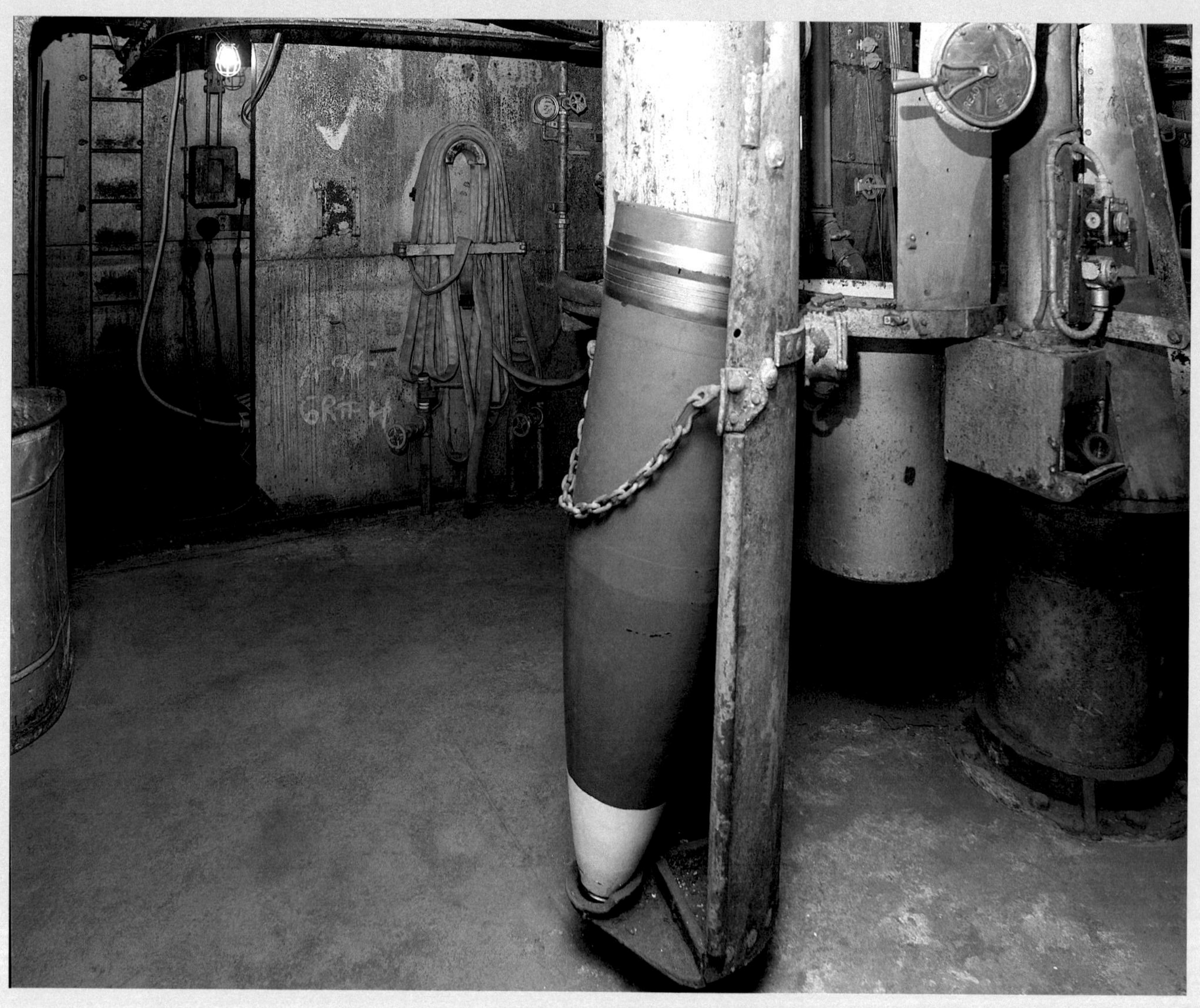

Projectile elevator, USS Texas 1914 The 14-inch projectiles came in three models with explosive weights varying from 23 to 104 pounds.<13> They were stored nose down in the magazine and sent aloft to the gun room in the turret above in the same pose. *At Battleship State Historic Site, Houston, TX*

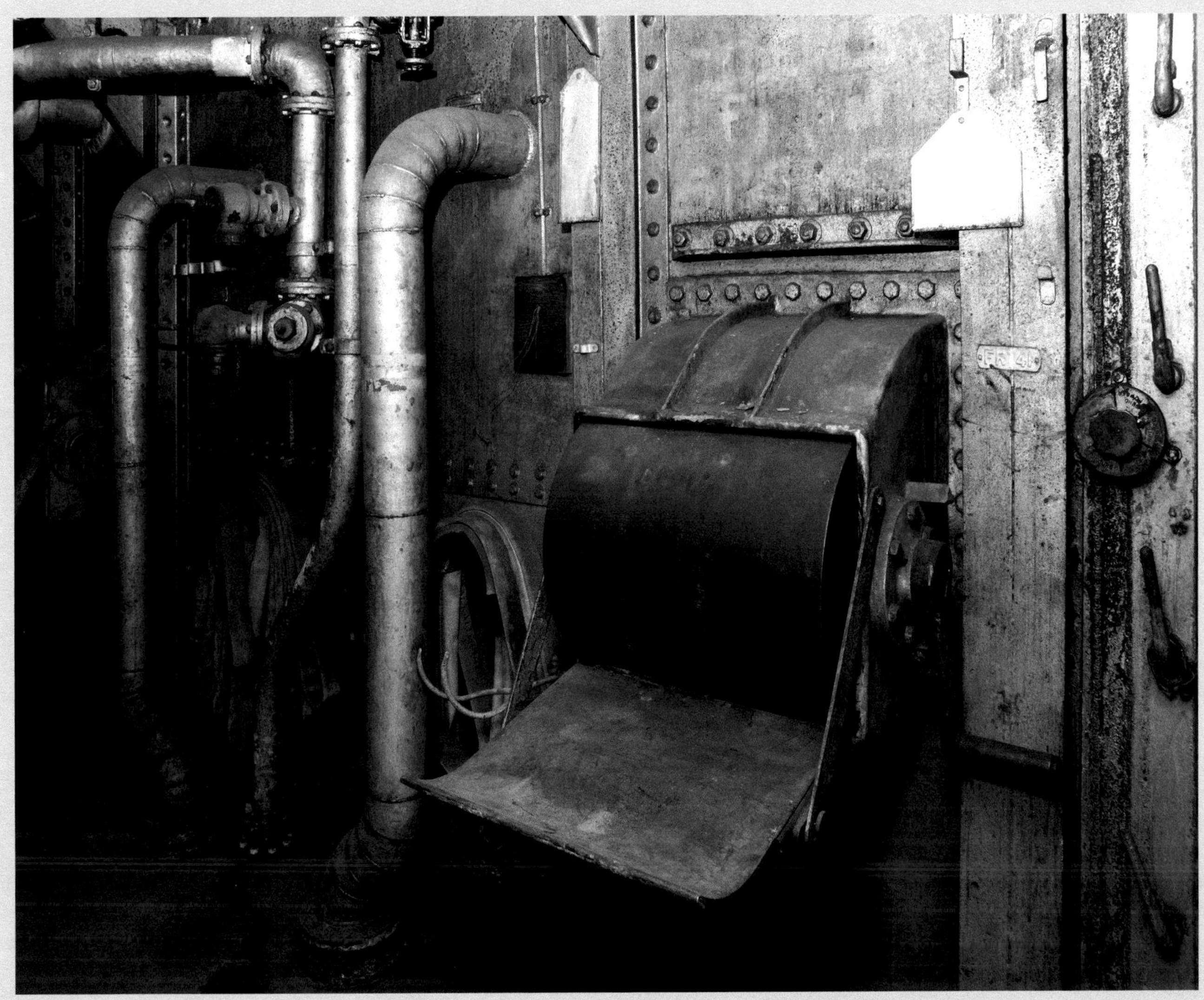

Gun Powder Flash Baffle, USS *Texas* 1914 One of the most serious shipboard disasters was for the powder magazines to catch fire and explode. This could lead to detonation of the projectile explosive filler and quickly sink the ship. To prevent fire from propagating to the magazine after hits to the turret above, these revolving drum baffles were used in the wall between the powder magazine and shell-handling room where powder bags and projectiles were sent by hoist to the guns in the turret above. The hoist tunnels were themselves fitted with baffles at both ends and closed automatically after the sensitive materials were sent aloft.<14> *At Battleship State Historic Site, Houston, TX*

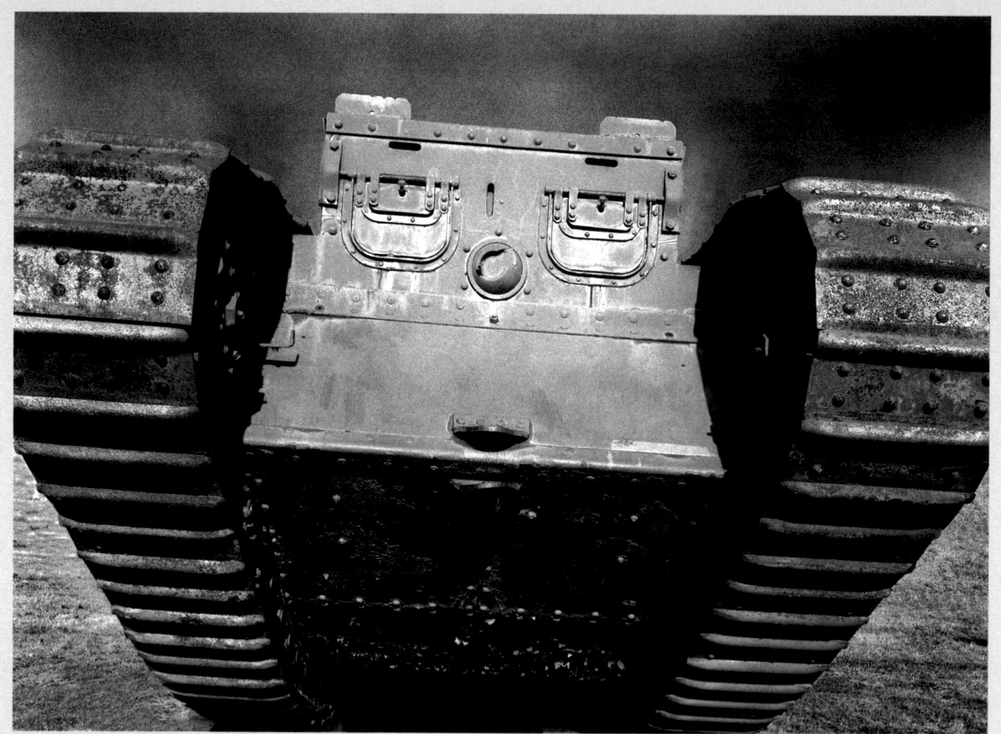

British tank, Mk IV "female," 30.2 tons combat weight<15> 1917 The first successful breach of German lines by tanks, including the Mark IV, was at Cambrai on 20 November 1917. The attack involved more than 400 tanks and gained some seven miles beyond the Hindenburg Line of defenses. However, insufficient preparations for exploiting the breakthrough led to the Germans recapturing the ground in a counterattack ten days later. Nonetheless, the tank's potential for breaking the stalemate was demonstrated.<16> *Former Collection of the US Army Ordnance Museum, Aberdeen Proving Ground, MD*

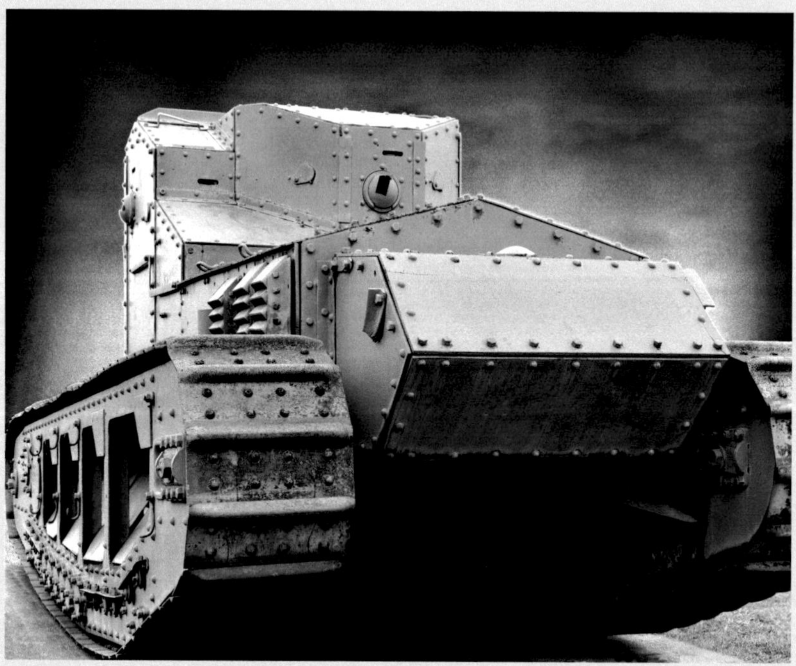

British medium tank, Mk A Whippet 1917 Armed with five .303-inch Lewis Model 1917 machine guns and weighing 15.7 tons,<17> the Whippet more than doubled the speed of the Mark IV. It first appeared in battle on 26 March 1918. However, it was not until the battle at Amiens on 8 August 1918 that the use of tanks led to a major breakthrough of the German line. German commander General Erich von Ludendorff called it "the black day of the German Army."<18> *Former Collection of the US Army Ordnance Museum, Aberdeen Proving Ground, MD*

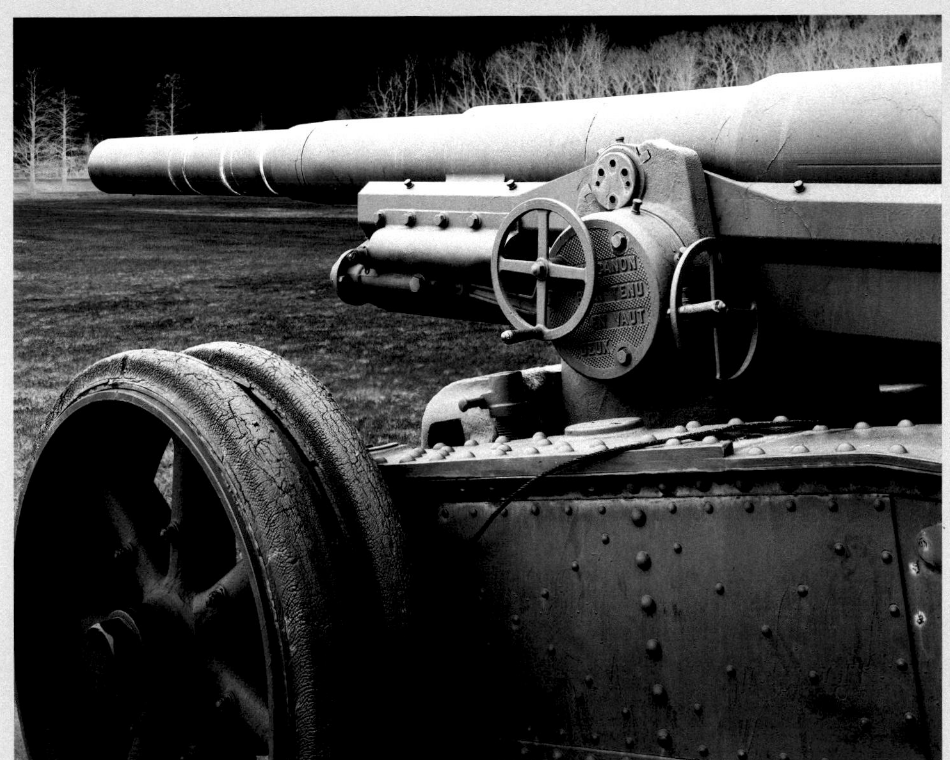

French 155-mm GPF cannon 1917 This long-range heavy artillery piece was designed by French Colonel L.J.F. Filloux (GPF is the abbreviation for Grande Puissance Filloux, or Great Power Filloux). It was made by a French government arsenal and under contract by Renault for both French and American forces. Capable of hurling a 95 pound shell to a distance of 10 miles, it became a mainstay of both armies from WWI into WWII.<19> *Former Collection of the US Army Ordnance Museum, Aberdeen Proving Ground, MD*

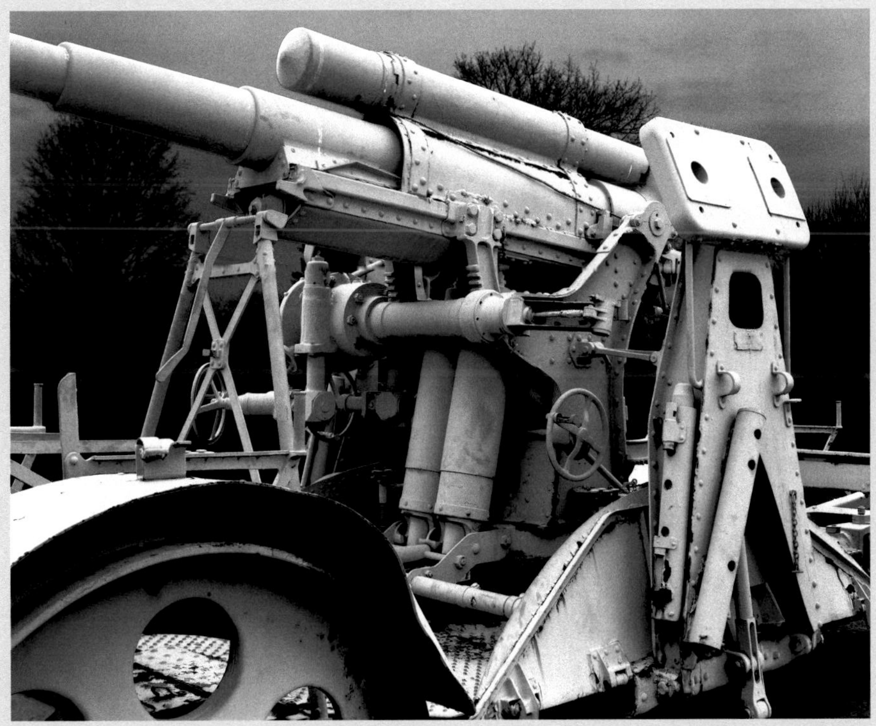

German 88-mm Flak gun 1917 When the German High Command determined that the new aircraft could be a serious threat, an 88-mm naval gun was adapted to fire at high elevation angles. This gun was the result. By the end of the war, some 1200 Allied airplanes had been shot down by Flak guns.<20> *Former Collection of the US Army Ordnance Museum, Aberdeen Proving Ground, MD*

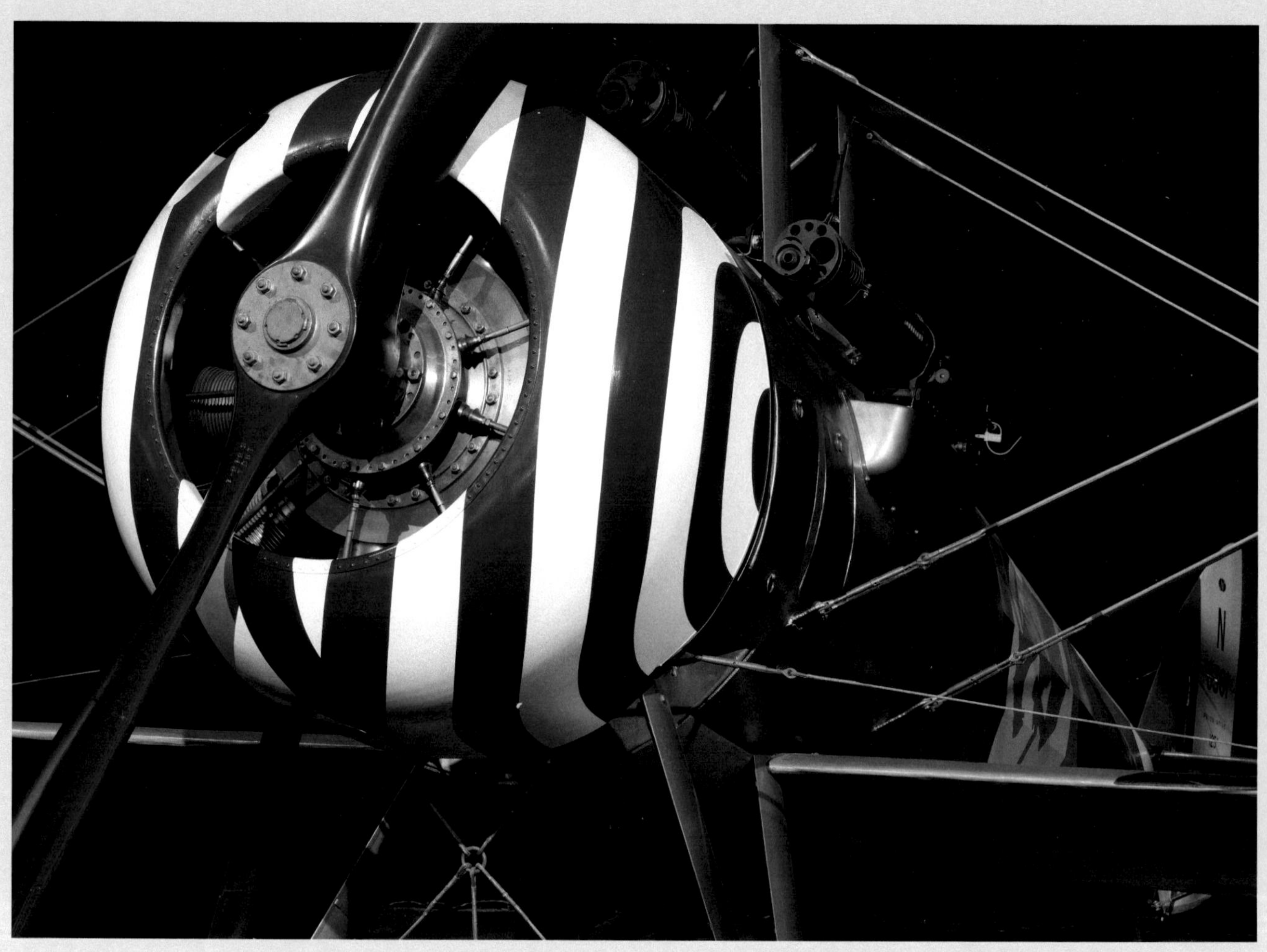

French Nieuport 28 1918 The Nieuport 28 was developed in the latter half of 1917 and put into production in early 1918. It did not prove to be a reliable fighter and the French quickly adopted the SPAD XIII instead. The newly arrived American Expeditionary Force, on the other hand, were desperate for aircraft and bought 297 Nieuport 28s. This particular plane was rebuilt from some original parts by Air Force museum restorers and painted in the colors of the 95th Aero Squadron, one of the first American units to fly them. Emblematic of the new paradigm of rapid obsolescence, these fighters were replaced by the newer SPAD XIII within months.<21> *Collection of the National Museum of the US Air Force*

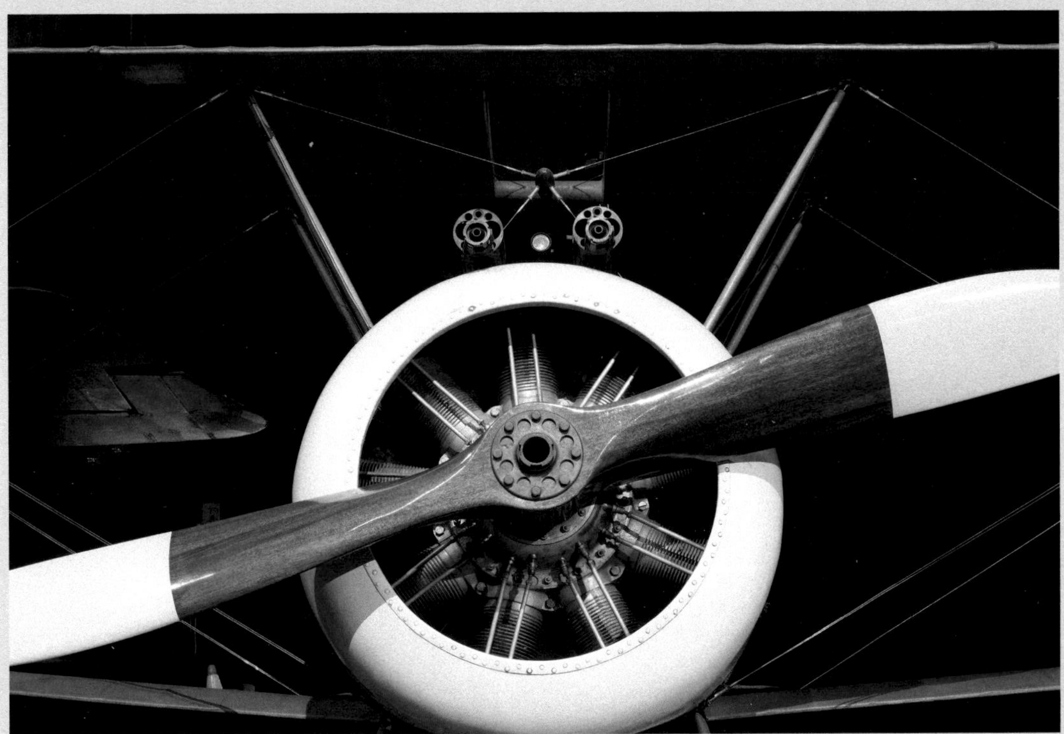

British Sopwith F.1 Camel 1917 This legendary plane was responsible for more enemy aircraft kills than any other in the Allied fleet. Its high maneuverability came at a cost, however. Because of its challenging handling characteristics, almost as many pilots died in training as in actual combat. It was armed with two Vickers .303-inch-caliber machine guns, which are visible just above the engine cowling. This particular aircraft is a replica completed in 1974 by US Air Force personnel from original factory drawings. Almost 5,500 were originally built but few are extant today.<22> *Collection of the National Museum of the US Air Force*

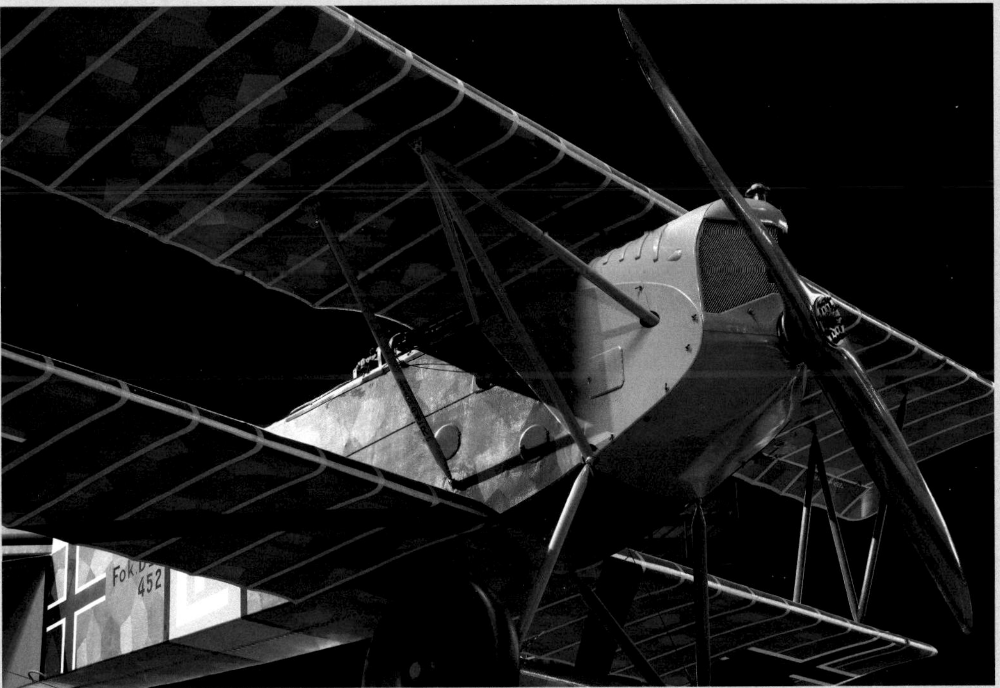

German Fokker D.VII 1918 First appearing in late spring 1918, the D.VII was chosen over thirty competing designs. It proved to be Germany's ablest fighter right up to the end of the war. Its easy handling qualities facilitated the rushed training of new pilots in the late stages of the war. More than 1,720 were produced compared with only several hundred of its predecessor, the Fokker Dr. I triplane.<23> *Collection of the National Museum of the US Air Force*

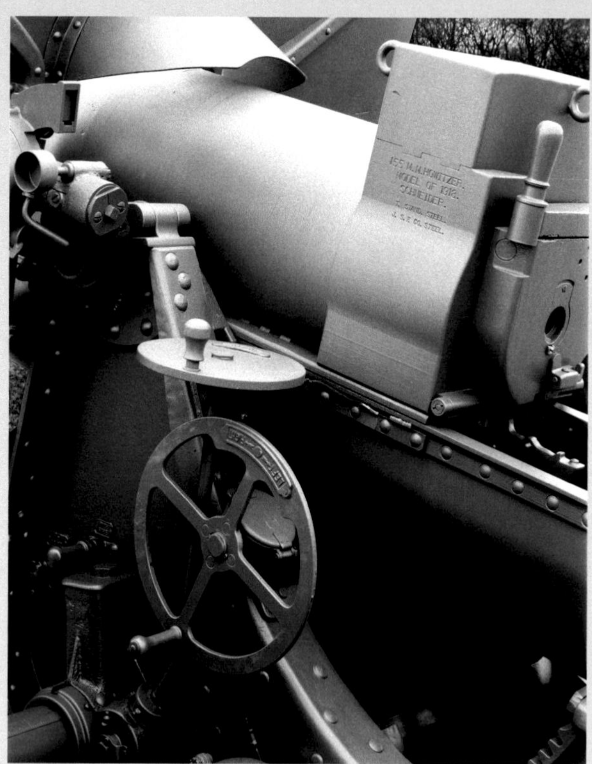

US 155-mm gun (Schneider) 1918 French design adopted by the US as M1918. Some, including this specimen, were made in the US late in the war. Shells for this gun and the 155-mm GPF were made in Baltimore, Maryland by the Bartlett-Hayward Corporation as well as in France. Due to its late start, the Baltimore plant had produced only 135,600 rounds by the end of the war.<24> *Former Collection of the US Army Ordnance Museum, Aberdeen Proving Ground, MD*

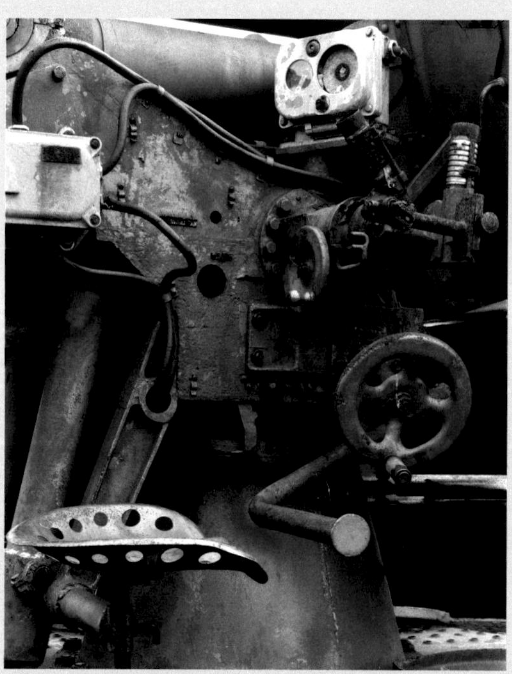

US 3-inch field gun 1918 When the US entered the war in April 1917, it had less than seven hundred 3-inch field guns. This was enough for an army of 250,000 but not for the expected two million man army to be sent to France. New designs were tried resulting in the M1916 and M1917 models but problems with inaccuracy and lack of robustness caused delays due to redesigns and retooling. The Ordnance Department also tried manufacturing the French 75mm gun but only a little over a hundred were completed by the end of the war.<25> This model of 1918 never entered service before war's end. *Former Collection of the US Army Ordnance Museum, Aberdeen Proving Ground, MD*

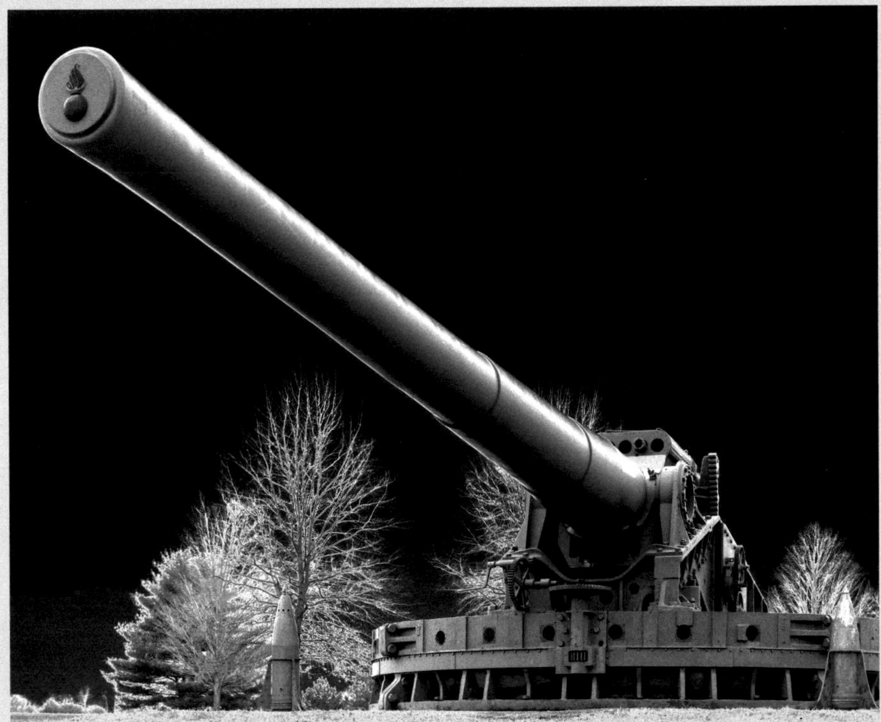

US 16-inch coastal defense gun 1922 As a result of the 1922 Washington Naval Treaty limiting the size of warships and their armament, 16" guns of 50-caliber length (66.6 feet) intended for next-generation battleships and battlecruisers were rendered unusable by the US Navy and some twenty were transferred to the Army for use as coastal defense guns.<26> Using a powder charge up to 700 pounds, the gun could throw a one-ton projectile some 25 miles.<27> *Former Collection of the US Army Ordnance Museum, Aberdeen Proving Ground, MD*

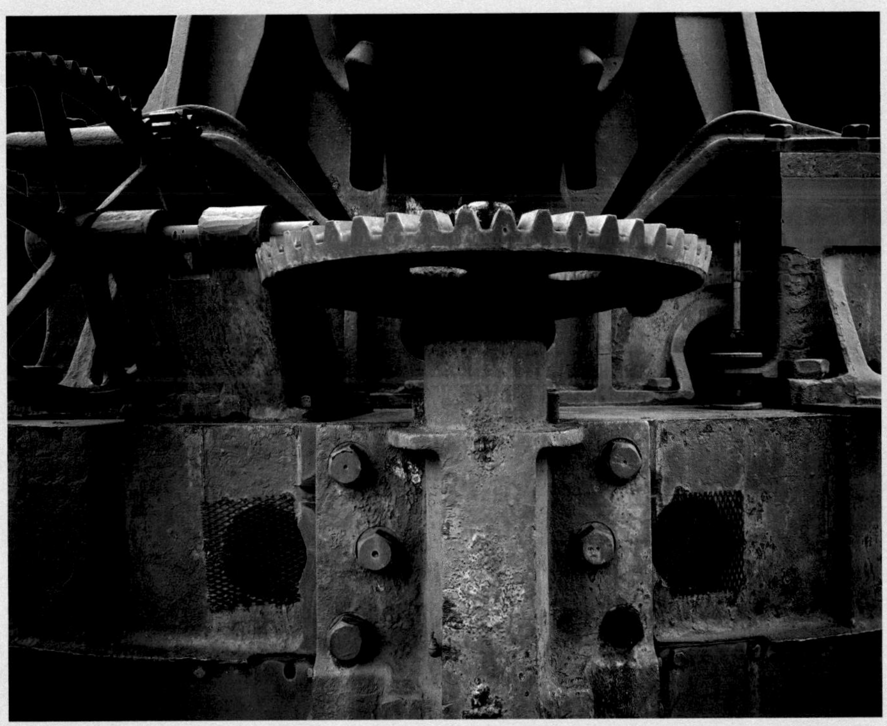

Traverse gear, US 16-inch coastal defense gun 1922 The barrel of this gun is so massive that a special technique was used to fabricate it. A 1-inch x 1-inch steel wire was wound under tension around an inner sleeve. Then steel hoops were then expanded by heating and shrunk about the windings.<28> *Former Collectin of the US Army Ordnance Museum, Aberdeen Proving Ground, MD*

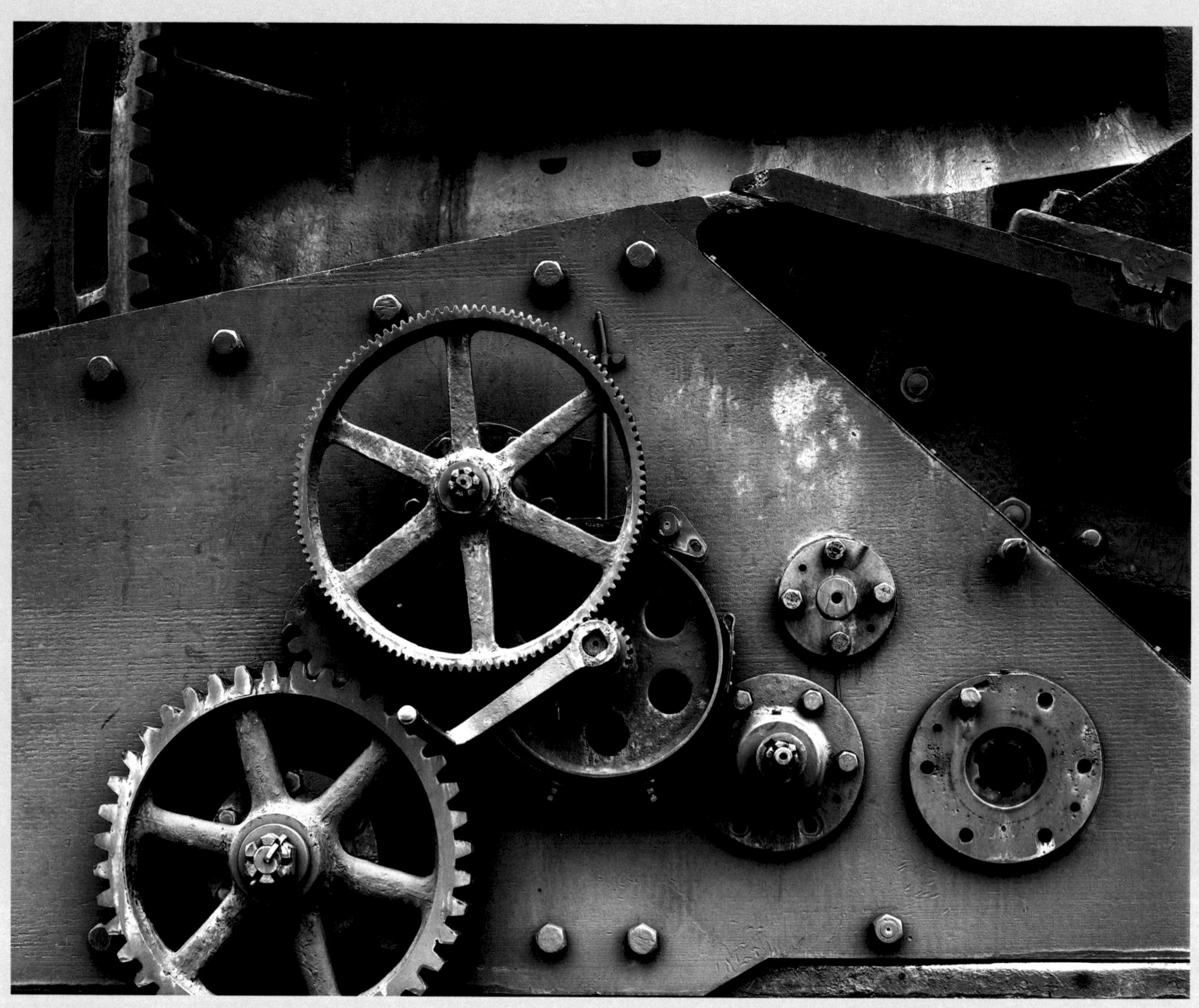

Elevation mechanism, US 16-inch coastal defense gun 1922 A special breech block prevented the escape of the propellant gases without the need for a brass shell casing. The projectile was first loaded, then up to eight bags of smokeless powder were inserted behind the shell. A primer then ignited black powder which, in turn, ignited the propelling charge.<29> Black powder continued to be used over most of the 20th century to ignite the smokeless-powder charges in artillery ammunition. *Former Collection of the US Army Ordnance Museum, Aberdeen Proving Ground, MD*

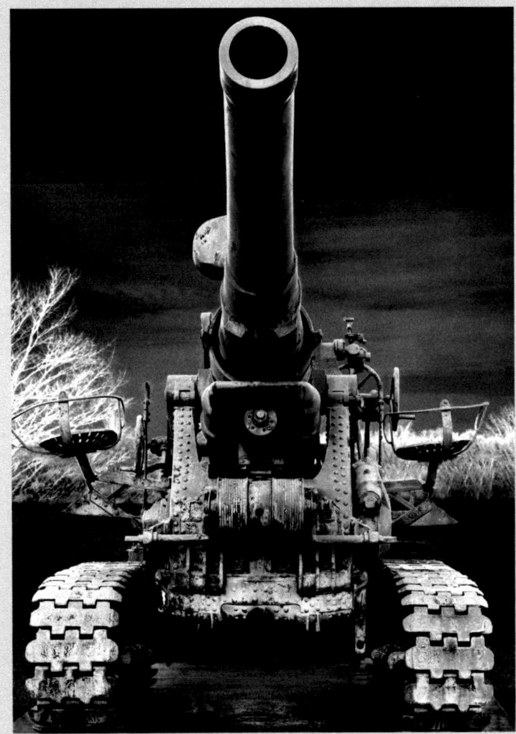

Soviet 203-mm self-propelled howitzer 1932 This artillery piece is a product of Stalin's early industrialization program to make the Soviet Union competitive in the modern world. Tractor chassis were becoming increasingly common and one was adapted to make the 203mm heavy howitzer self-propelled. It proved to be devastatingly effective in breeching German fortifications during the Soviet drive to Berlin near the end of the war. German soldiers called it "Stalin's Sledgehammer" and even pressed captured weapons into service against the Soviets.<30> *Former Collection of the US Army Ordnance Museum, Aberdeen Proving Ground, MD*

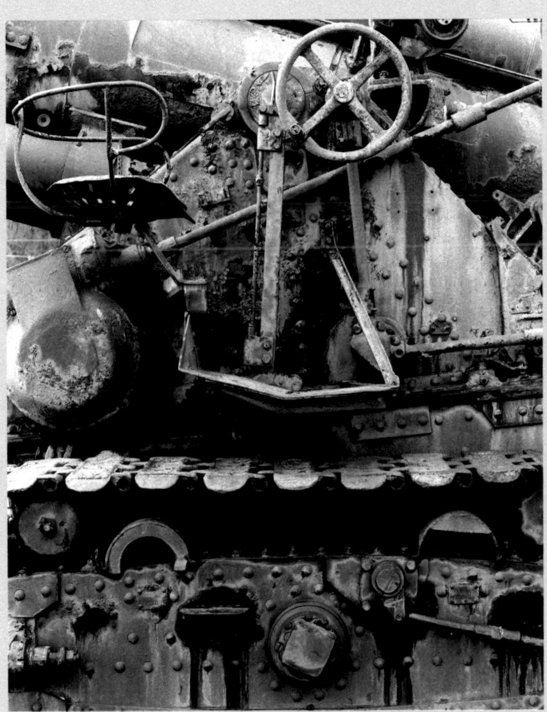

Elevation mechanism, Soviet 203-mm self-propelled howitzer 1932 The combat weight of the M1931 was about 20 tons and required a crew of 10. It could engage targets up to 11 miles. Beginning in 1932, the Soviets produced almost 900 of these weapons, more than any other heavy artillery piece.<31> *Former Collection of the US Army Ordnance Museum, Aberdeen Proving Ground, MD*

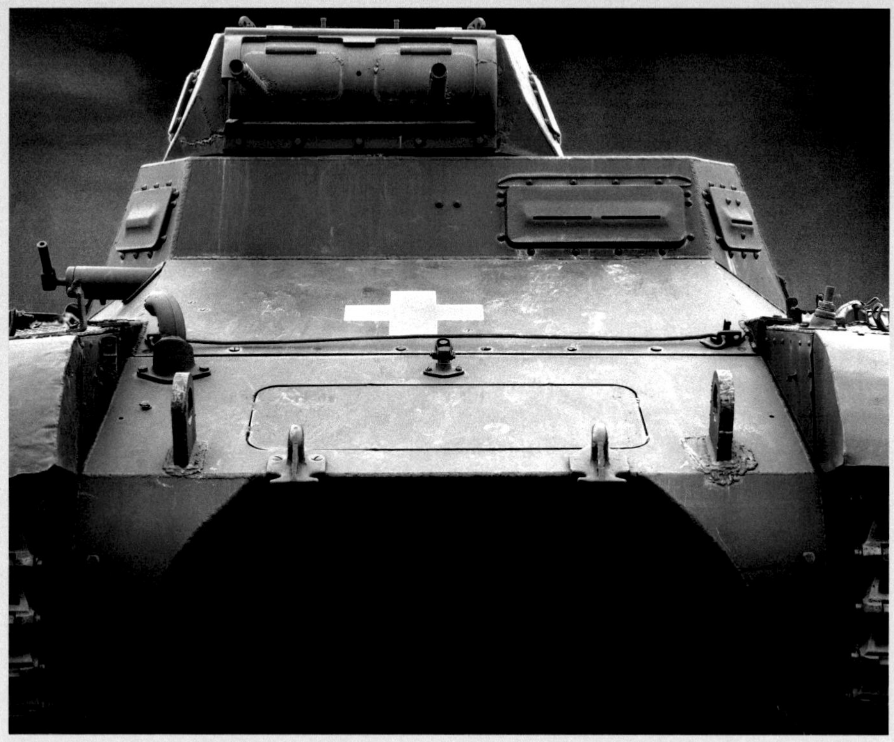

German light tank, *Pzkpfw* I, Model B with twin 7.9-mm machine guns, 6.4 tons combat weight<32> **1934** Nazi Germany's first tank, the *Pzkpfw* I was designed by Krupp to be primarily a training tank. Until Hitler renounced the Versailles Treaty in 1935, maneuvers with this tank took place in the Soviet Union. The two-man tank was underpowered, under-armed, and under-armored but still was involved in combat operations from 1936 in the Spanish Civil War to the Soviet Union in 1941.<33> *Former Collection of the US Army Ordnance Museum, Aberdeen Proving Ground, MD*

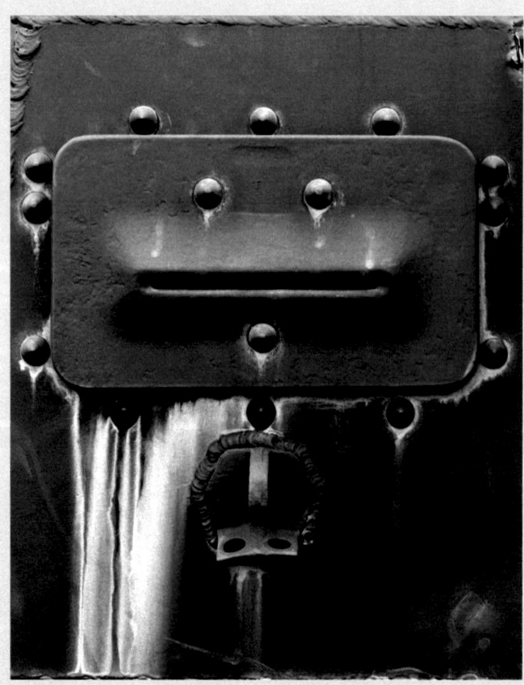

Viewport, German tank, *Pzkpfw* I 1934 Heinz Guderian, who developed Blitzkrieg tactics for Hitler between the wars, was a radio communications officer in WWI, and he realized how vital communications could be among his armored forces. He made sure that his tanks, beginning with the *Pzkpfw* I, were equipped with radio, and he oversaw the development of throat microphones that allowed speech to be understood over the engine and battle din. *Former Collection of the US Army Ordnance Museum, Aberdeen Proving Ground, MD*

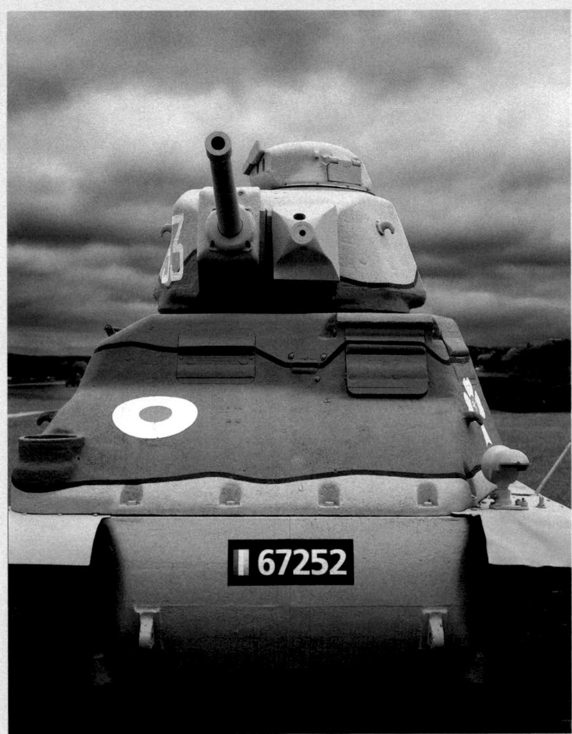

French medium tank, Somua with 47-mm gun, 22.1 tons combat weight<34> **1935** This tank is credited with having excellent firepower, mobility, and armor. Some believe it superior overall to the German *Pzkpfw* II and even IV against which it fought in the opening weeks of the battle for France in WWII. They were outclassed, however, by Guderian's tactics. About 500 of the 3-man tanks had been produced at the time of France's defeat, and they were even appropriated for training and some combat by Germany as the *Pzkpfw* 35C 739(f).<35> *Former Collection of the US Army Ordnance Museum, Aberdeen Proving Ground, MD*

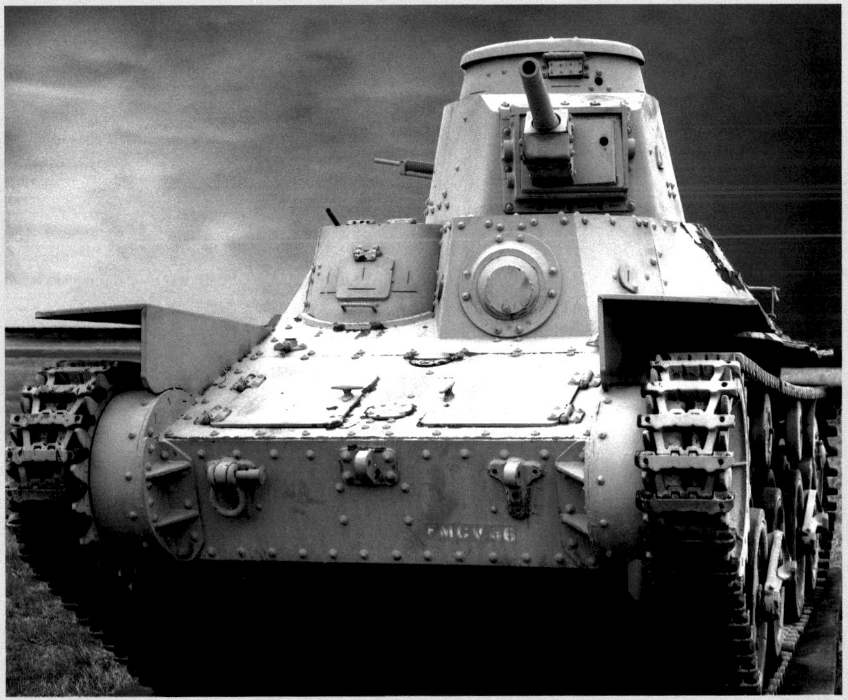

Japanese Type 95 light tank with 37-mm gun, 8.5 tons combat weight<36> **1935** About 1,250 of these 3-man tanks were produced by 1943, most built by Mitsibishi. The Type 95 was most active in the early part of the war in the Pacific, being overtaken in the cycles of obsolescence. Some were used as fixed fortifications on islands defended by the Japanese.<37> *Former Collection of the US Army Ordnance Museum, Aberdeen Proving Ground, MD*

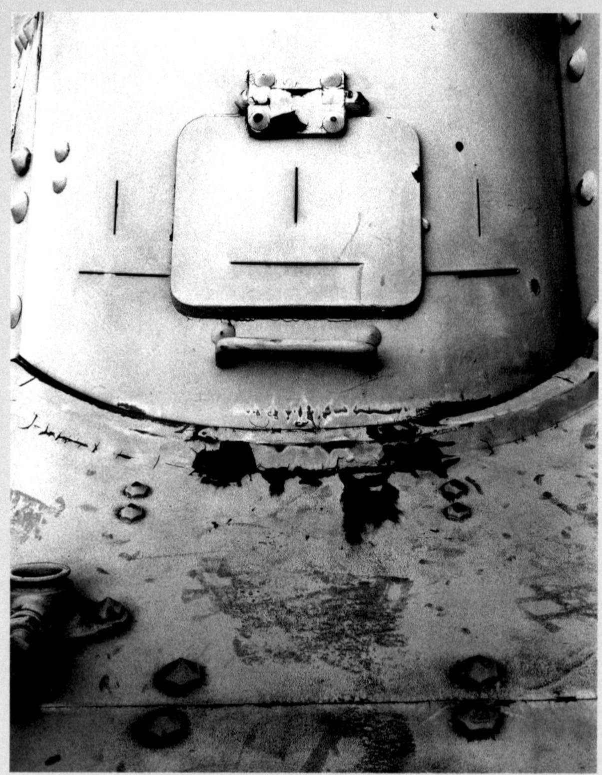

Viewport detail, Japanese Type 95 light tank 1935 The Type 95 employed a 6-cylinder diesel engine which performed better in cold weather was less vulnerable to fuel fires, but its armor was so thin that a 75mm shell could penetrate its hull on both sides. Visibility through these narrow viewing slits was also poor.<38> *Former Collection of the US Army Ordnance Museum, Aberdeen Proving Ground, MD*

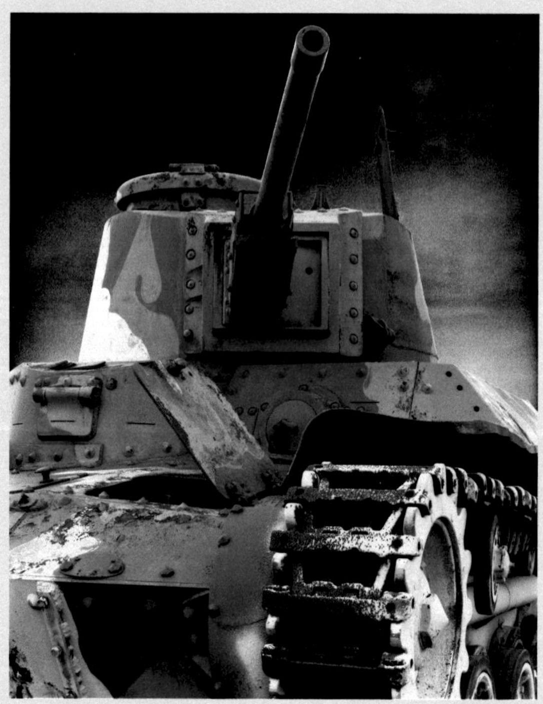

Japanese Type 97 Chi-Ha medium tank with 47-mm gun, 15 tons combat weight<39> **1937** The Type 97 was the best of the Japanese tanks and was in production until the end of the war. It utilized a crew of four and was powered by a 12-cylinder diesel engine. It was built by Mitsubishi, which had problems with mass production. As a result only about 3,000 were built.<40> *Former Collection of the US Army Ordnance Museum, Aberdeen Proving Ground, MD*

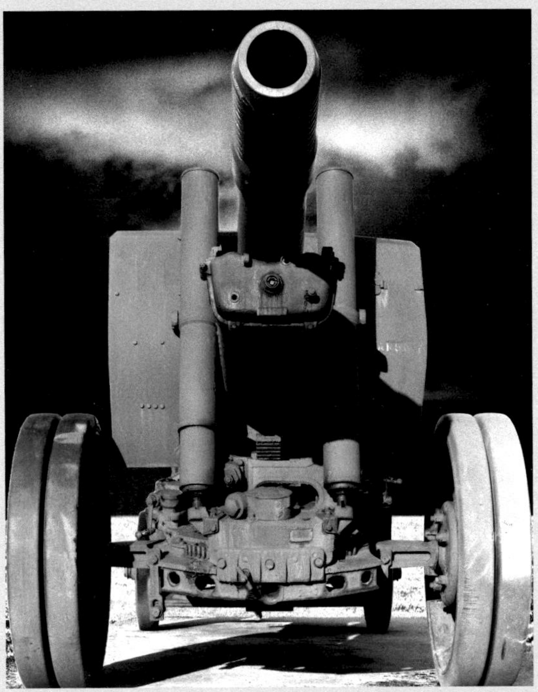

Soviet 152-mm howitzer/gun ML-20 1937 This towed artillery piece was intended to serve both the role of a high-velocity, flat-trajectory gun and a low-velocity, high-trajectory howitzer. It grew out of modifications to a 1910 French Schneider-built gun harking back to the time before the First World War when France was helping to rebuild the Russian military after the disastrous Russo-Japanese war of 1904–1905. Ammunition existed for multiple tactical purposes such as high-explosive anti-fortification, anti-personnel fragmentation, and high-explosive anti-tank roles. Almost 7,000 were built from 1937–1947.<41> And, as was often the pattern, the Germans thought enough of the piece to place captured weapons into action against their producers.<42> *Former Collection of the US Army Ordnance Museum, Aberdeen Proving Ground, MD*

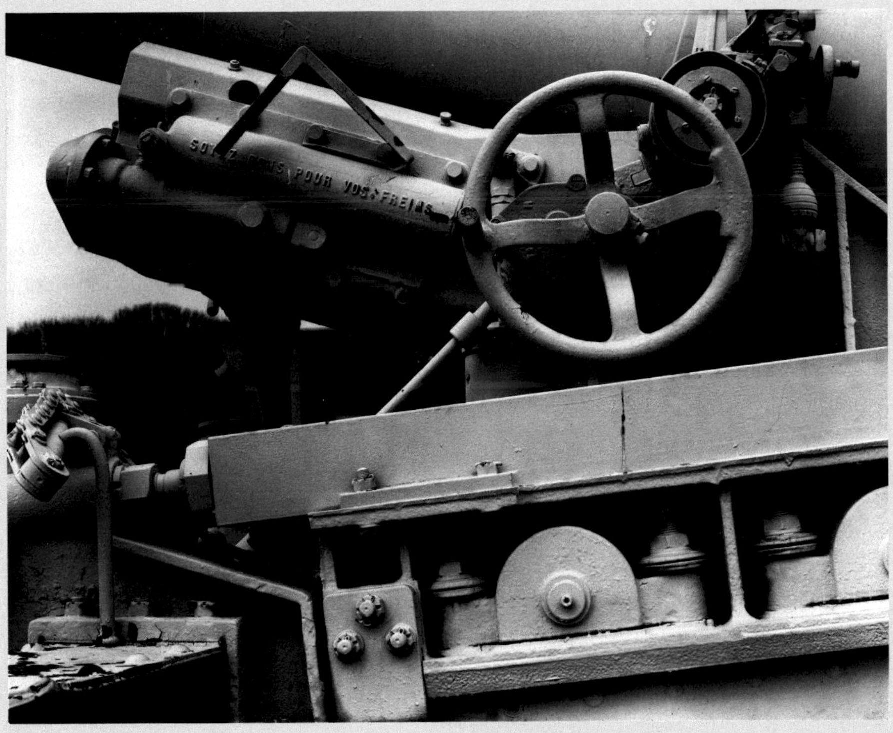

Elevation mechanism, French self-propelled 194-mm cannon GPF 1938 Another artillery piece designed by French Colonel L.J.F. Filloux in a larger caliber and this one with a tracked carriage. It could hurl a 178-pound, high-explosive shell to a distance of 11 miles.<43> *Former Collection of the US Army Ordnance Museum, Aberdeen Proving Ground, MD*

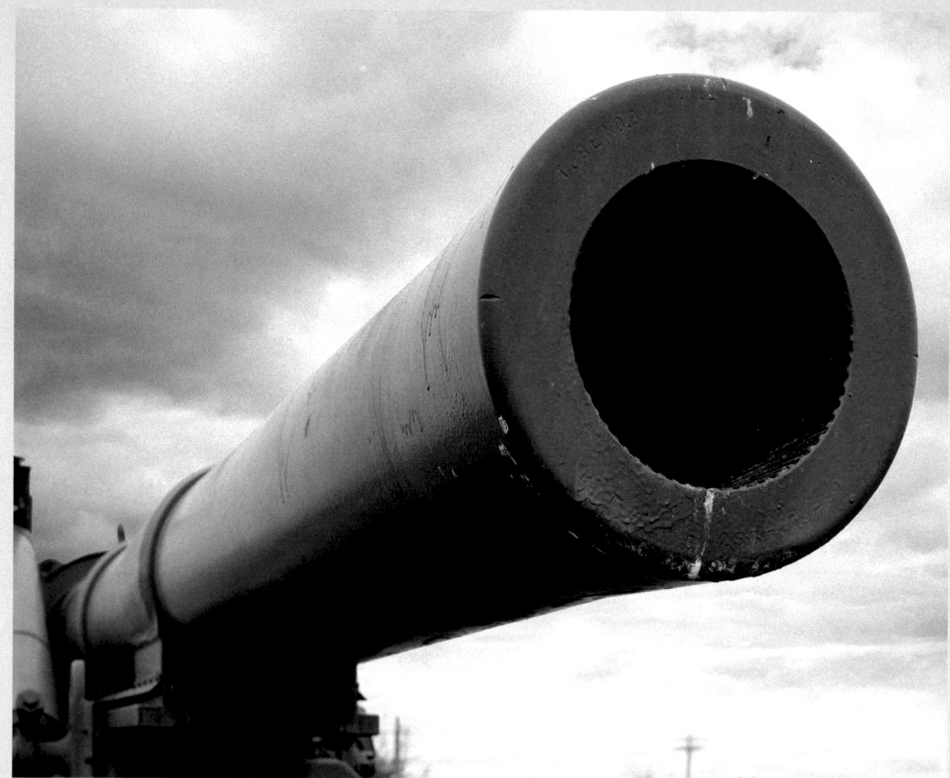

US 155mm M1 "Long Tom" field gun 1938 Between the wars the US Army Ordnance Department redesigned the French 155-mm GPF canon used by the US in WWI to include a new breech of British origin. The result was the Long Tom with a nearly 23-foot barrel. It could throw a 95-pound shell at targets 15 miles away with high accuracy. It was used in WWII, Korea, and in Vietnam.<44> *Former Collection of the US Army Ordnance Museum, Aberdeen Proving Ground, MD*

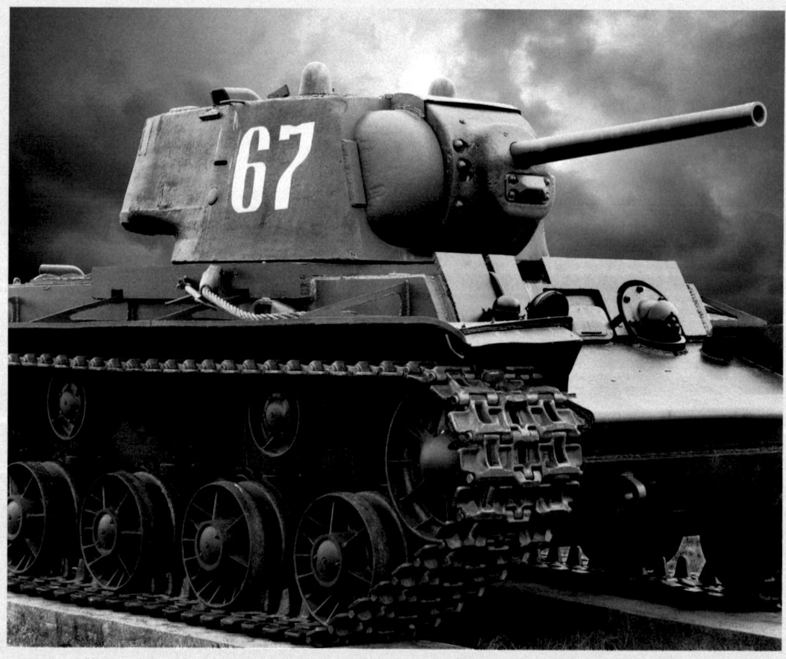

Soviet heavy tank, KV-1C with 76.2-mm gun 52 tons combat weight<45> 1939 The KV-1 was the only heavy tank in the world at the outset of WWII and the Nazi guns could not penetrate its thick armor early on. Once again superior German tactics and training overcame opposition from what was an arguably better tank than any possessed by Germany. When the Wehrmacht advanced on Leningrad the factory building the KV-1 had to be relocated to Chelyabinsk, just east of the Ural Mountains. Ultimately some 13,500 of these five-man tanks and guns using the same chassis were built.<46> *Former Collection of the US Army Ordnance Museum, Aberdeen Proving Ground, MD*

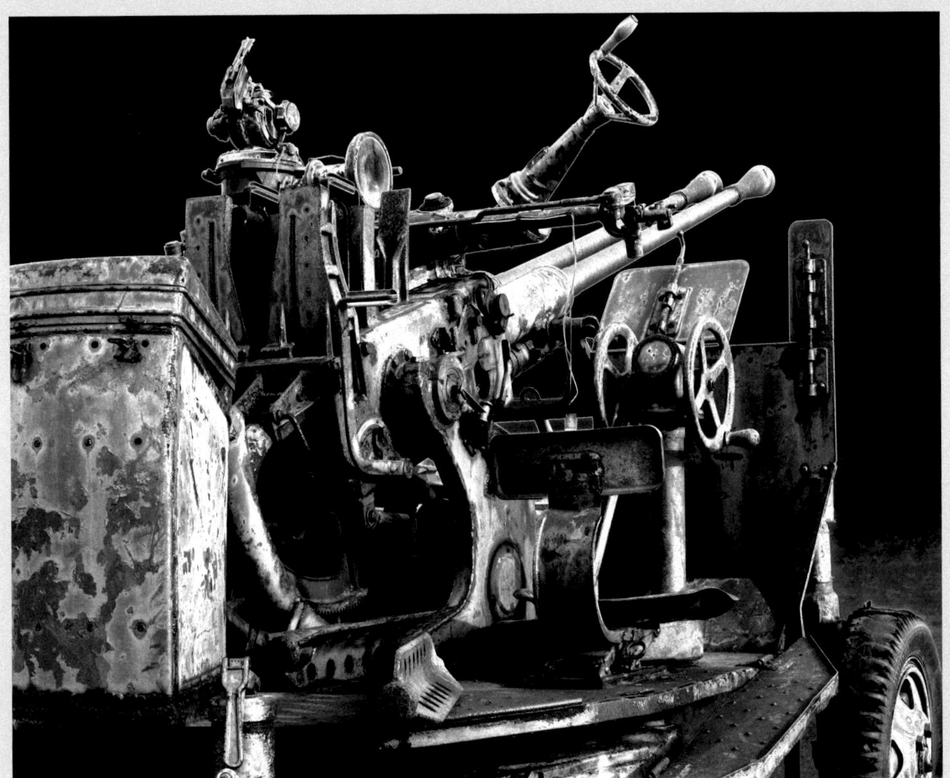

Chinese Type 65 twin-barreled 37-mm anti-aircraft gun 1939 This is a Chinese copy of a Soviet 37-mm AA gun that was, in turn, a knockoff of the 40-mm Swedish Bofors gun. The Bofors gun was manufactured under license by both the UK and US and was copied by the Germans as well as the Soviets without permission.<47> This use by all sides is reminiscent of the universality of the Maxim machine gun in WWI. *Former Collection of the US Army Ordnance Museum, Aberdeen Proving Ground, MD*

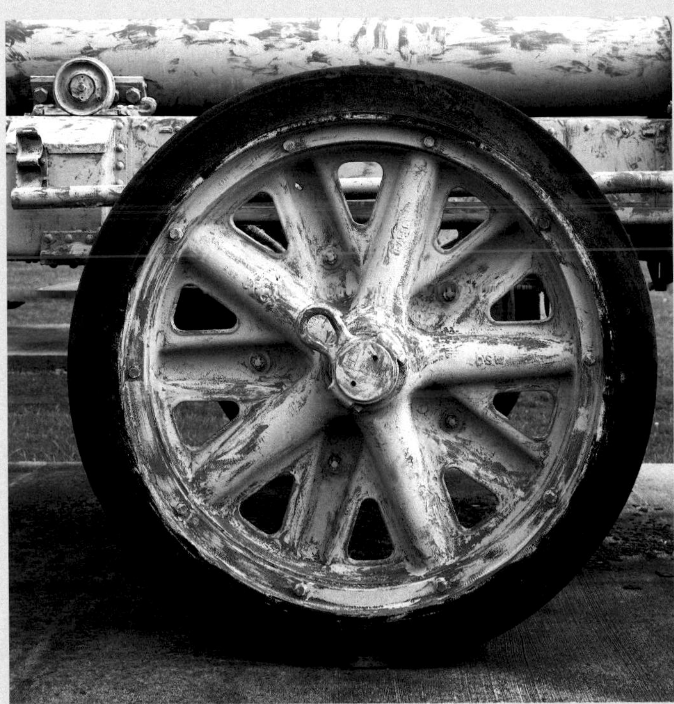

Wheel, German 210-mm heavy howitzer, Mörser 18 1939 Krupp made a little over 700 of these howitzers during the course of the war. It could send a 250-pound projectile to an effective range of 9 miles. Its predecessor, the *Mörser* 16 of the same caliber, was used in WWI. Because of its high angle of fire, the German Army classed it as a mortar (in German, *Mörser*).<48> *Former Collection of the US Army Ordnance Museum, Aberdeen Proving Ground, MD*

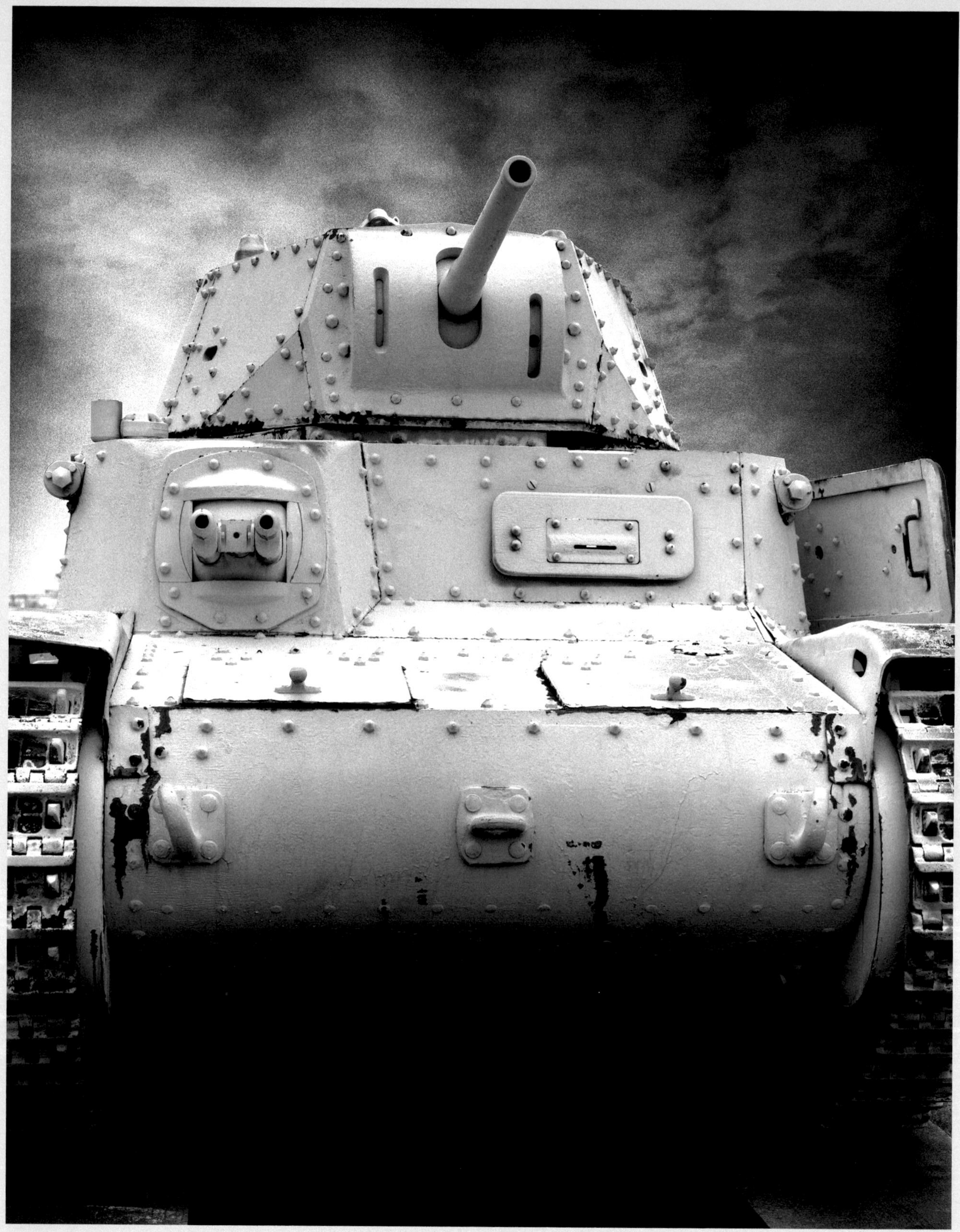

Italian medium tank, Carro Armato M13-40 with 47-mm gun, 15.4 tons combat weight<49> **1940** The M13 proved to be undergunned and its gasoline engine underpowered and prone to breakdown in the deserts of North Africa because it lacked a sand filter. Many were abandoned intact by their four-man crews and appropriated for use by the Allies. Some 800 were produced.<50>
Former Collection of the US Army Ordnance Museum, Aberdeen Proving Ground, MD

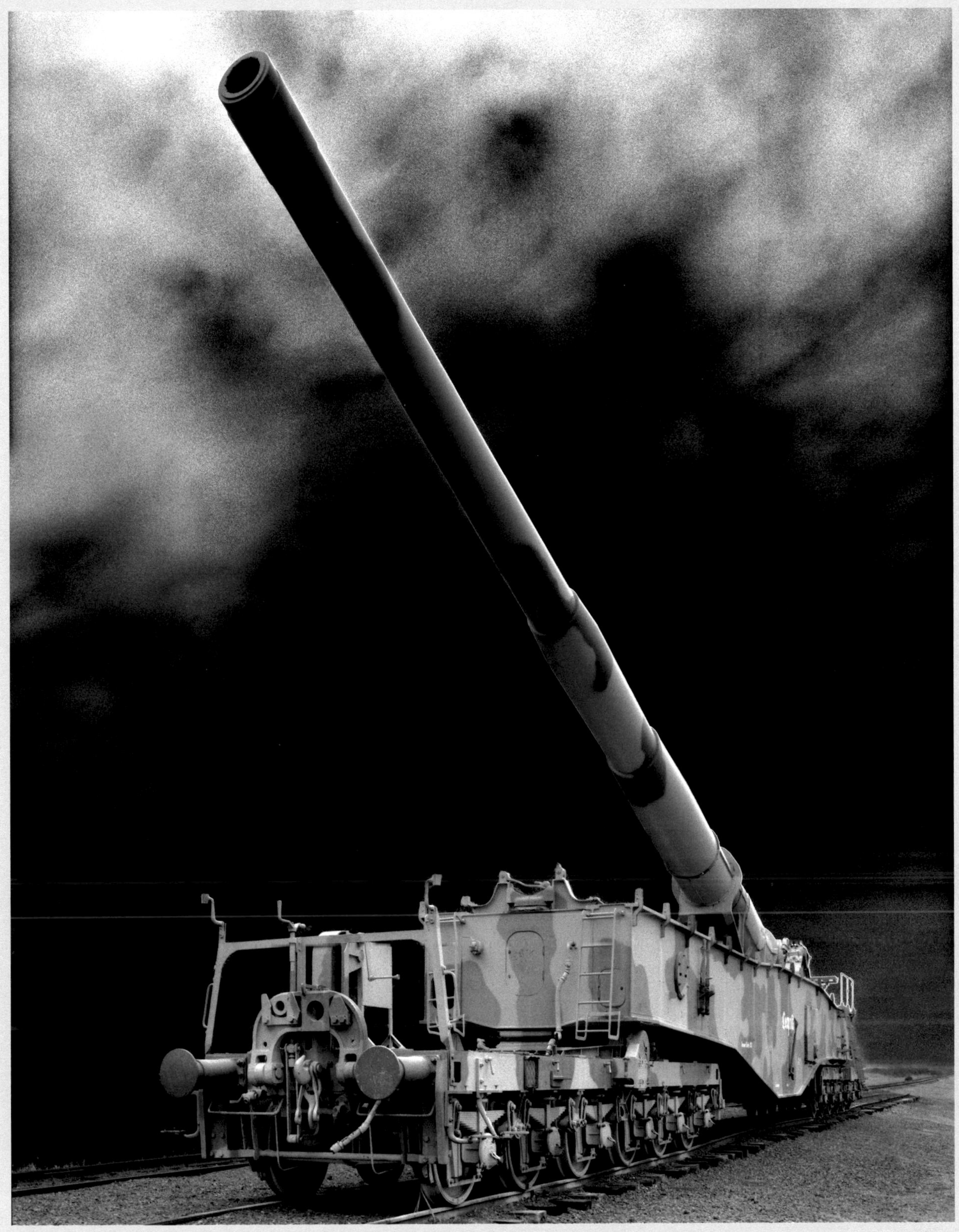

German 280-mm K5 railway gun, Leopold ("Anzio Annie") 1940 The K5 was designed to provide a range of 50 kilometers (31 miles). It was jointly developed by Krupp and Hanomag of Hannover. Nicknamed *schlanke Bertha* (slender Bertha), the K5 was the most successful of the rail guns and ultimately 22 were produced.<51> In the Allied assault at Anzio in January 1944 the German sent two K5 railway guns to fire on the beach heads, one named Robert and the other Leopold. They both used tunnels to hide within when not shooting. By the end of April, they had fired more than 500 rounds.<52> The K5 guns at Anzio are often known as Anzio Annies. *Former Collection of the US Army Ordnance Museum, Aberdeen Proving Ground, MD*

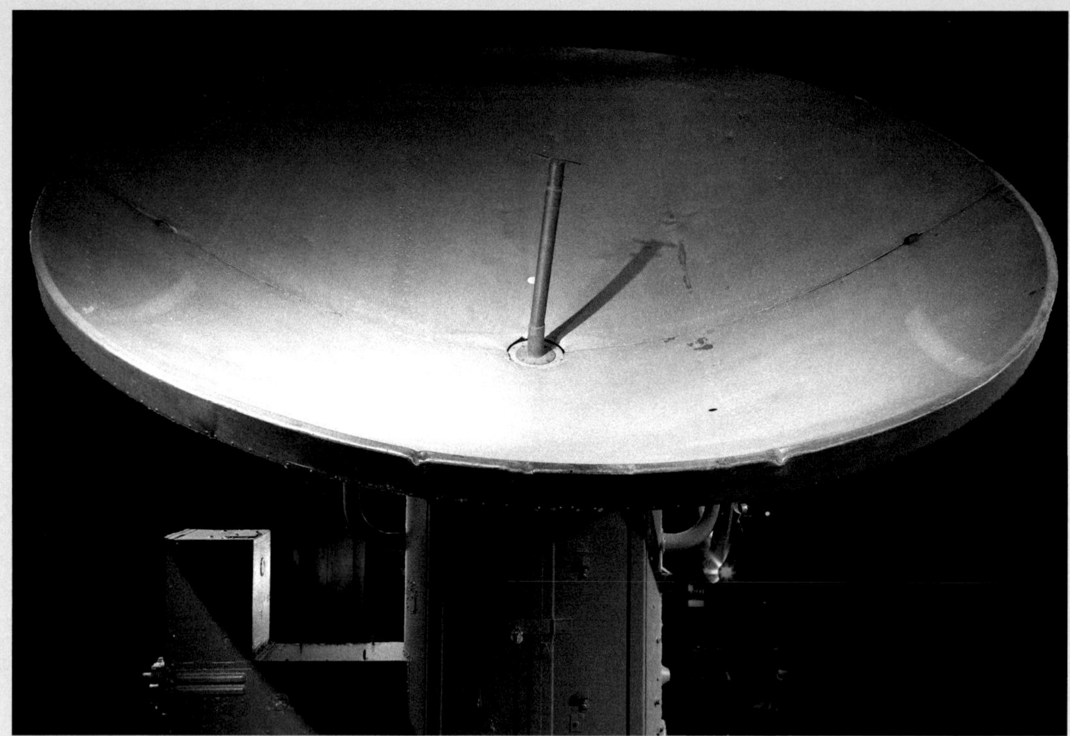

German Würzburg-A gun-laying radar antenna 1940 It was only in 1887 that German scientist Heinrich Hertz discovered a way to generate radio waves. He also showed that they could be reflected from objects. This discovery unleashed a flurry of wireless applications including radar, which was pursued independently in many countries. In Germany these investigations led to the Würzburg-A, fielded in 1940, which could direct gunfire towards enemy aircraft even through clouds or fog.<53> *Collection of the British Imperial War Museum*

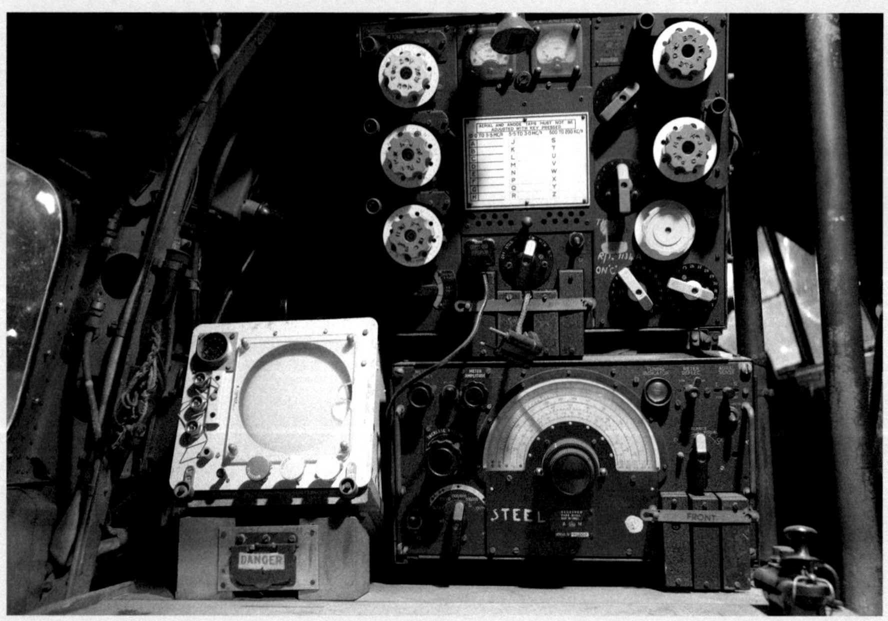

Radio Operator's station in a British Avro Lancaster Bomber 1943 This is a view from the seat of the bomber's wireless operator. The top box at right is the transmitter and below it is the receiver. The Morse-code key is at lower right. The device with circular display at the left of the radio stack is code-named Fishpond, part of the H2S ground-mapping radar system. Rather than mapping the distant landscape below, Fishpond showed the presence of aircraft in a five-mile radius below the bomber. In addition to his other duties, the wireless operator monitored Fishpond and warned the crew of the presence of enemy fighters.<54> The navigator sat around the corner to the right and forward of this station. One can almost feel the stress of operating these complex devices in the terrors of battle at night high in the sky over enemy territory. *Collection of the British Imperial War Museum*

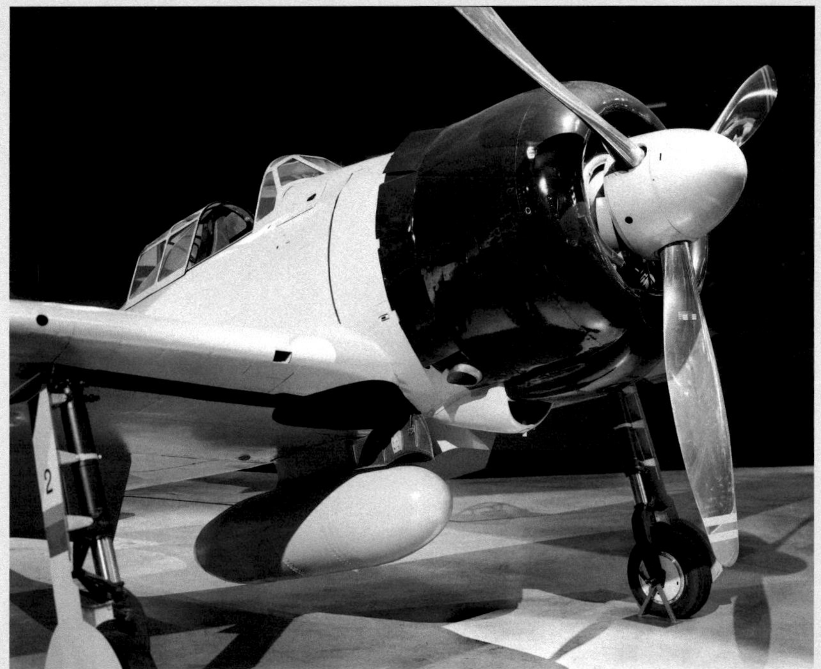

Japanese fighter aircraft, Mitsubishi A6M2 Zero 1940 The A6M was developed as a long-range fighter aircraft suitable for carrier basing. This model had a truly astonishing range of 1,930 miles. In the attack on Pearl Harbor 7 December 1941 six carriers launched 353 Zeros. Early in the war, the Zero's combat reputation achieved mythical proportions, and some of it was deserved. In mid-1942 a mostly intact plane was discovered ditched on an Aleutian island after a raid on Dutch Harbor. After a thorough examination, several weaknesses were found including no armor protection for the pilot or self-sealing fuel tanks, and this analysis helped in the design of the next generation F6F Hellcat. Special tactics such as the "Thach Weave" also helped even the odds as the war heated up in the Pacific with the Battle of Midway in June 1942.<55> *Collection of the National Museum of the US Air Force*

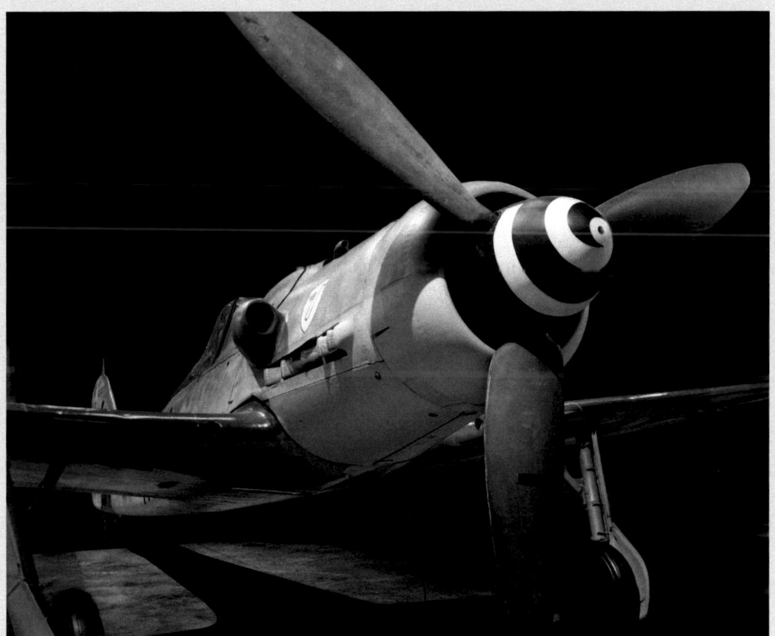

German fighter aircraft, Focke-Wulf Fw 190D 1940 The Fw 190 earned a reputation of being one of the best fighters of WWII. Over the course of the war over 20,000 were produced. In the hands of highly trained pilots they were nearly unbeatable, but, as the war dragged on, Germany was not able to maintain the quality of its training in the face of its losses. One of their main missions was to stop the Allied bombers that pushed deeper and deeper into the German heartland. A successful tactic was to form a linear formation behind the bombers and peel off in different directions once within range of the bomber's tail gunner.<56> The multiple targets that suddenly burst out in all directions challenged even the best tail gunners. *Collection of the National Museum of the US Air Force*

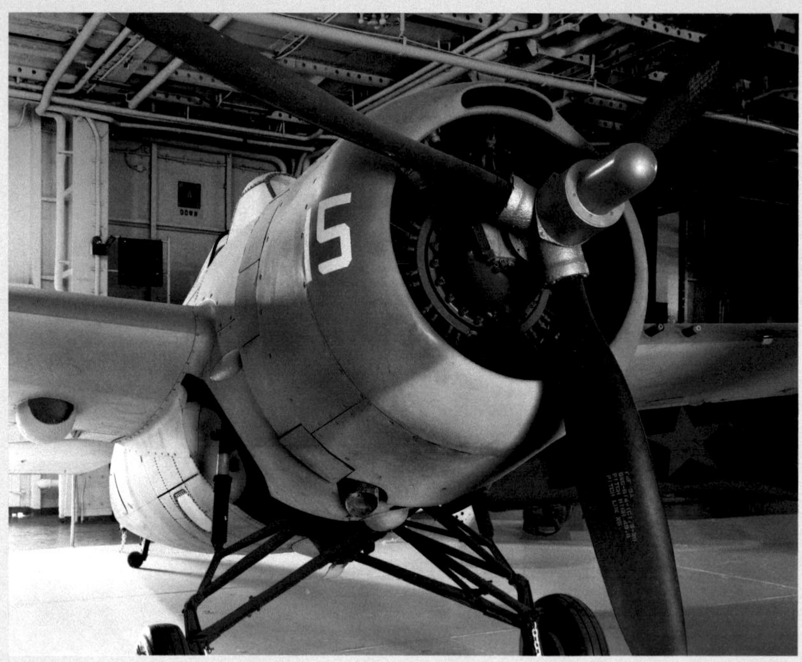

US carrier-based fighter aircraft, F4F-3A Wildcat 1941 The range of the Grumman F4F Wildcat was less than half that of the Zero, but the F4F compensated with self-sealing fuel tanks, armor protection for the pilot, rugged construction, and the superior firepower of its 0.50 caliber machine gun. John Thach developed his famous Thach Weave tactic for the F4F. Four Wildcats would fly in pairs several hundred yards apart. When one pair spotted a Zero after the other pair, the two pairs would turn toward each other and the first pair would engage the Zero head-on. The two pairs would then cross one another's path and back into a parallel formation. The Zero proved very vulnerable to the 0.50 caliber machine gun of the F4F.<57> *At Patriots Point Naval & Maritime Museum, Mt. Pleasant, SC*

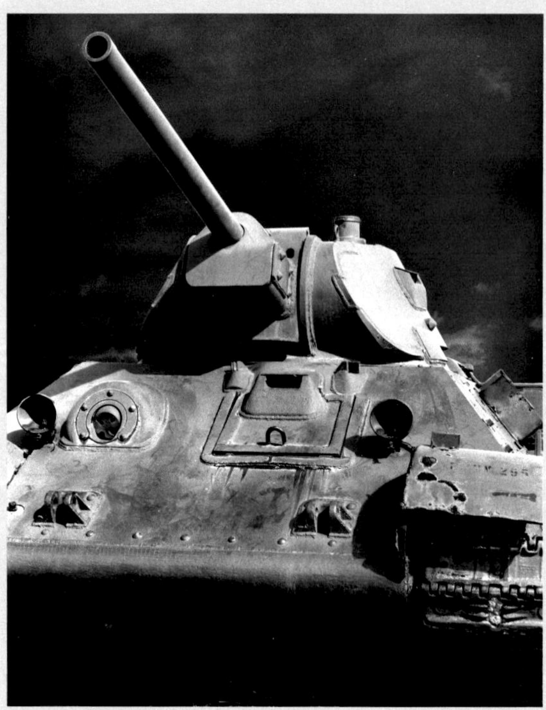

Soviet medium tank, T34 with 76.2-mm gun, 32 tons combat weight<58> **1940** By the time of the June 1941 Nazi attack on the Soviet Union, code named Barbarossa, some 1,200 T34s had been produced. But not until September 1941 were sufficient numbers deployed to gain the Wehrmacht's respect as a fighting machine superior to any the Germans possessed. No less a figure than General Guderian recommended that Germany reverse engineer the T34 and produce their own to deal with this unexpected threat.<59> That was not about to happen, but many elements of the T34 were incorporated into the next generation of German tanks, the Panthers and Tigers.<60> *Former Collection of the US Army Ordnance Museum, Aberdeen Proving Ground, MD*

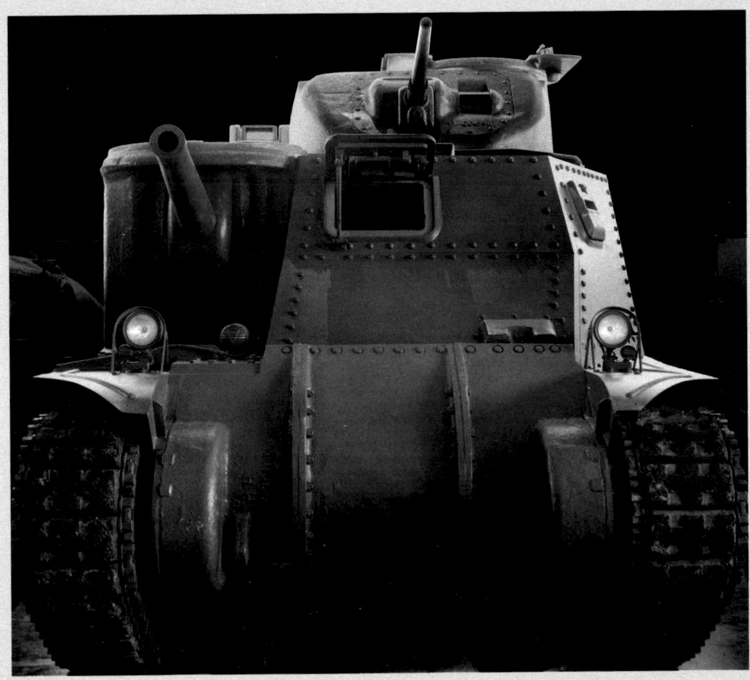

US medium tank, M3 Lee with 75-mm gun, 32 tons combat weight 1941 The M3 was the first American tank placed in mass production, which began in August 1941. The British wished for a tank of their design to be produced in the US, but the American government offered the M3 instead to be made under the Lend-Lease Act. These tanks were designated the M3 Lee after American Civil War General Robert E. Lee. Production ceased at the end of 1942 by which time a total of 6,258 had been produced. Five manufacturers had been involved: Chrysler, American Locomotive (ALCO), Baldwin Locomotive Works, Pressed Steel Car Co., and Pullman Car Co. The list is a microcosm of the wartime mobilization of the civilian economy.<61> *Collection of the Canadian War Museum*

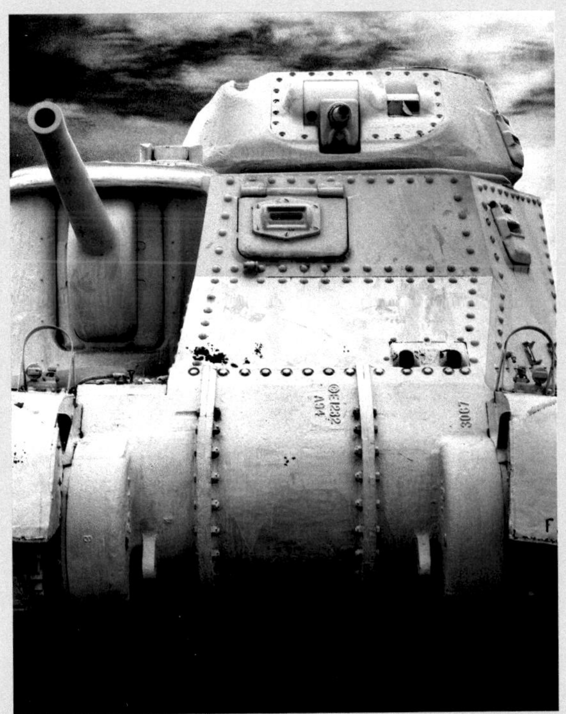

British medium tank, M3 Grant, 31.1 tons combat weight<62> 1941 The US did agree to make a variant of the M3 with no commander cupola at British request, and this model was designated the M3 Grant after the Civil War General Ulysses S. Grant. The M3 tanks worked well in the deserts of North Africa, impressing even German General Irwin Rommel with their performance. The North African Campaign ended in Allied victory about the middle of May 1943 and with it the virtual end of the combat life of the M3 as production was ramping up for the M4 Sherman.<63,64> *Former Collection of the US Army Ordnance Museum, Aberdeen Proving Ground, MD*

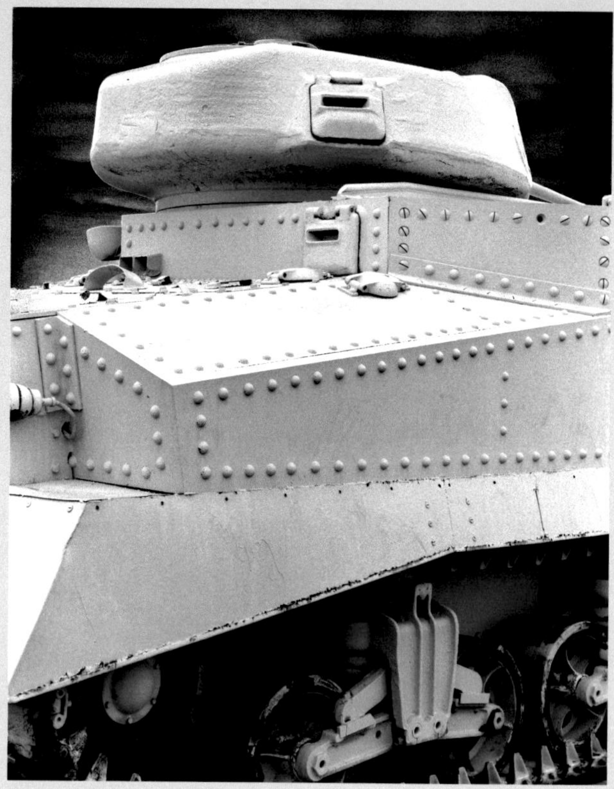

Armor plate detail, British medium tank, M3 Grant 1941 The M3 was America's first tank built using the mass production methods of the automotive industry. Early versions had riveted hulls like this one that later gave way to cast or welded hulls. *Former Collection of the US Army Ordnance Museum, Aberdeen Proving Ground, MD*

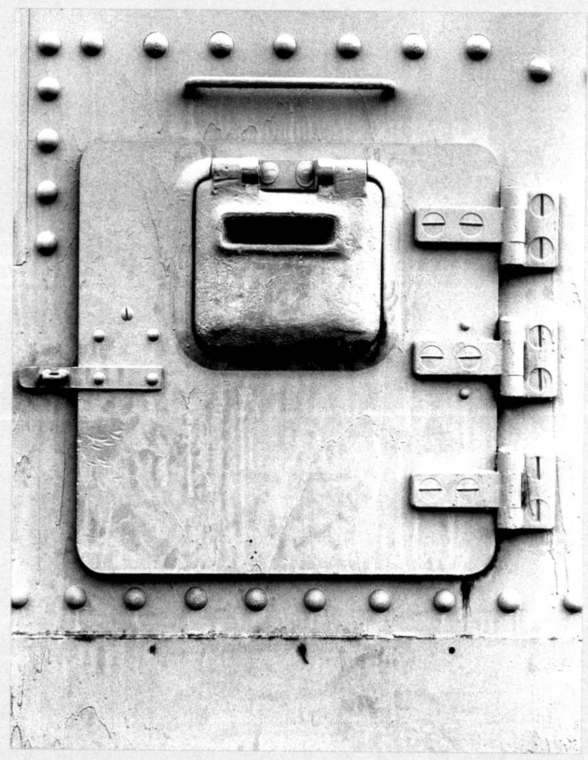

Viewport, British medium tank, M3 Grant 1941 Some 6,300 M3s were built before production ended near the end of 1942.<65> *Collection of the US Army Ordnance Museum, Aberdeen Proving Ground, MD*

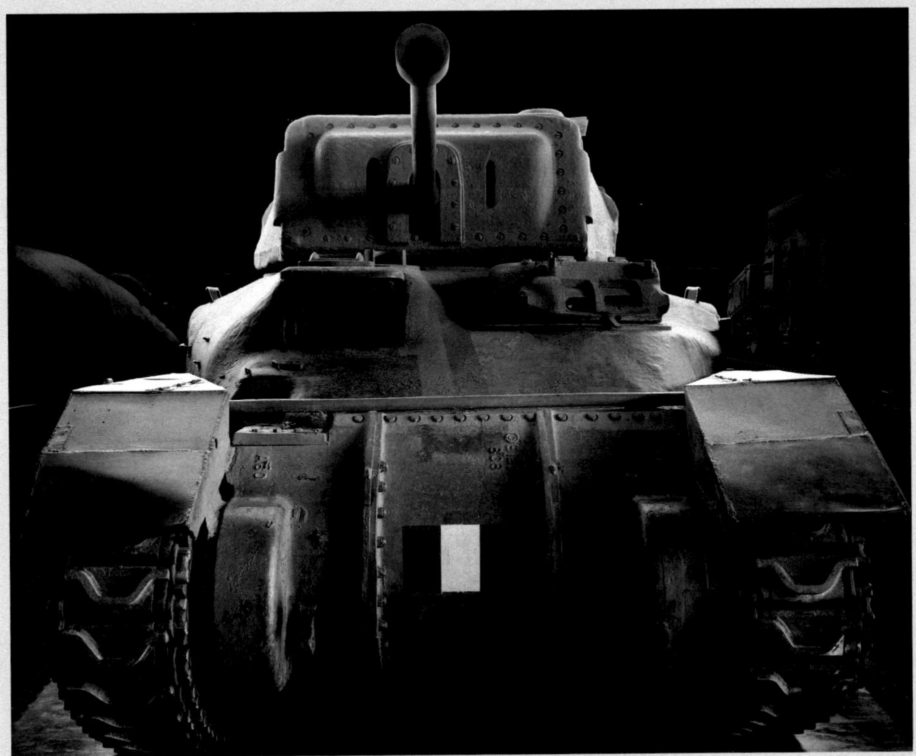

Canadian cruiser tank, Ram II with 57-mm gun, 29 tons combat weight 1942 The Ram was built by the Montreal Locomotive Works to improve on the M3 tank. It dispensed with the sponson-mounted gun in favor of a revolving turret mounted gun. The Ram I was armed with a 2-pounder gun that was updated to a 6-pounder (57mm) in the Ram II. In July 1943 production was switched over to the Canadian version of the M4 Sherman, the Grizzly.<66> *Collection of the Canadian War Museum*

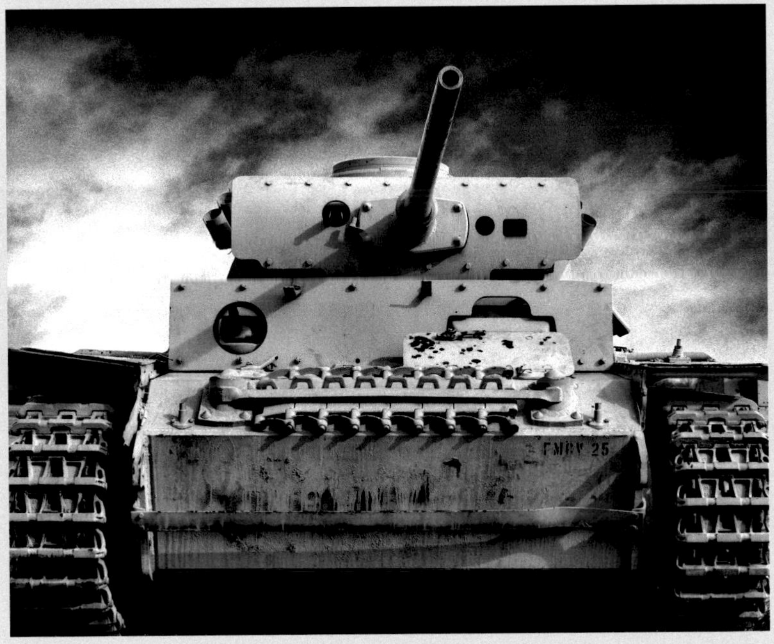

German medium tank, *Pzkpfw* III Model L with long 50-mm gun, 25.4 tons combat weight<67> **1942** The *Pzkpfw* III, with turret by Krupp and chassis by Daimler-Benz, was ultimately produced in 14 versions, illustrating the many modifications necessitated by combat experience. The five-man tank was used in Poland, France, and the Soviet Union.<68> Some 168 were converted to the *Tauchpanzer* III (diving tank) in preparation for Operation Sea Lion, the invasion of Britain that never was implemented. These tanks were sealed and fitted with snorkels attached to floats and expected to travel along the seabed at depths up to 50 feet.<69> *Former Collection of the US Army Ordnance Museum, Aberdeen Proving Ground, MD*

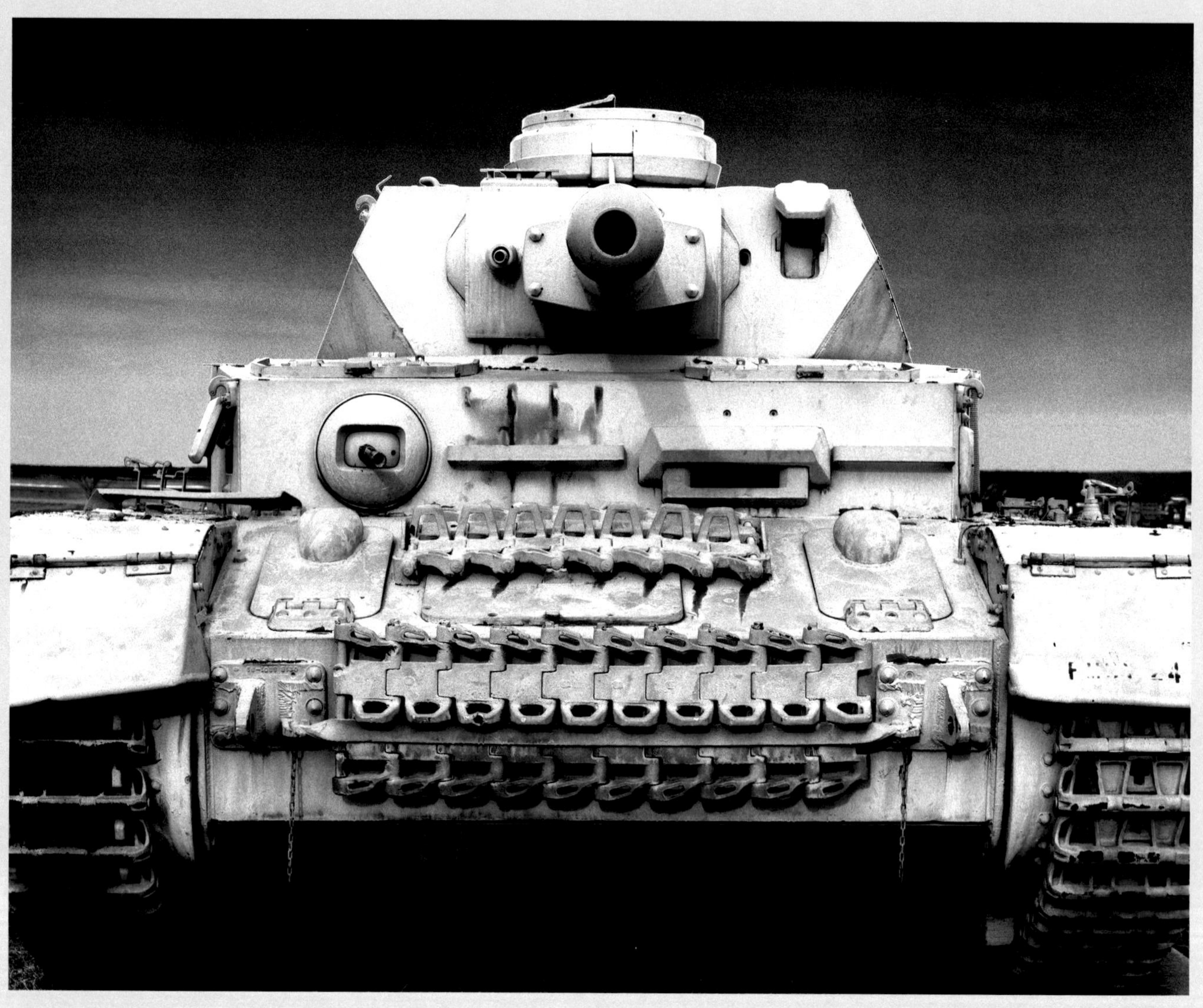

German Medium Tank, *Pzkpfw* IV Model F2 (renamed G) with long 75-mm gun, 25.8 tons combat weight<70> **1942** In the mid-1930s early German tank-combat doctrine called for two types of tank: a smaller caliber, high-velocity gun to take on other tanks and one with a larger-caliber, low-velocity gun to engage troop formations and fortifications with high-explosive rounds. The *Pzkpfw* III was to serve the former role and the *Pzkpfw IV* the latter. As tank capabilities escalated during the war the *Pzkpfw* IV was up-gunned to serve both roles, and it did so until the end of the war. This was the only tank in production throughout the war in Germany with a total of 8485 in ten versions produced.<71,72> In one version or another it served in every German theater of the war. *Former Collection of the US Army Ordnance Museum, Aberdeen Proving Ground, MD*

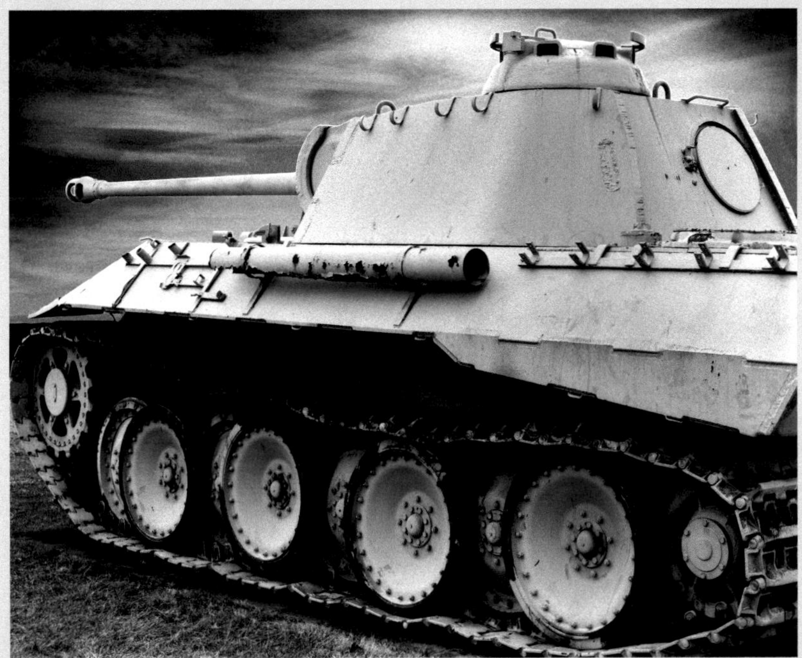

German heavy tank, *Pzkpfw* V, Panther Model IA, 50.2 tons combat weight<73> **1942** The success and versatility of the *Pzkpfw* IV prior to the invasion of the Soviet Union in June 1941 had slowed the development of Germany's next generation tank. However, its shocking inferiority to the Soviet T34 reenergized the program and redirected it to include those features that made the T34 such an effective tank, powerful gun, sloping armor, and wide tracks. The Panther was the result of these efforts, and through a succession of versions, ultimately earned a reputation as one of the best tanks of WWII.<74> However, the first production models such as this one were prone to mechanical failures because of the rush to get them into the fight with the Soviets. At their combat debut in the massive tank battle of Kursk in July 1943, most of them broke down before they could be deployed in the battle.<75> *Former Collection of the US Army Ordnance Museum, Aberdeen Proving Ground, MD*

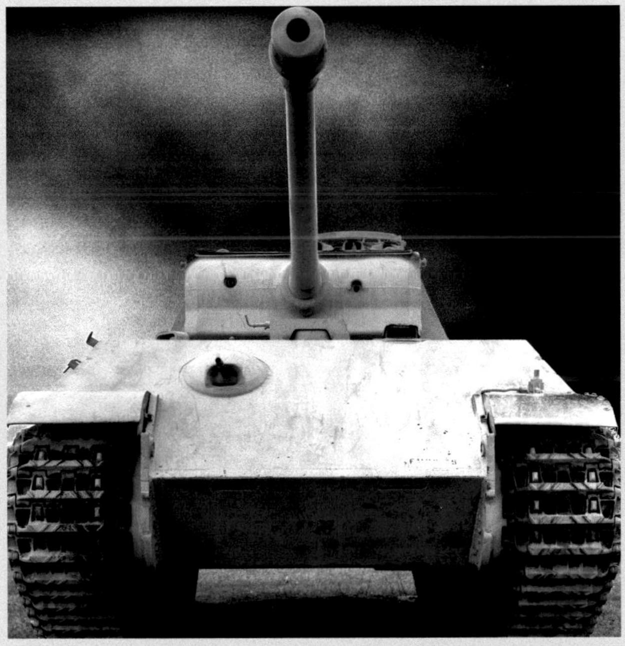

German heavy tank, *Pzkpfw* V, Panther Model IG, 50.2 tons combat weight<76> **1943** Two important changes in the Panther from Version IA to IG was an increase in frontal armor from 60 mm to 80 mm and the addition of a skirt on the armored mantlet from which the main gun protrudes to prevent incoming rounds from being deflected downward into the crew compartment.<77> The Panther successfully adapted all of the innovations of the T34 except one—simplicity of manufacture and maintenance. As the Allies squeeze the Wehrmacht from all sides, Germany's failure to manufacture enough Panthers hastened their eventual defeat. *Former Collection of the US Army Ordnance Museum, Aberdeen Proving Ground, MD*

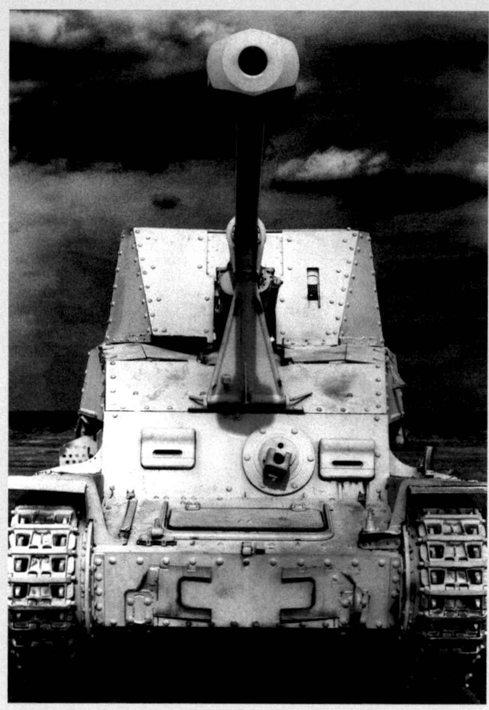

German tank destroyer, *Marder Panzerjäger* (Marten tank hunter) III with Czech 38 chassis and captured Soviet 76.2-mm gun, 11.6 tons combat weight<78> **1941** The *Marder* III is a perfect example of wartime resourcefulness. When Czechoslovakia was absorbed into the Third Reich in March 1939, all of its armaments works became available to arm the Wehrmacht. Tracked chassis for the Czech TNH light tank were adapted to the tank hunter role by mounting later captured Soviet 76.2-mm anti-tank guns that were rechambered for a German cartridge case. The combination proved very effective early in the war.<79> *Former Collection of the US Army Ordnance Museum, Aberdeen Proving Ground, MD*

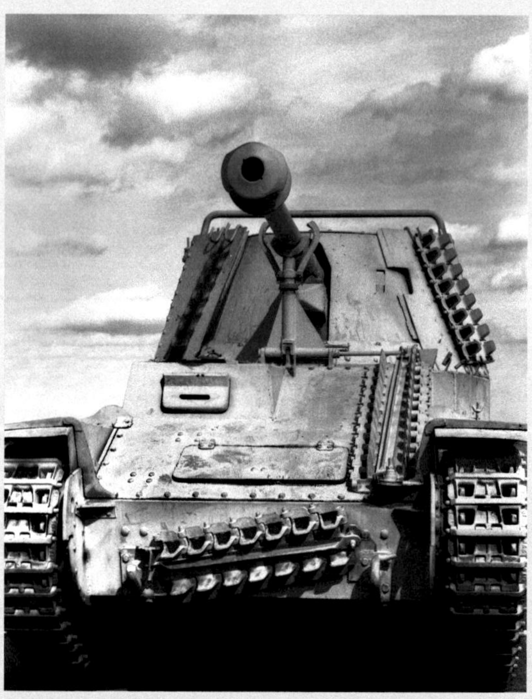

German Tank Destroyer, *Marder Panzerjäger* (Marten tank hunter) III with Czech 38 chassis and German 75-mm Pak 40 gun, 11.6 tons combat weight<80> **1943** This is the final configuration of the Marder III. The profile was lowered and the German 75-mm Pak 40 anti-tank gun installed in place of the Soviet 76.2 mm. Almost a thousand were built between April 1943 and March 1944.<81> *Former Collection of the US Army Ordnance Museum, Aberdeen Proving Ground, MD*

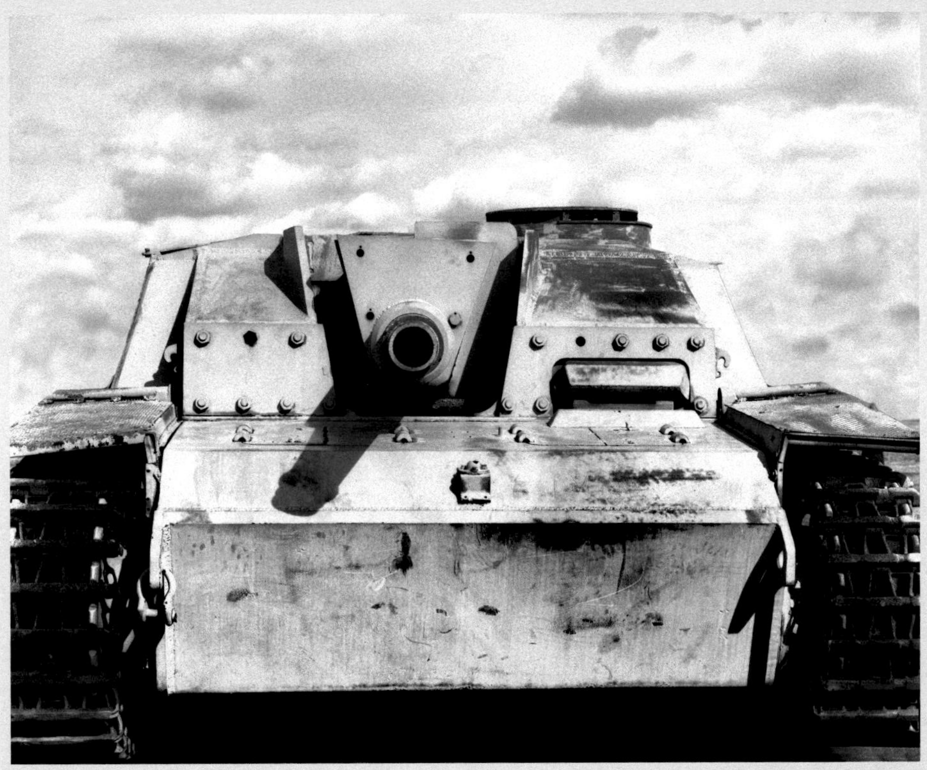

German *Sturmhaubitze* 42 (assault howitzer) III, with 105-mm howitzer, 26.4 tons combat weight<82> **1942** The first *Sturmgeshütz or StuG* III dated from 1937 with a low-velocity 75-mm gun mounted on a *Pzkpfw* III chassis. Its purpose was to provide direct fire support for infantry. Later a high-velocity gun was mounted, giving it an anti-tank role as well. This variant gave it the ability to destroy fortifications as well.<83> *Former Collection of the US Army Ordnance Museum, Aberdeen Proving Ground, MD*

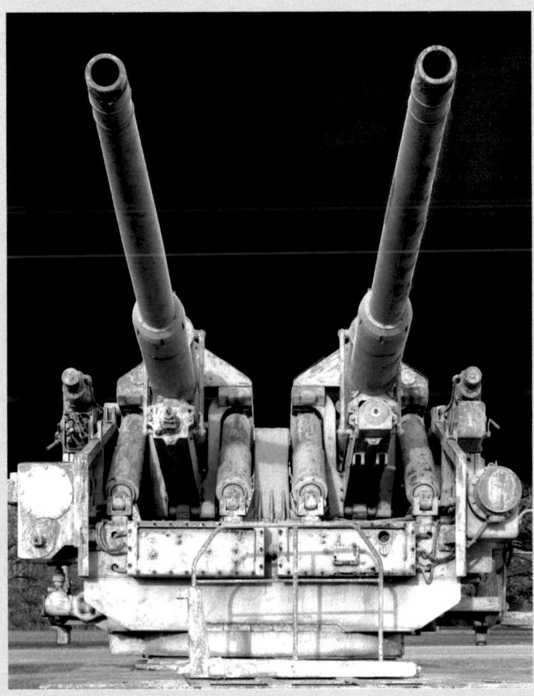

German dual 128mm anti-aircraft guns, *Flakzwilling* 40 1942 Like the 88-mm gun, the 128-mm gun was originally designed as an anti-tank gun, and, like the 88, it proved to be a potent anti-aircraft weapon. The two-gun version shown here proved too heavy for mobile use and was mostly used at the top of the enormous Flak Towers that protected cities like Berlin, Hamburg, and Vienna.<84> *Former Collection of the US Army Ordnance Museum, Aberdeen Proving Ground, MD*

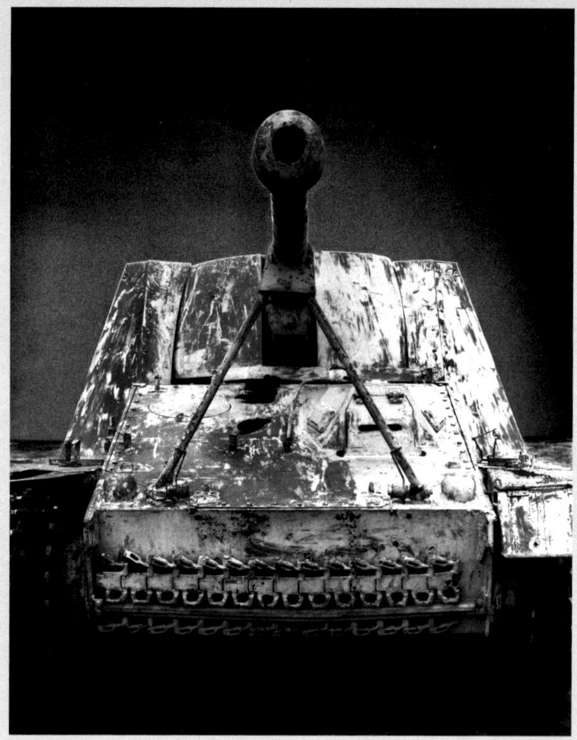

German Tank Destroyer, *Nashorn* (Rhinoceros) Model GW IV with 88-mm gun, 26.5 tons combat weight<85> **1942** The *Nashorn* mated the potent 88-mm gun to a modified *Pzkpfw* IV chassis. Though having little protection for its crew of four especially with its open top, it could knock out Allied tanks by standing off at long range. Nearly 500 were produced and they served predominantly in special detachments in the Soviet Union and Italy.<86> *Former Collection of the US Army Ordnance Museum, Aberdeen Proving Ground, MD*

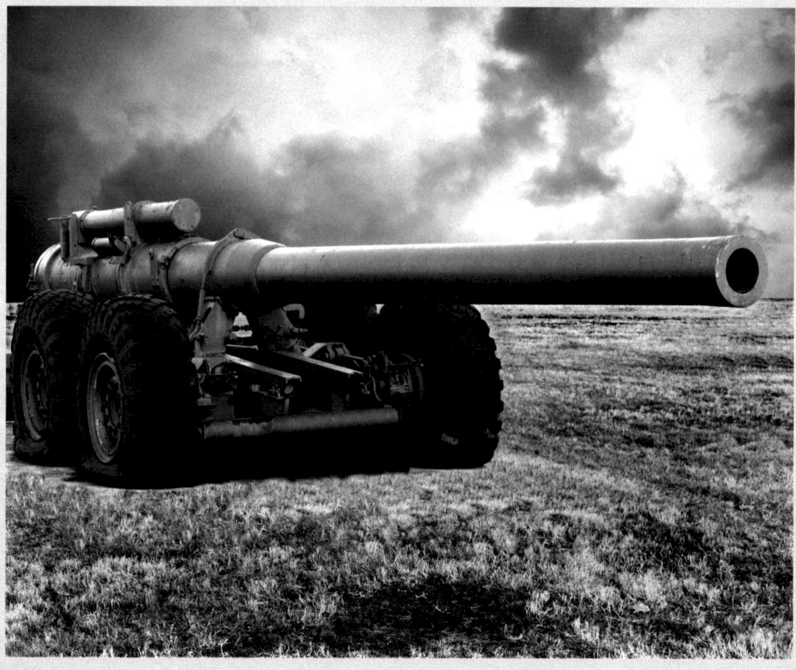

US 8-inch field gun, M1 1942 The 8-inch M1 had the longest range of any gun in the US inventory during WWII—20 miles. It was so large and heavy that its barrel had to be transported separately from its carriage and mated using a portable crane onsite. The barrel is shown here in its road transport trailer. It proved valuable in the mountain warfare in Italy, being able to engage targets on the opposite side of a mountain. In the winter of 1944 when bad weather prevented Allied air support, the 8-inch M1 was also used in blasting through German fortifications.<87> *Former Collection of the US Army Ordnance Museum, Aberdeen Proving Ground, MD*

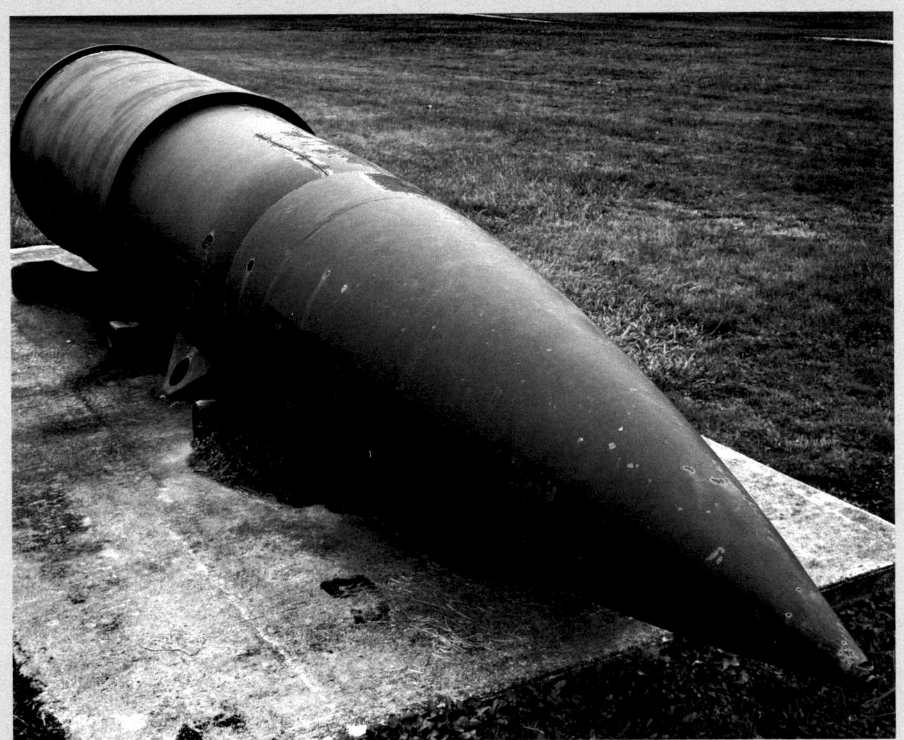

German 800-mm railway artillery shell, Heavy Gustav 1942 This massive shell, filled with 1,540 pounds of high explosive and weighing a total of 10,560 pounds, could be projected by the Heavy Gustav rail gun to a distance of 30 miles.<88> Only once was this weapon used—in the destruction of the Soviet fortress at Sevastapol. Otherwise, at the same cost as some 28 Tiger tanks, it was a frivolous waste of resources. Neither of the two guns that were built survive. *Former Collection of the US Army Ordnance Museum, Aberdeen Proving Ground, MD*

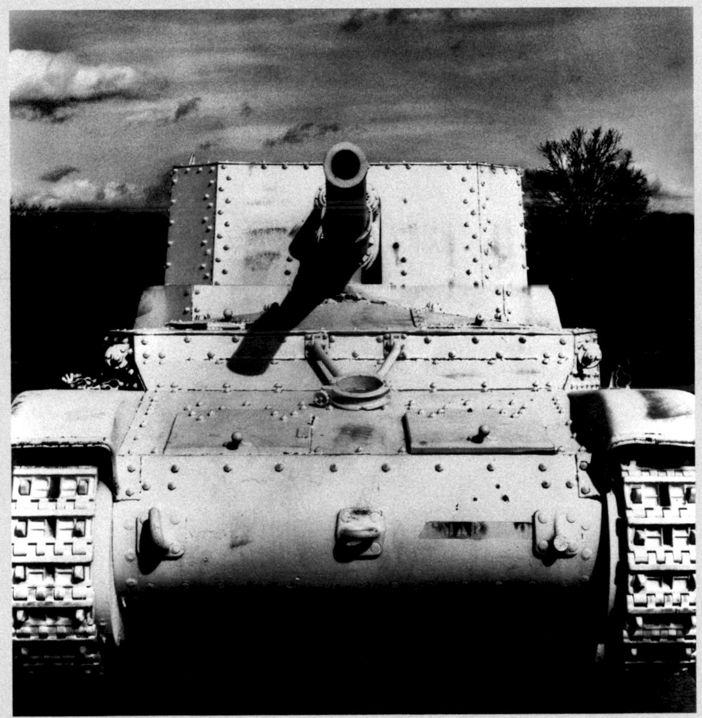

Italian Semovente M41 90-mm self-propelled gun<89> 1942 Although intended for use an anti-tank gun against Soviet forces, it came under operational command of the Wehrmacht and used in Sicily mostly as long-range artillery due to its poor armor protection.<90> *Former Collection of the US Army Ordnance Museum, Aberdeen Proving Ground, MD*

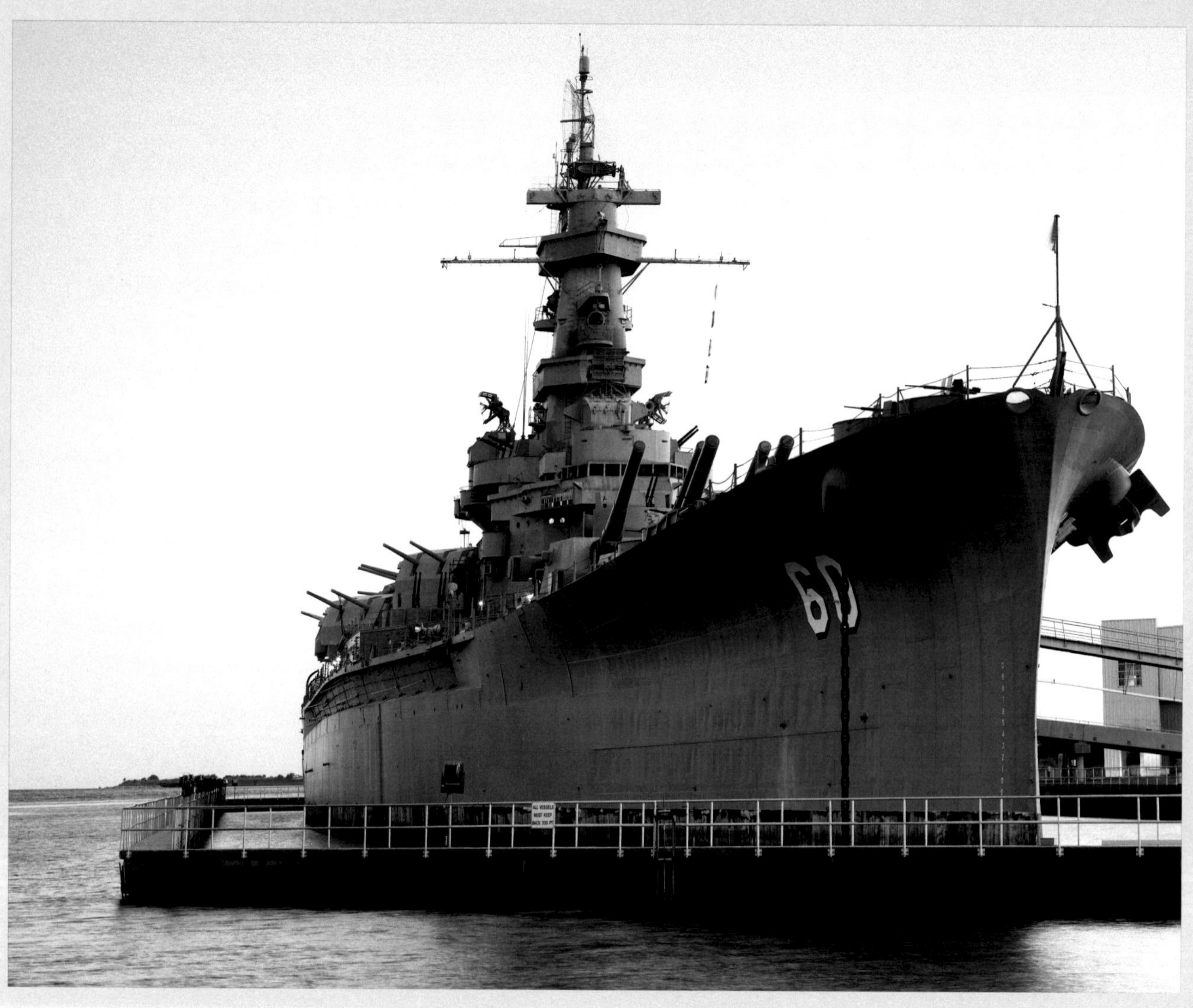

USS *Alabama* battleship at dusk, BB-60 1942 The *Alabama*, a *South Dakota* class battleship, was commissioned on 16 August 1942 at 44,519 tons displacement fully loaded. This was America's second largest battleship class and was about three quarters the size of the *Iowa* in both displacement and overall length.<91> In mid-1943 she spent a few months on duty in the North Atlantic before deploying to the Pacific Theater by September 1943. Here she would stay virtually in continuous action until the end of the war.<92> *At Battleship Memorial Park, Mobile, AL*

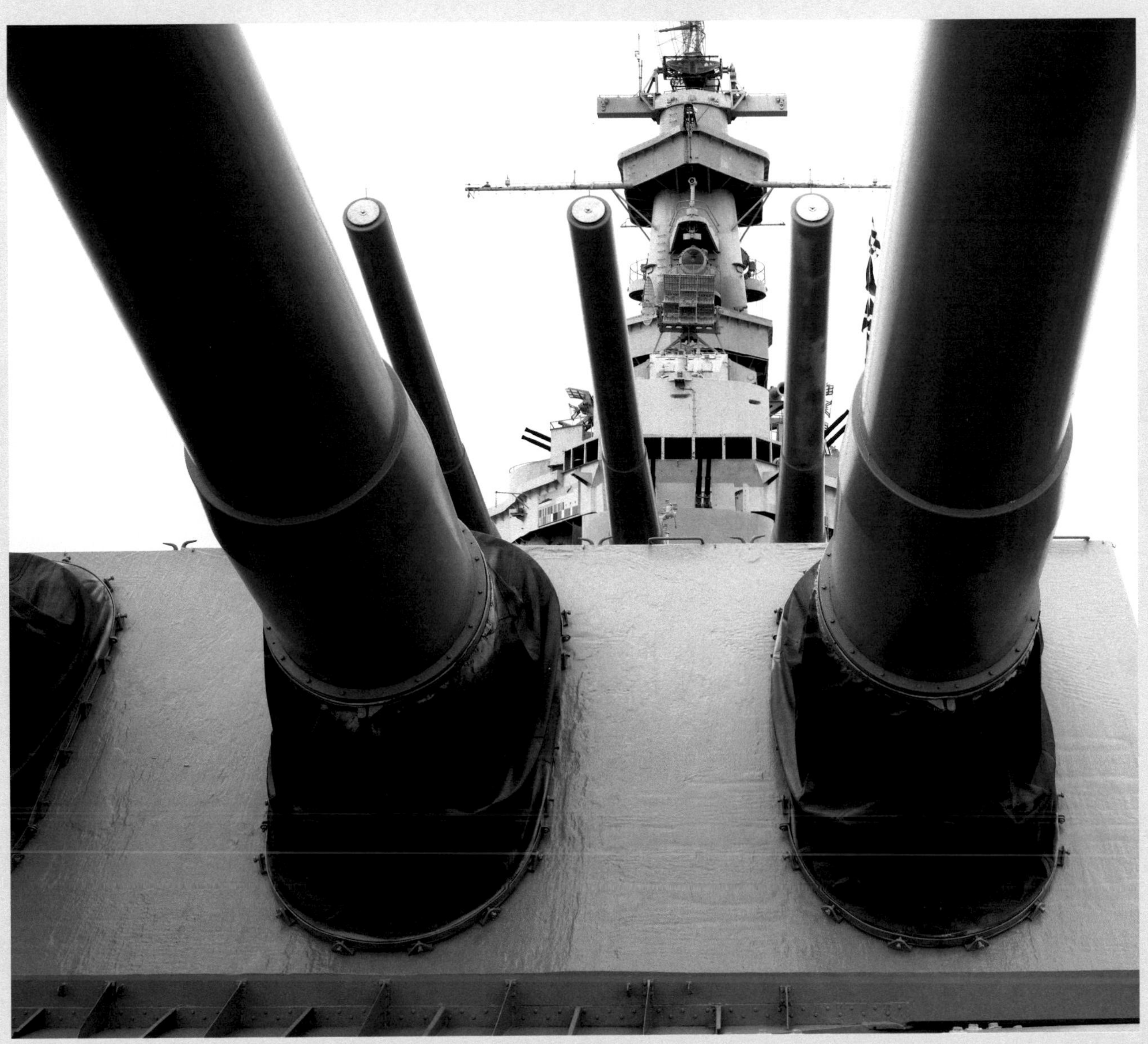

16-inch guns, USS *Alabama* battleship 1942 The *Alabama's* main armament consists of three turrets, two forward and one aft, with three 16-inch x 45-caliber long guns. Each round weighed 2,700 pounds and used up to 540 pounds of smokeless power in six separate bags. One hundred sailors toiled in each turret to fire a round every 30 seconds. At a distance of six miles, a penetration of nearly 22 inches of steel could be achieved. The maximum range of the gun was over 20 miles.<93> *At Battleship Memorial Park, Mobile, AL*

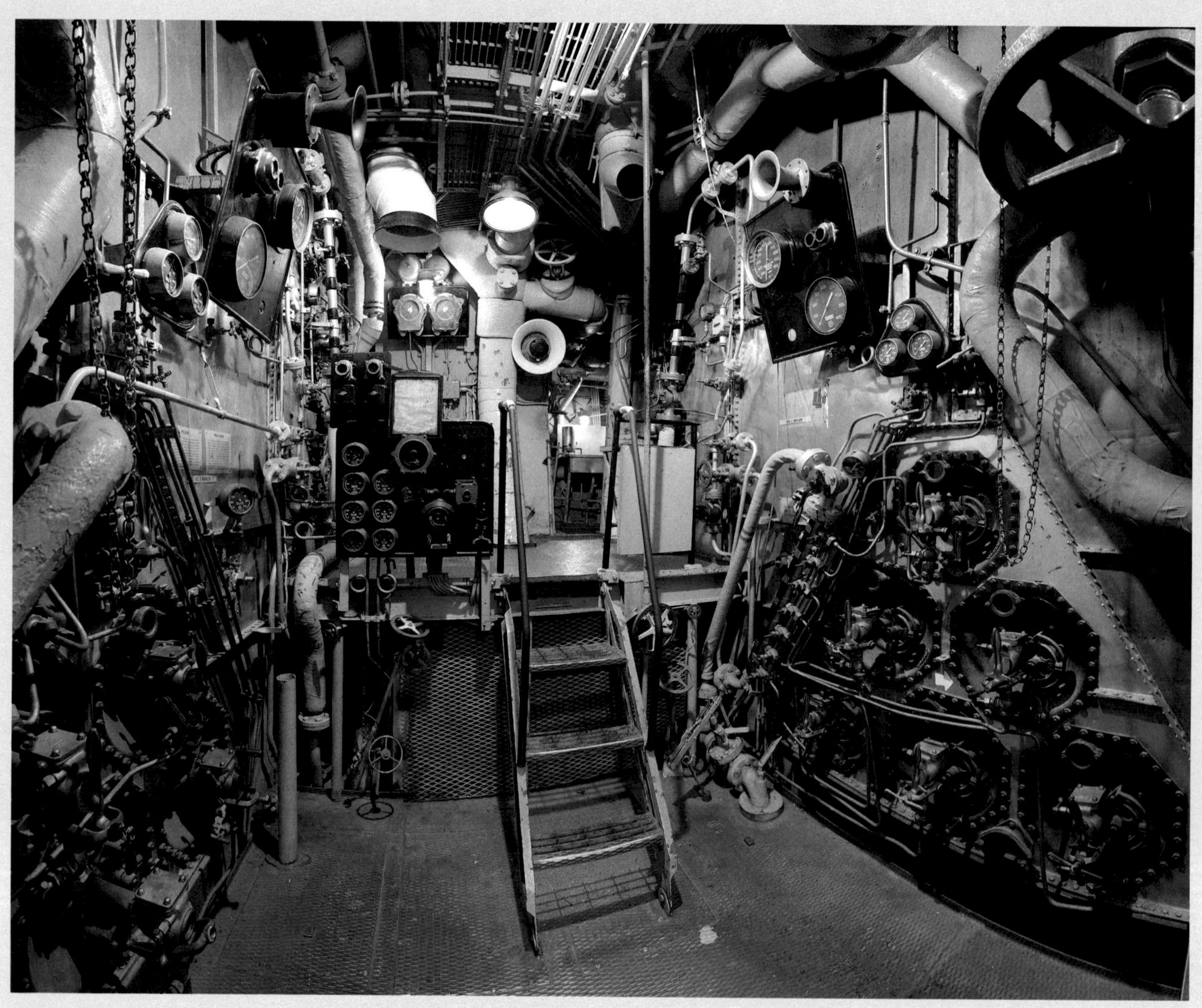

Boilers for No. 1 Engine Room, USS *Alabama* battleship 1942 Four turbines powered the Alabama, and these were operated by the steam from eight boilers. In the South Pacific sailors would have to endure temperatures up to 120 °F in these tight quarters. The turbines working together could deliver up to 130,000 horsepower to the four propellers, each of which were over 17 feet in diameter. At full steam the Alabama reached nearly 28 knots.<94> The complexity here is impressive. *At Battleship Memorial Park, Mobile, AL*

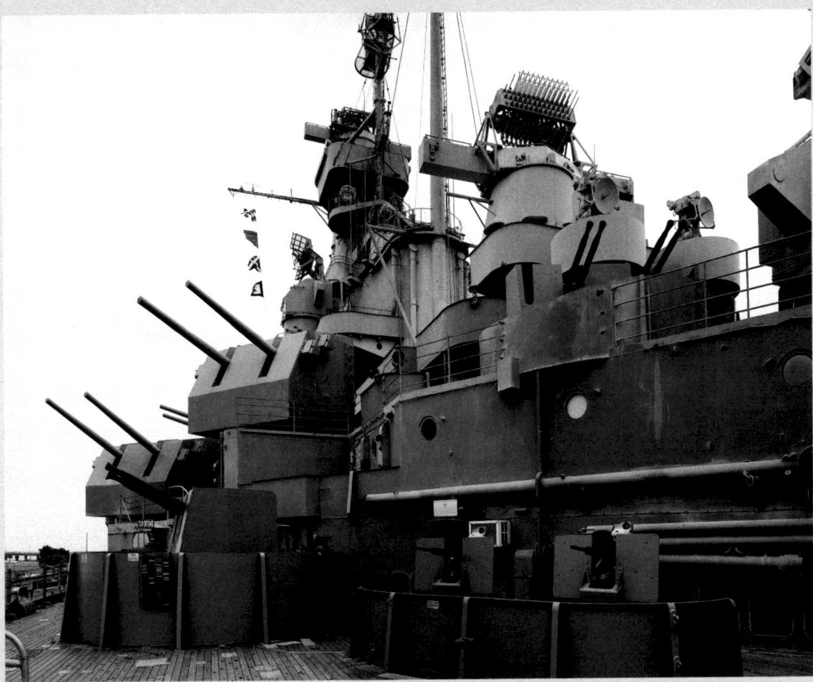

Port Side Armament, USS *Alabama* battleship 1942 This view shows some of the twenty 5-inch x 38-caliber guns and some hundred 20-mm and 40-mm anti-aircraft guns. The latter were put to good use during the first Kamikaze attacks in October 1944 and again in May 1945. Even as late as 9 August 1945, the day of the second atomic bomb attack, a nearby destroyer was hit by a Kamikaze pilot<95> *At Battleship Memorial Park, Mobile, AL*

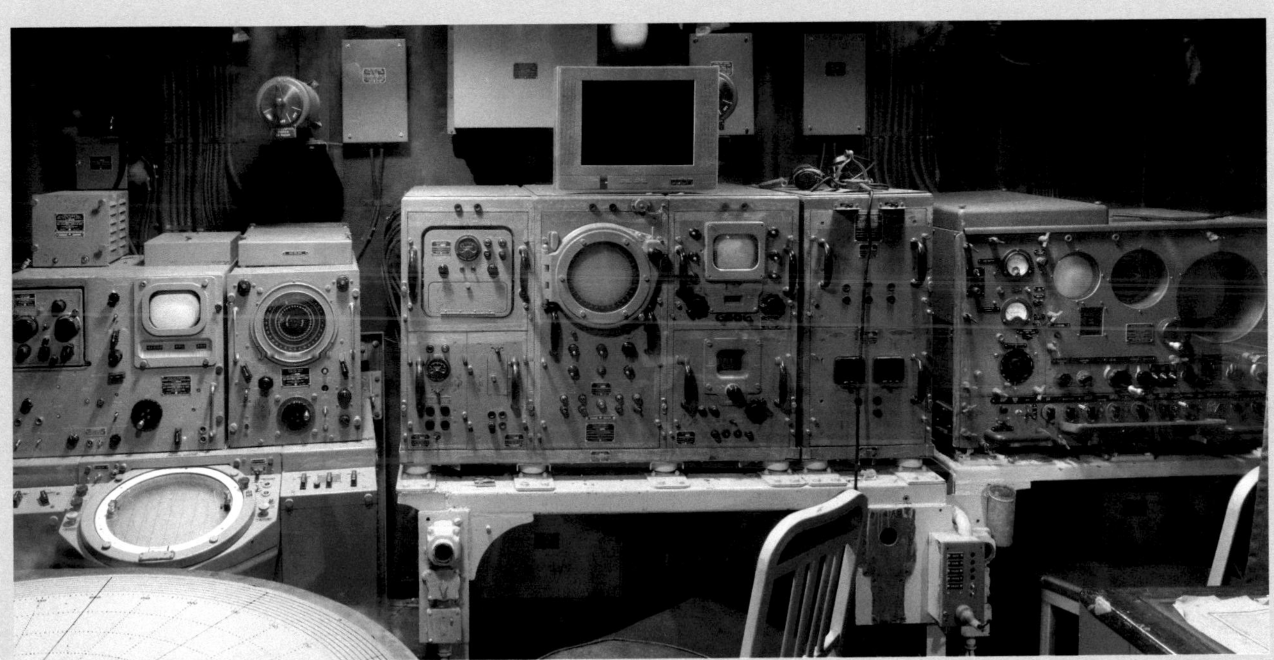

Combat Information Center, USS *Alabama* battleship 1942 Here is where the signals and radar information is gathered and organized for command decisions in time of combat. At the Battle of the Philippine Sea in June 1944 the approach of Japanese bomber at a range of 190 miles was first reported by the radar men of the *Alabama*.<96> The day I photographed the Combat Information Center there was an *Alabama* radar veteran visiting, and he said that they had received no training on the radar. It had just arrived one day in crates, and they had installed it and learned to use it from the manuals. *At Battleship Memorial Park, Mobile, AL*

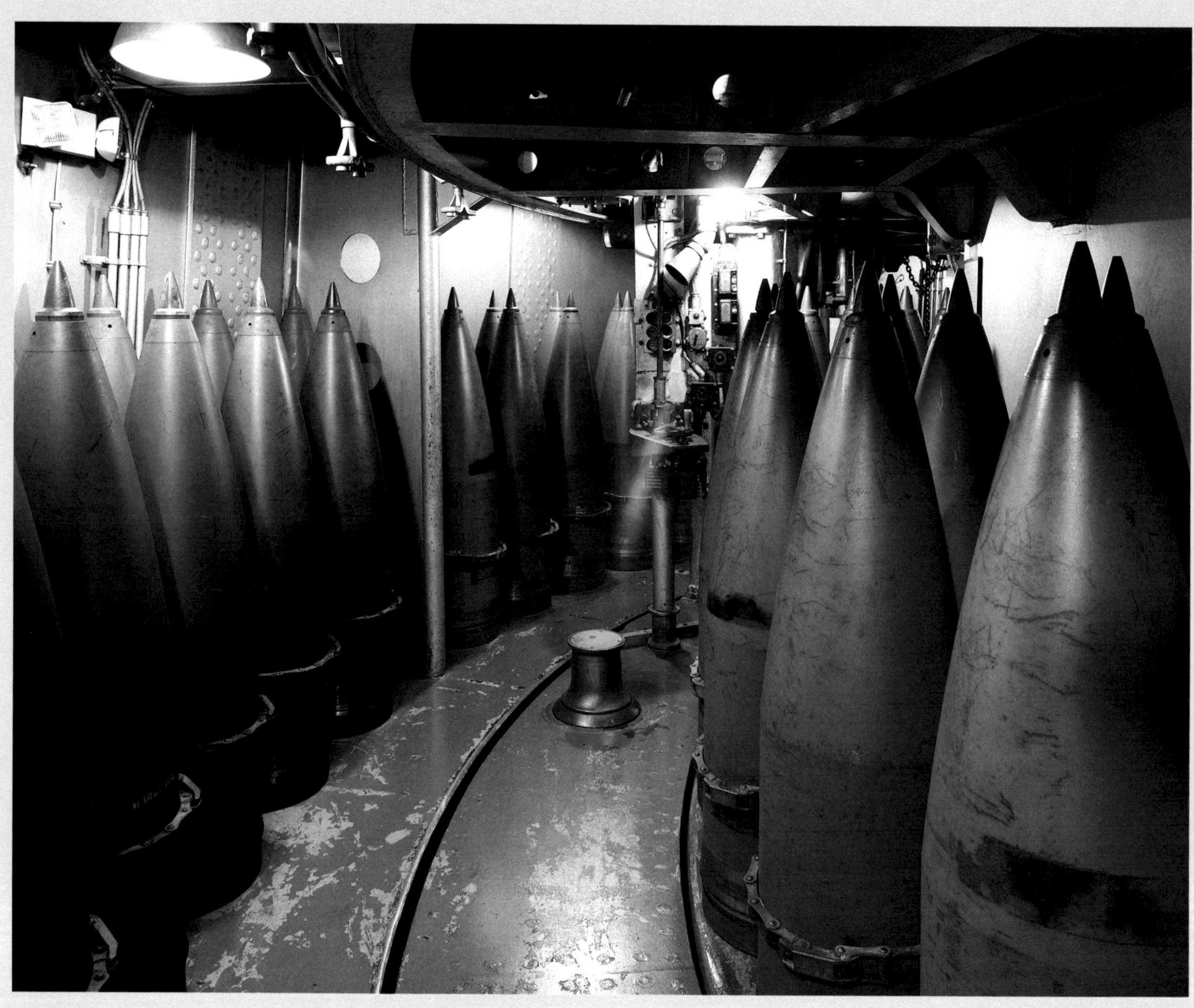

16-inch projectile handling platform, USS *Massachusetts* battleship, BB-59 1942 The *Massachusetts* was a sister ship of the *Alabama*, commissioned three months earlier. "Big Mamie," as she was known by her crew, had the improbable distinction of firing both the first 16-inch shell of the war at Casablanca on 8 November 1942 and the last 16-inch shell of the war at Kamaishi on Honshu Island on 9 August 1945.<97> *At Battleship Cove, Fall River, MA*

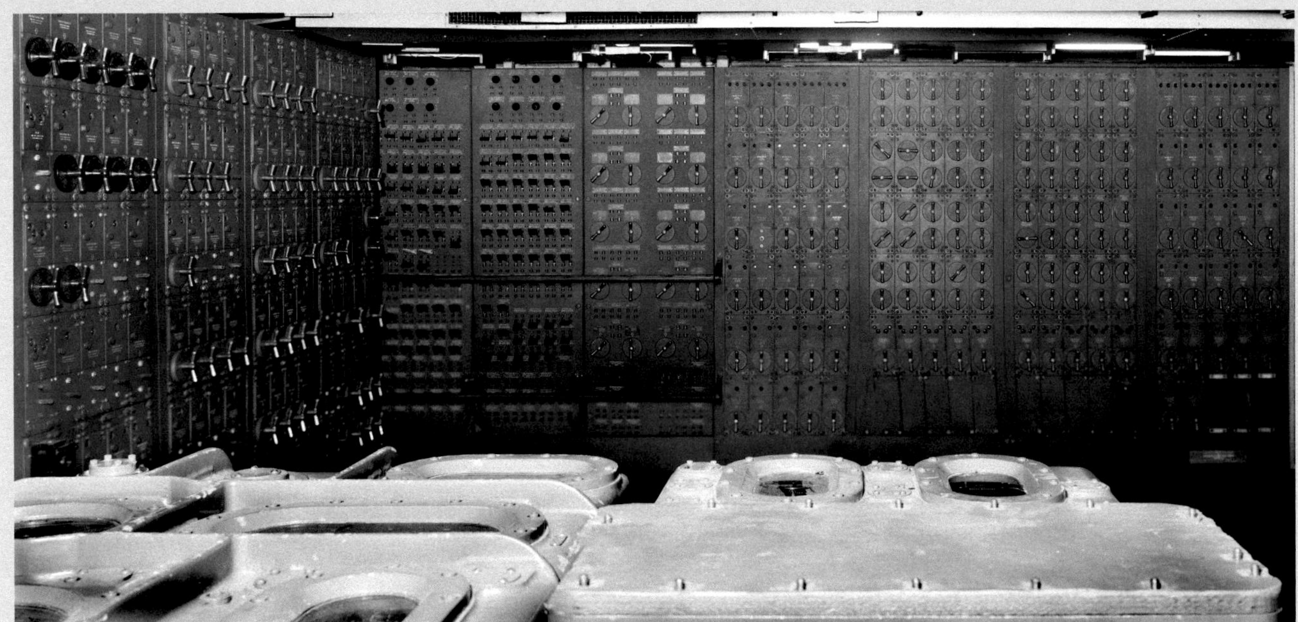

Main Battery Plotting Room, USS *Massachusetts* battleship 1942 The main battery fire-control switchboard is in the background. In the foreground are the mechanical analog computers used to calculate the gun aiming parameters and setting shell fuze times required to strike a given target. These complications became necessary when high energy, slow burning smokeless powders, coupled with rifled barrels, dramatically increased the ranges of naval guns, transforming them from direct-fire to indirect-fire weapons. The analog computer shown in the foreground here is a Ford Instrument Range Keeper Mark 8, Model 9 calibrated for the *Massachusetts'* 16-inch guns.<98> Hannibal Ford started his business of making fire-control systems for the US Navy in 1915. His first, the Rangekeeper Mark 1 was installed on the battleship USS *Texas* in 1917. Ford made versions of his ingenious devices for most of the Navy's larger gun systems in WWII. The accuracy could be stunning. In one instance at the Battle of Guadalcanal, the battleship *Washington* made nine hits of 75 fired at a Japanese battleship at a range of 11 miles.<99> *At Battleship Cove, Fall River, MA*

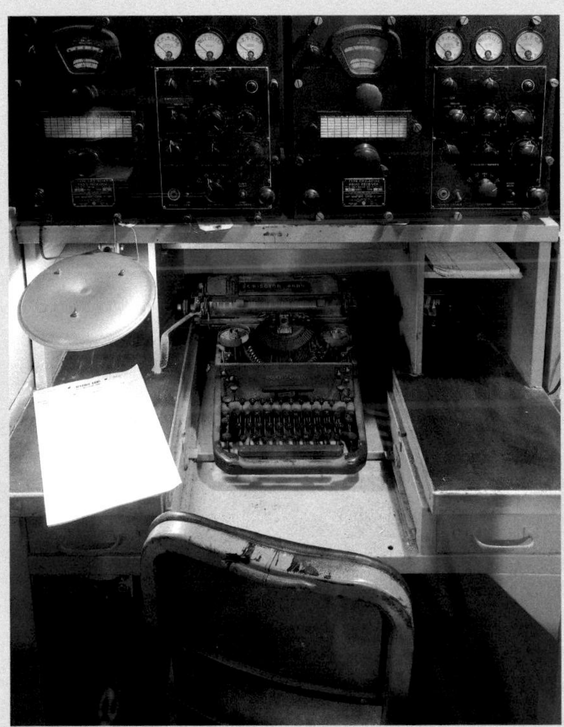

Radioman Station, USS *Massachusetts* battleship 1942 At these stations, and there were many such in the radio room, the radioman would listen at some frequency to Morse code communications and transcribe the message using the typewriter. This was obviously a highly skilled task as signals would often fade in and out and be obscured by static. *At Battleship Cove, Fall River, MA*

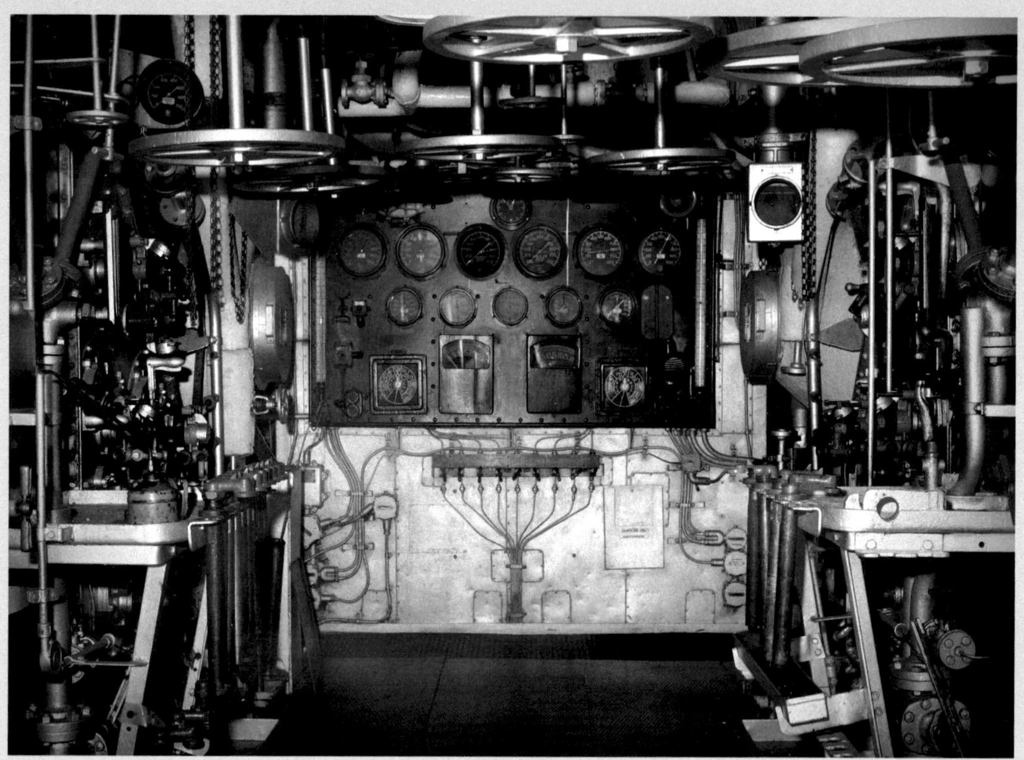

Engine Control Board, USS *Massachusetts* battleship 1942 Complexity was a new fact of life for military men of the twentieth century. However, like mass production itself, breaking any task down into a series of steps and imposing relentless regimens proved that mechanical complexity, even with highly dangerous machinery, can be mastered by ordinary individuals. *At Battleship Cove, Fall River, MA*

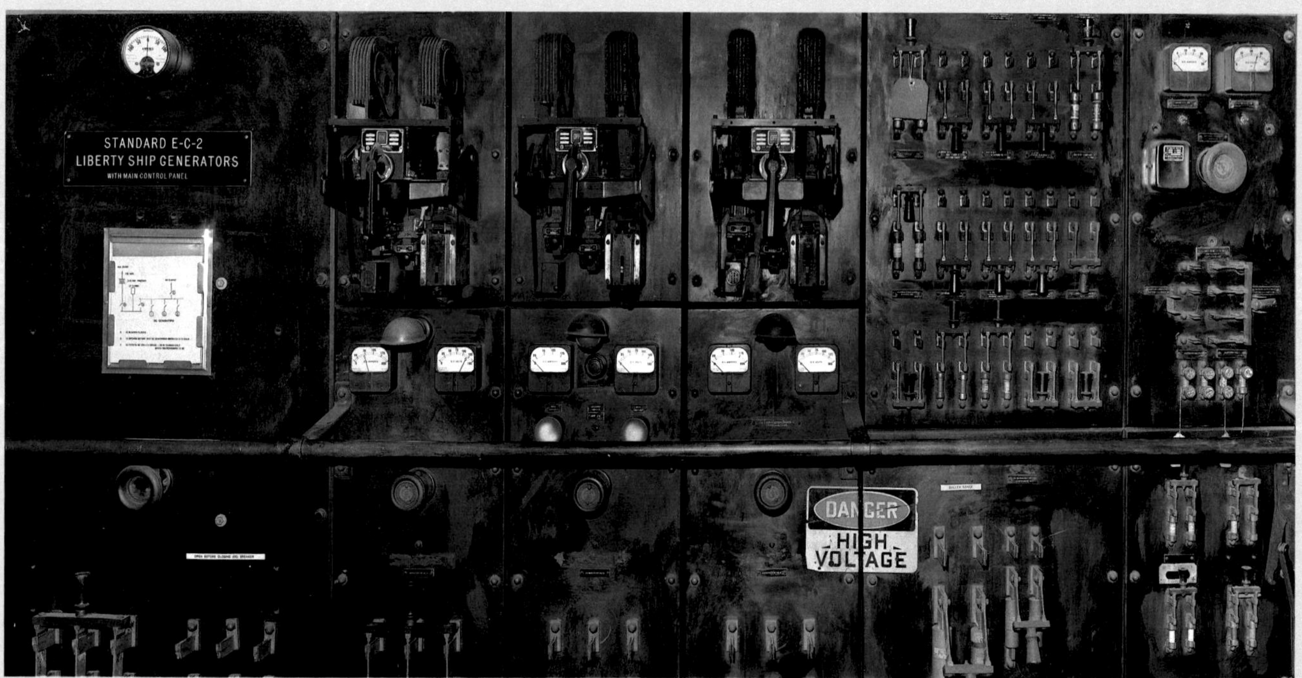

Electrical Panel, SS *John W. Brown* Liberty Ship 1942 Labor was a chronic problem in the shipyards and cooperation between labor unions and management was a continuing balancing act. A major influx of women into the workforce began in the fall of 1942, helping to relieve the shortages. Every trade was opened to them despite some concerns about the dangerous work in shipyards. Kaiser, in fact, found that women could be trained faster and did a better job in fabricating intricate electrical circuits requiring a high degree of dexterity.<100> *At Project Liberty Ship, Baltimore Harbor*

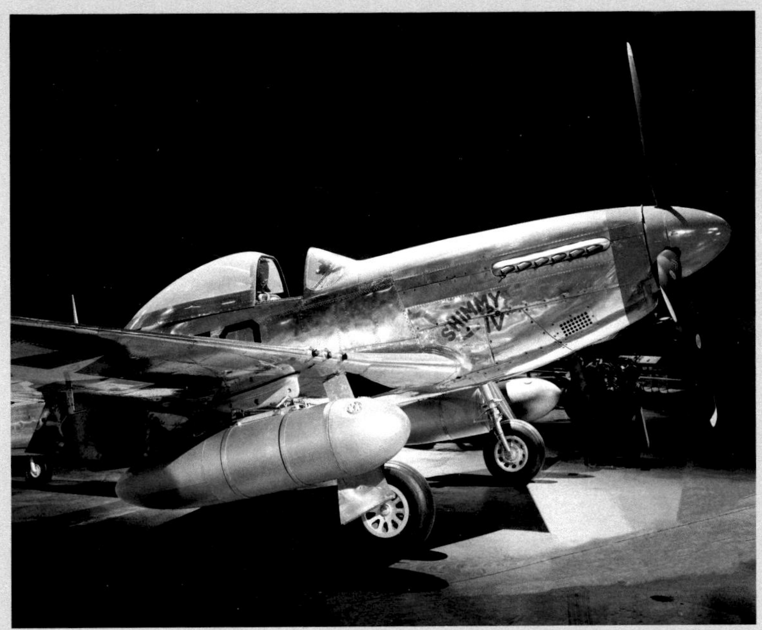

US long-range fighter aircraft, P-51D_Mustang 1942 In trying to carry the fight to Nazi Germany early in the war Allied bombers could only have fighter escort for part of the distance because of their limited range. This all changed with the development of the North American P-51 Mustang—but not right away. With its original Allison engine, the P-51 showed poor performance at high altitudes, but when the British fitted it with a Rolls Royce Merlin engine from a Spitfire, it excelled. The problem was that the Merlin would have to be made in large numbers and fast. American industrial troubleshooter Bill Knudsen convinced the Packard Motor Company to build the Merlin XX engine under license. The plans for this engine were so secret, however, that they were sent by battleship to the US. Rolls built all of their engines by hand; the American automobile industry did things differently. The engineers at Packard essentially had to redesign the engine to make mass production work. In six months the first of the new engines rolled off the assembly line. It would be followed by more than 55,000.<101> *Collection of the National Museum of the US Air Force*

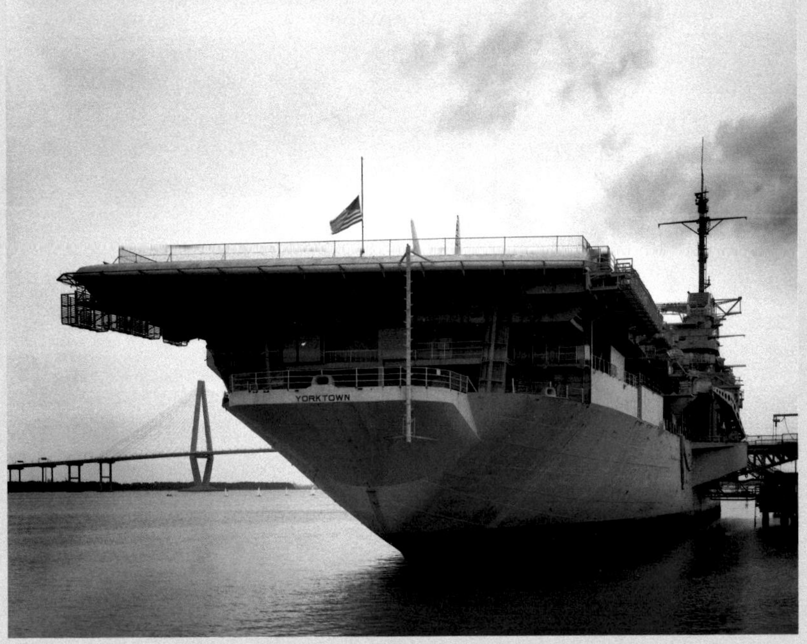

USS *Yorktown* aircraft carrier, CV10 1943 The first *Yorktown* carrier, CV5, was sunk at the Battle of Midway in July 1942. This *Yorktown*, some 40% larger in displacement, was a member of the *Essex* class with 24 sister ships.<102> Commissioned on 15 April 1943, the Yorktown arrived in the Pacific three months later carrying 360 officers, 3,000 enlisted men, 37 F6F Hellcat fighters, 36 SBD-5 Dauntless dive bombers, and 18 TBF Avenger torpedo bombers. After a storied combat history in WWII, she went on to support combat operations in both Korea and Vietnam.<103> *At Patriots Point Naval & Maritime Museum, Mt. Pleasant, SC*

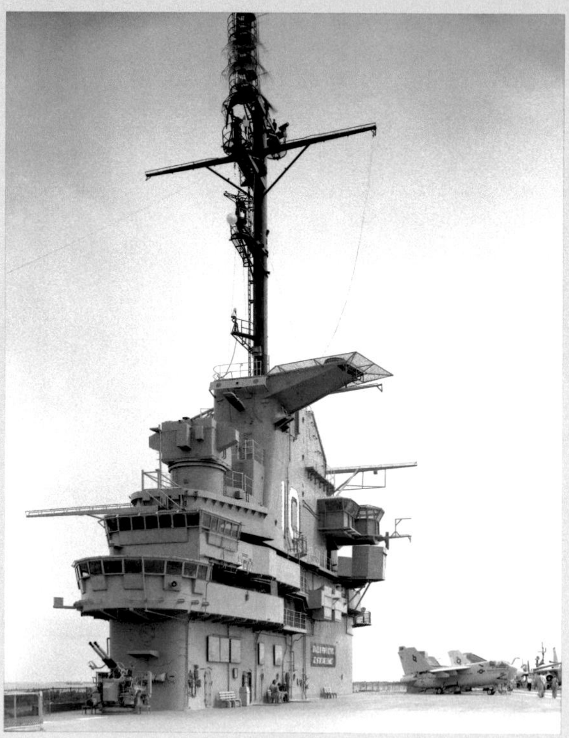

Control Island and part of the flight deck, USS *Yorktown* 1943 In March 1945 under heavy attack by Japanese dive bombers, a bomb succeeded in hitting the signal bridge and after a series of bounces exploded near the water line. Two large holes were blasted through the hull starting fires on the third deck. Heroic efforts quenched the blaze, but five men died and 26 more were wounded. Repairs were feverishly made overnight and by 0600 the next morning all fighters were launched to attack airfields on the main island of Honshu.<104>
At Patriots Point Naval & Maritime Museum, Mt. Pleasant, SC

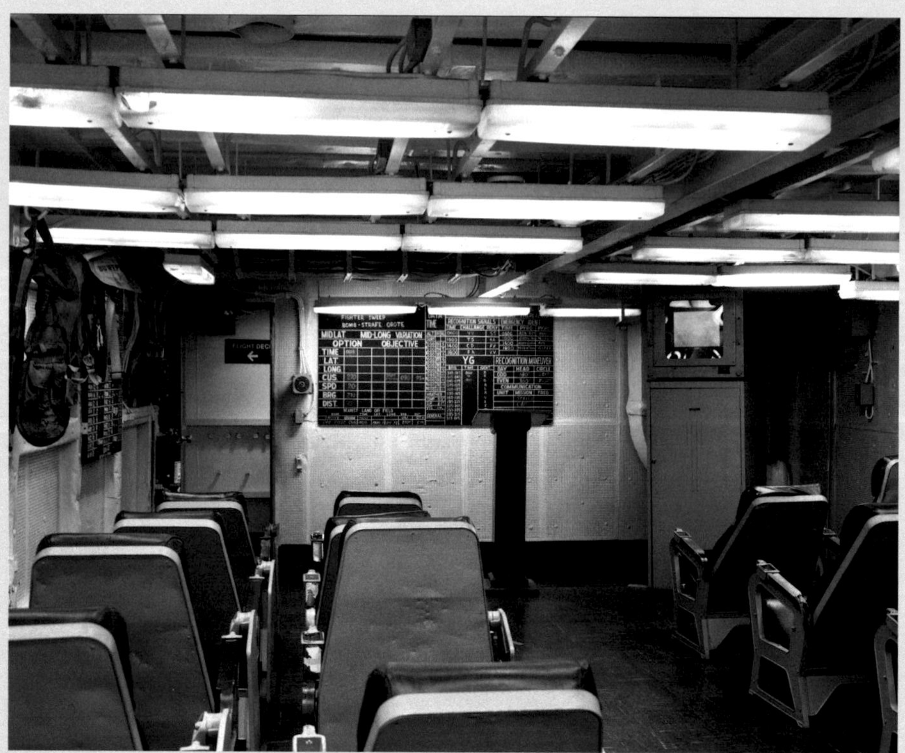

Pilots' ready room, USS *Yorktown* CV10 1943 Three generations of pilots have received their mission instructions in this room—World War II, Korea, and Vietnam. One can feel the tension still. *At Patriots Point Naval & Maritime Museum, Mt. Pleasant, SC*

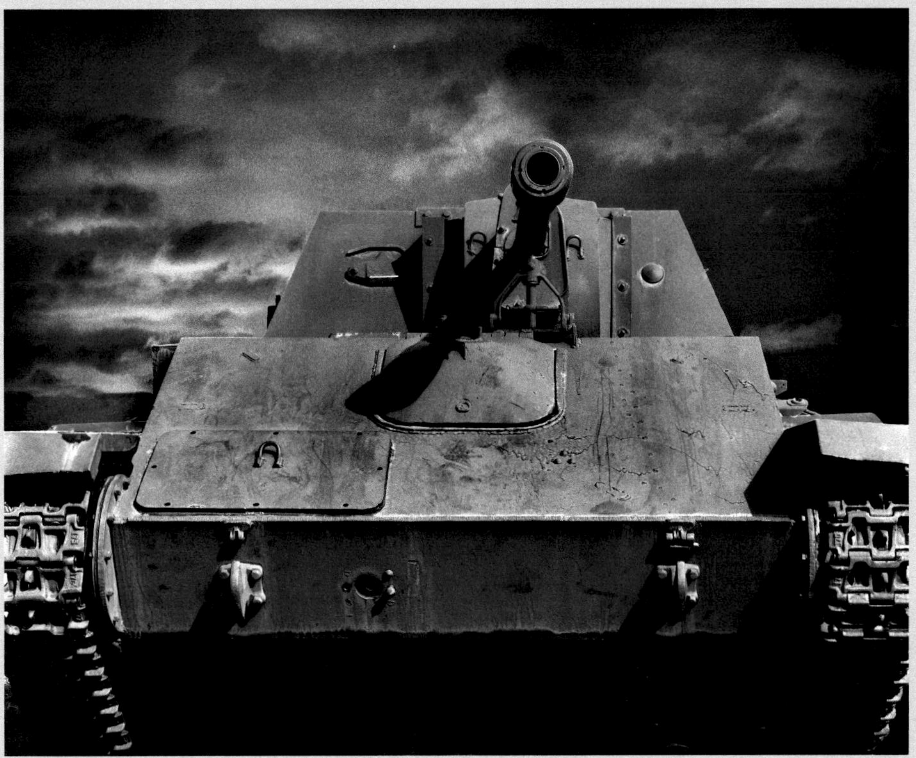

Soviet self-propelled 76.3-mm gun, SU-76, 12.3 tons combat weight<105> **1943** Originally designed for both the anti-tank role and close fire support of infantry, improvements in German armor obsoleted the former role, but it continued successfully in the latter until the end of the war. More Su-76s were produced than any other Soviet vehicle except the T-34.<106> *Former Collection of the US Army Ordnance Museum, Aberdeen Proving Ground, MD*

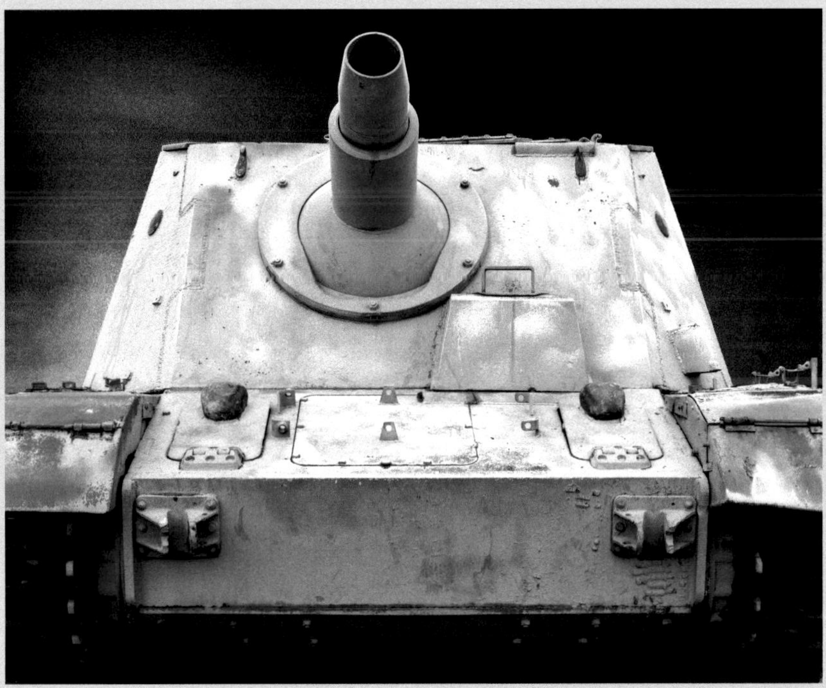

German *Sturmpanzer* (assault howitzer) IV *Brummbär* (Grizzly Bear) with 150mm gun, 30.4 tons combat weight<107> **1943** Yet another variant built on the *Pzkpfw* IV chassis, the *Brummbär* used its 150-mm gun for close infantry support and to reduce fortifications. It was first used at Kursk and heavily relied upon afterward until the end of the war. Some 300 were built.<108> *Former Collection of the US Army Ordnance Museum, Aberdeen Proving Ground, MD*

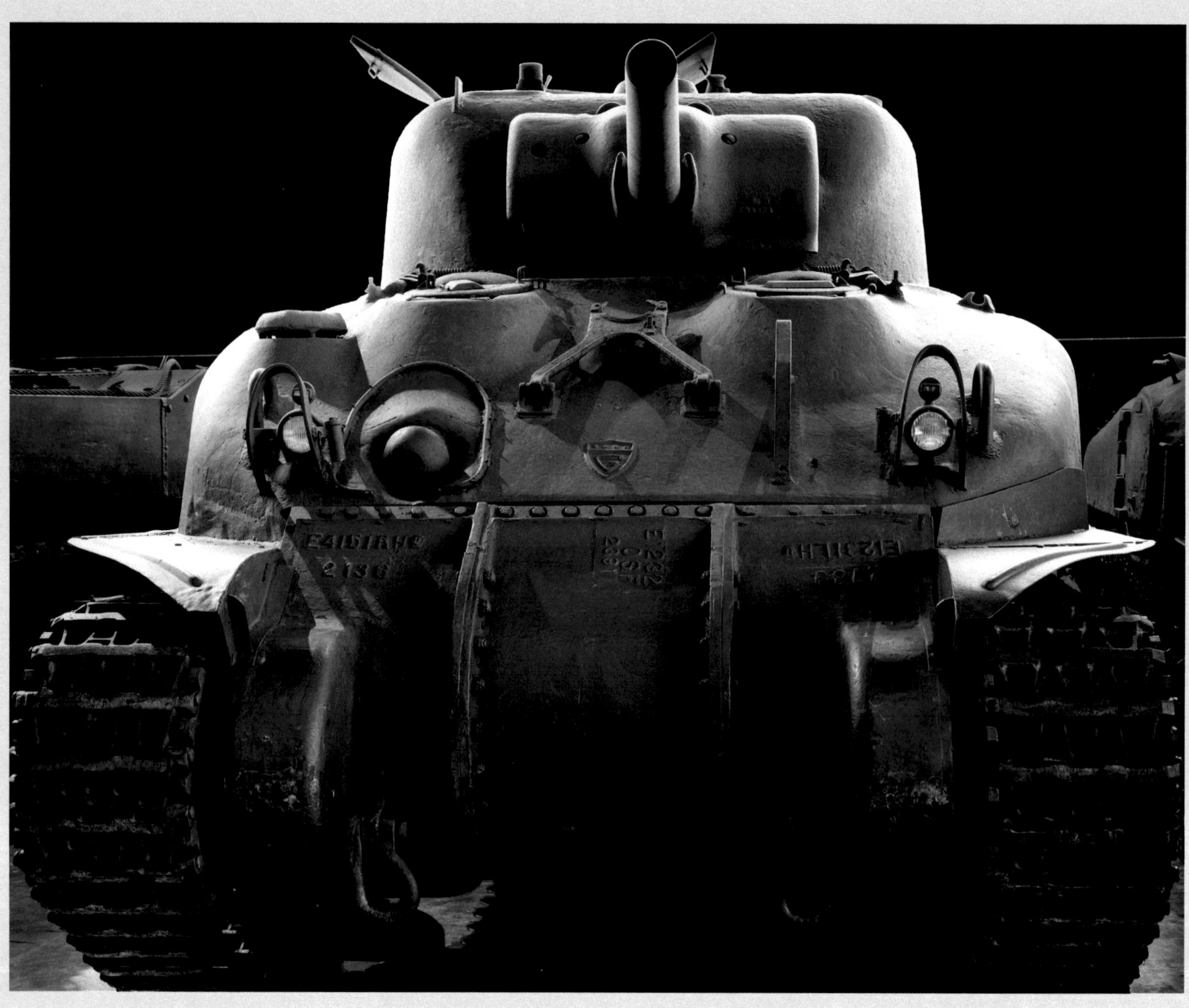

Canadian tank, M4A1 Grizzly I with 75-mm gun, 33.5 tons combat weight 1943 The Canadian version of the M4 Sherman tank, built by the Montreal Locomotive Works, incorporated a cast hull in place of a welded-plate hull. Production started in July 1943 but was cancelled five months later.<109> *Collection of the Canadian War Museum*

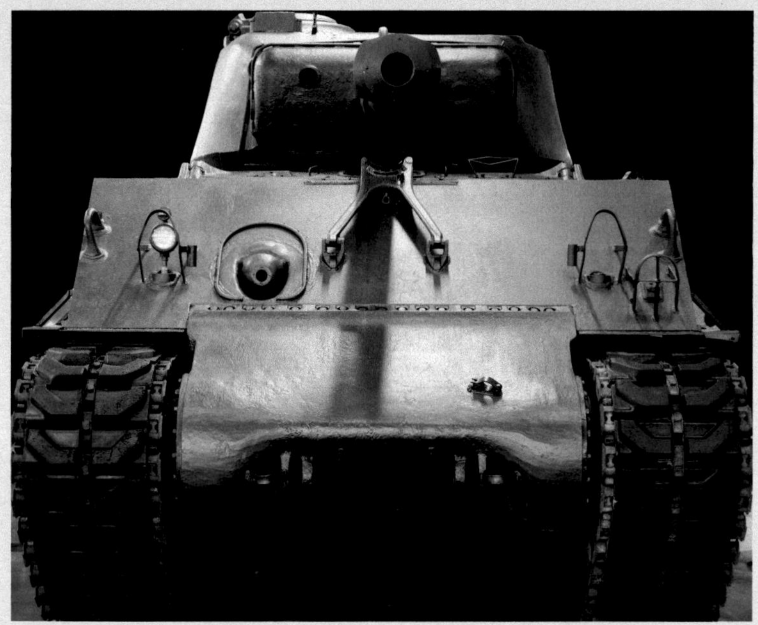

Canadian medium tank, Sherman M4A2E8 with 76-mm gun, 35 tons combat weight 1943 This tank may have been used by Canadian forces, but it was not built in Canada. Only the Grizzly I in the previous plate was built there. There were eleven companies in all that produced the Sherman along with countless subcontractors making component parts. They were: American Locomotive (ALCO), Baldwin Locomotive Works, Chrysler Defense Arsenal (CDA), Federal Machine & Welder, Fisher Tank Arsenal, Ford Motor Company, Lima Locomotive Works, Montreal Locomotive Works, Pacific Car and Foundry, Pressed Steel Car, and Pullman Standard. Fisher, General Motors' body subsidiary, received government commitment to fund a new plant in mid-November 1941. They began construction of the plant in January 1942, and completed their first tank on 7 March 1942 by impressing existing plant floors into operation. What an impressive achievement!<110> By war's end some 50,000 Shermans had been produced.<111> *Collection of the Canadian War Museum*

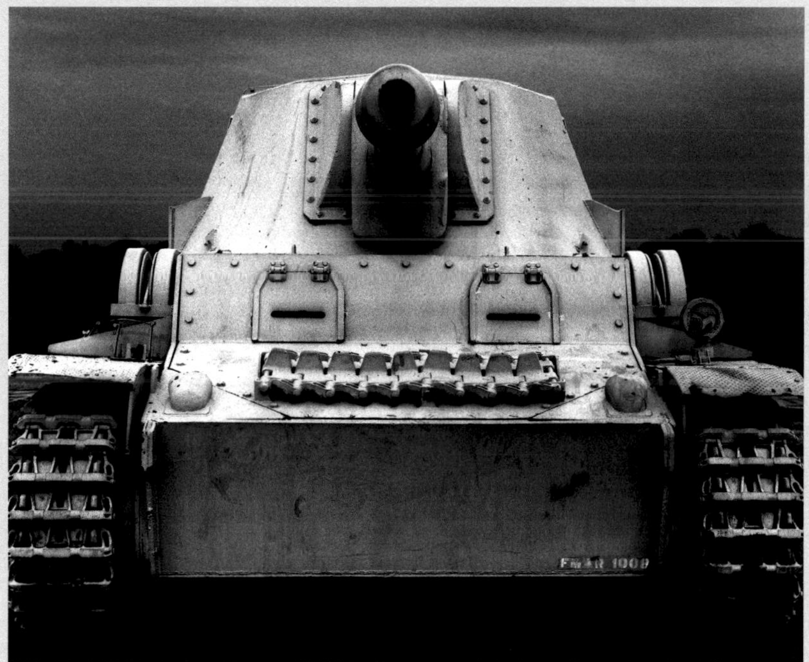

German self-propelled howitzer, *Heuschrecke* (Grasshopper) IV with 105-mm gun<112> 1943 One of the more bizarre variants of Germany's self-propelled artillery, the Krupp-built Heuschrecke is equipped with a removable turret. The drums at either side of the hull are part of a hull-mounted parallelogram lifting frame that allows the gun turret to be removed to a wheeled carriage and used as towed artillery. The contraption was never put into production.<113> *Former Collection of the US Army Ordnance Museum, Aberdeen Proving Ground, MD*

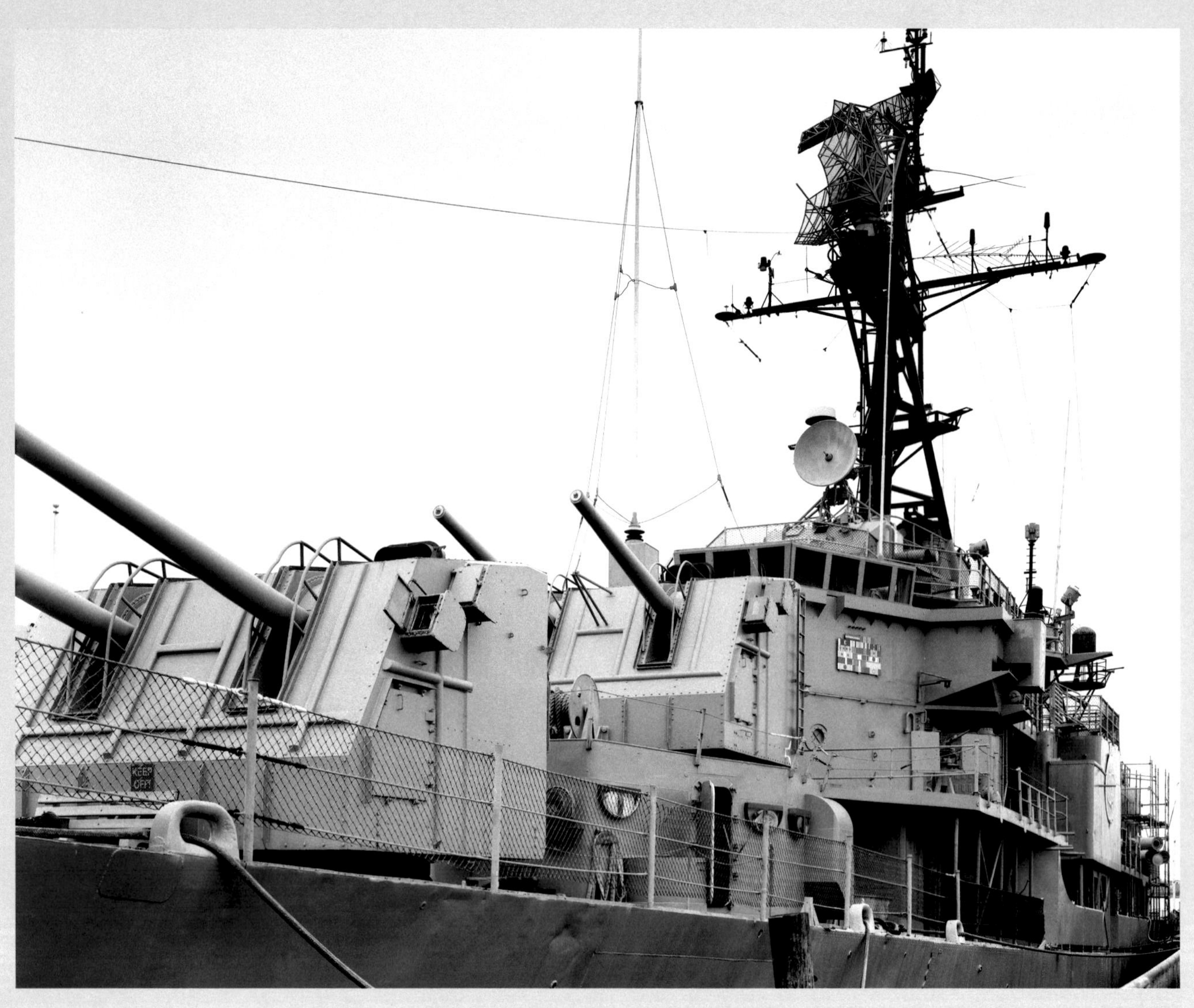

USS *Laffey* destroyer, DD-724 1944 The *Laffey* was a celebrated Kamikaze combatant of the Second World War. On 16 April 1945 it was on radar picket duty off Okinawa when it was attacked by 22 Japanese bombers and Kamikazes over the space of a harrowing hour and a half. According to the ship's action report, she was hit by seven Kamikazes, a low-flying bomber, and four bombs, but managed to shoot down nine planes. Five of the planes did extensive damage. Her captain, Commander F. Julian Becton heroically kept the ship afloat despite her rudder being disabled, 32 crewmen killed, and 71 wounded.<114> *At Patriots Point Naval & Maritime Museum, Mt. Pleasant, SC*

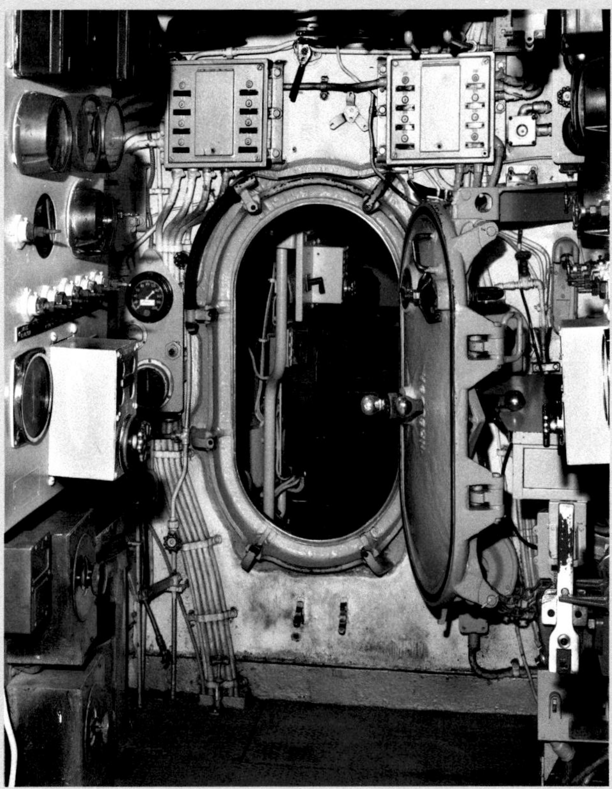

Hatchway, USS *Lionfish* submarine, SS-298 1944 *Lionfish*, a Balao-class sub, was commission on 13 October 1944 and spent WWII on war patrol in the Pacific. Her first captain was Edward Spruance, son of Admiral Raymond Spruance, illustrious commander of US naval forces at the Battle of Midway and the Battle of the Philippine Sea.<115> *At Battleship Cove, Fall River, MA*

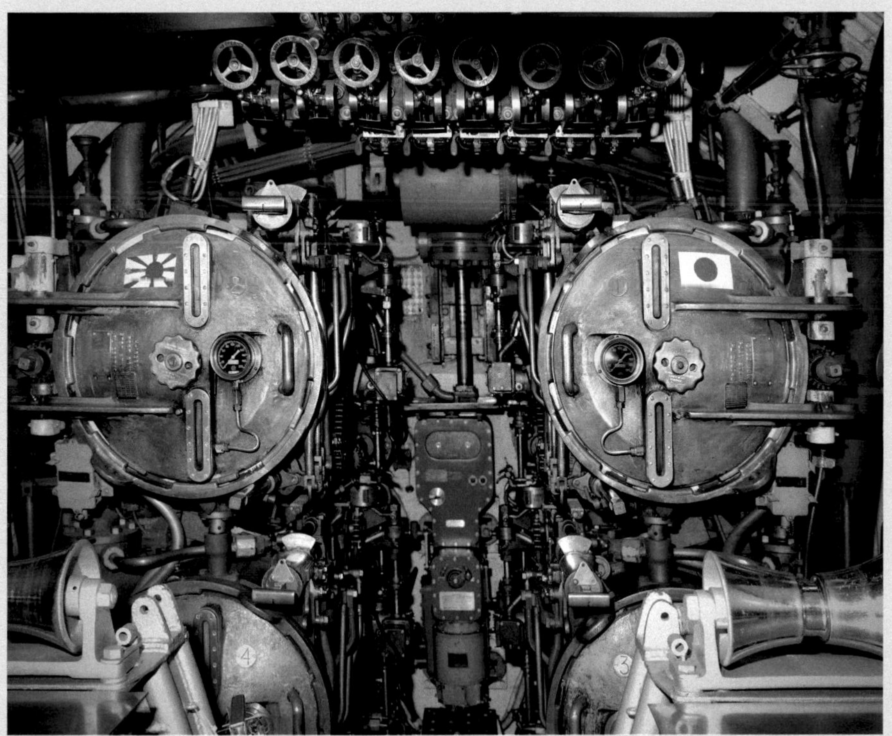

Aft Torpedo Tubes, USS *Lionfish* submarine, SS-298 1944 In addition to these four aft torpedo tubes, the Lionfish had six forward tubes. Every crew member aboard the submarine had to know how to perform every other job in case of emergency. This required intensive and recurring training regimens.<116> *At Battleship Cove, Fall River, MA*

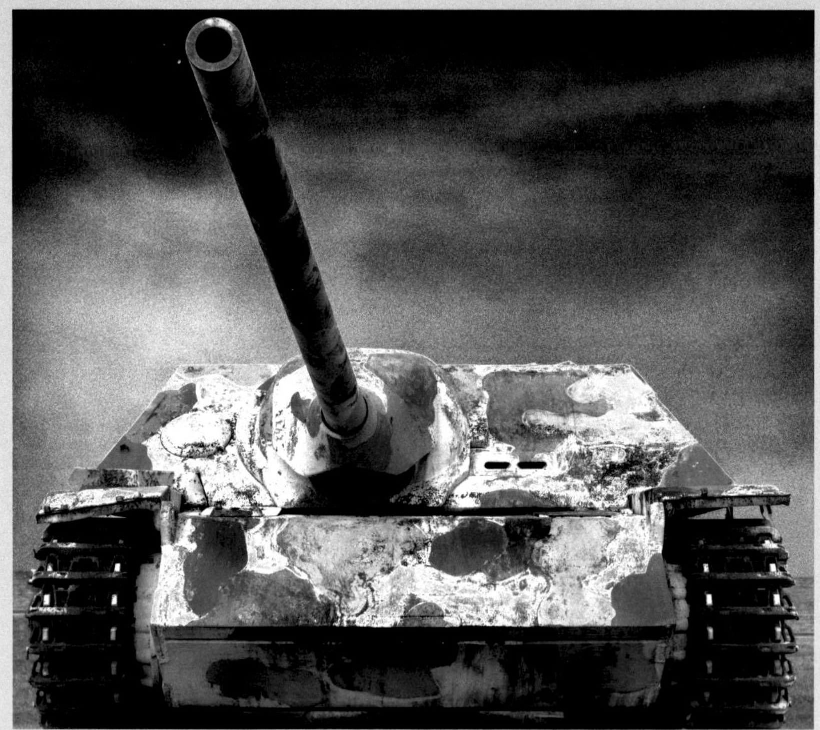

German *Jagdpanzer* (tank hunter) IV with 75-mm gun, 26.9 tons combat weight<117> **1944** The earlier *Sturmgeschutz* IV had gradually shifted in its role of close infantry support to that of tank hunter, but a dedicated ground-up design was found to be desirable to counter the heavy Soviet tanks. So, once again, the *Pzkpfw* IV chassis was adapted to a potent 75-mm anti-tank gun. The result proved very effective and some 930 were built before war's end.<118> *Former Collection of the US Army Ordnance Museum, Aberdeen Proving Ground, MD*

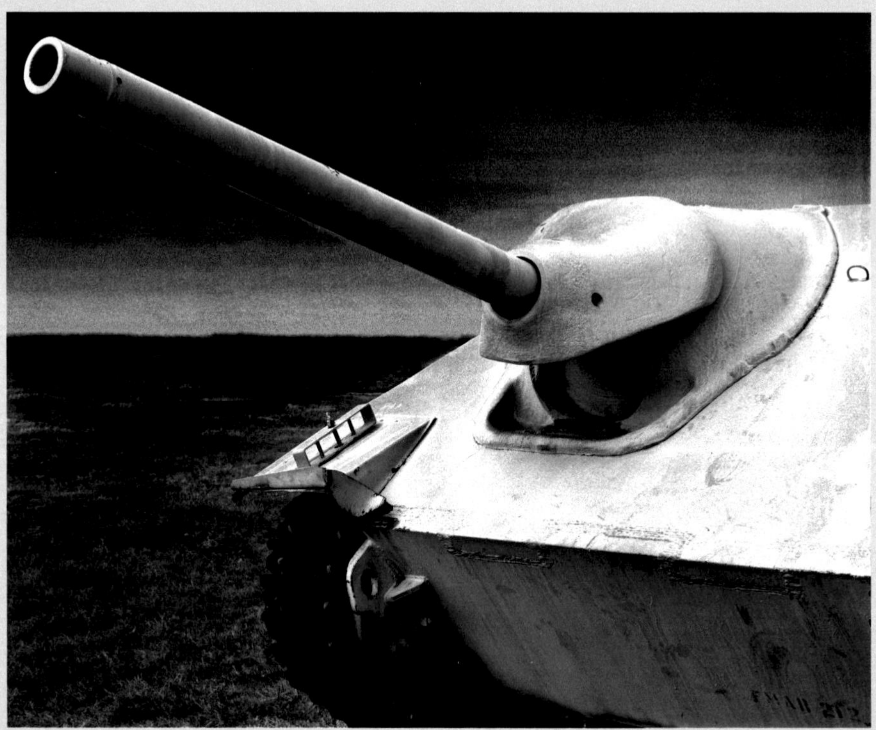

German tank destroyer, Panzerjäger *Hetzer* (rabble-rouser) with 75-mm gun, 17.6 tons combat weight<119> **1944** A light tank hunter built on the Czech 38 chassis, the *Hetzer* proved so effective in the latter stages of the war that the Skoda Works was given over to its exclusive production. Its low profile led to deadly stealth. Over 2500 were built.<120> *Former Collection of the US Army Ordnance Museum, Aberdeen Proving Ground, MD*

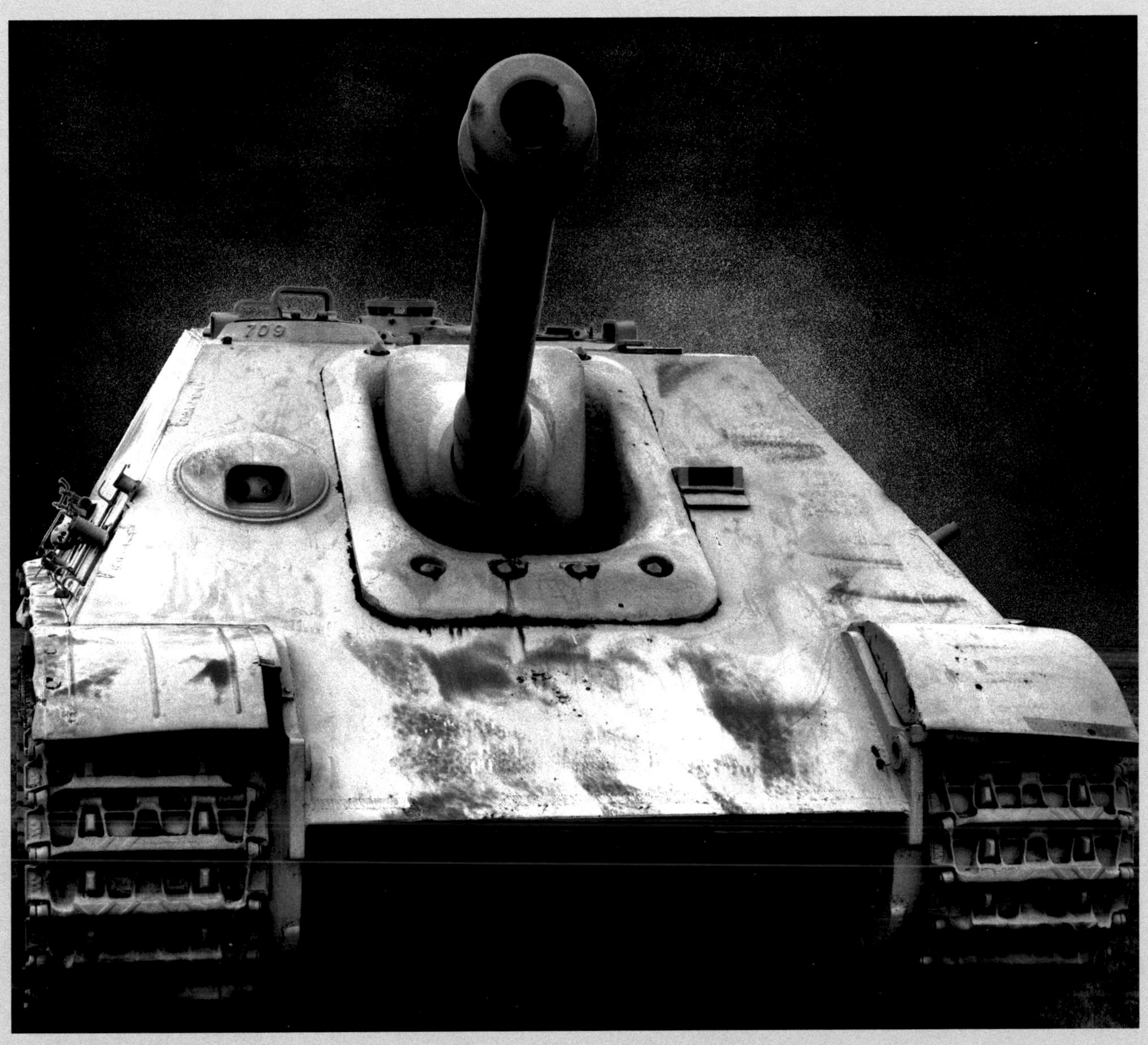

German tank destroyer, Panzerjäger Panther "Jagdpanther" with 88-mm Pak 43 gun, 51.3 tons combat weight<121> 1944 The Jagdpanther was originally designed by Krupp but pressures of other work moved its production to other manufacturers. It had good armor, good mobility, and a proven anti-tank gun; its effectiveness against all Allied tanks was superb, particularly in engaging them at long distance. Production did not begin until 1944 and by then was constantly being interrupted by Allied bombing. A little less than 400 were built.<122> *Former Collection of the US Army Ordnance Museum, Aberdeen Proving Ground, MD*

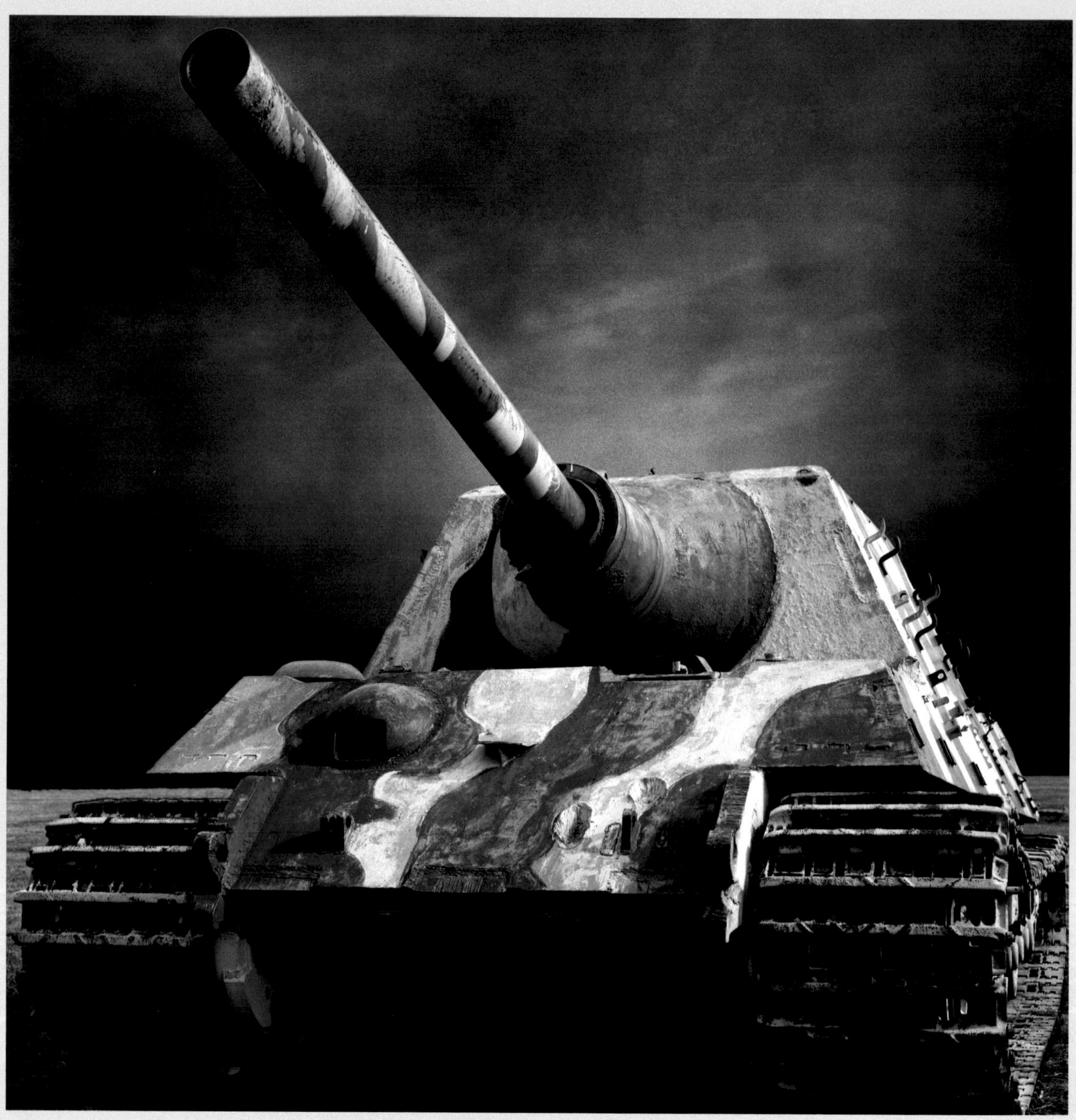

German *Jagdtiger (hunting tiger)* Model B with 128-mm PJK 80 gun, 79 tons combat weight<123> **1944** This was probably the most powerful armored vehicle of the war. Built on the *Pzkpfw* VI chassis and armed with the massive 128-mm gun originally developed for the next-generation, super-heavy tank *Maus* (mouse), its enormous weight and ponderous size limited its mobility and versatility, but it did see action in the Battle of the Bulge. Some 85 were produced before the war ended.<124> *Former Collection of the US Army Ordnance Museum, Aberdeen Proving Ground, MD*

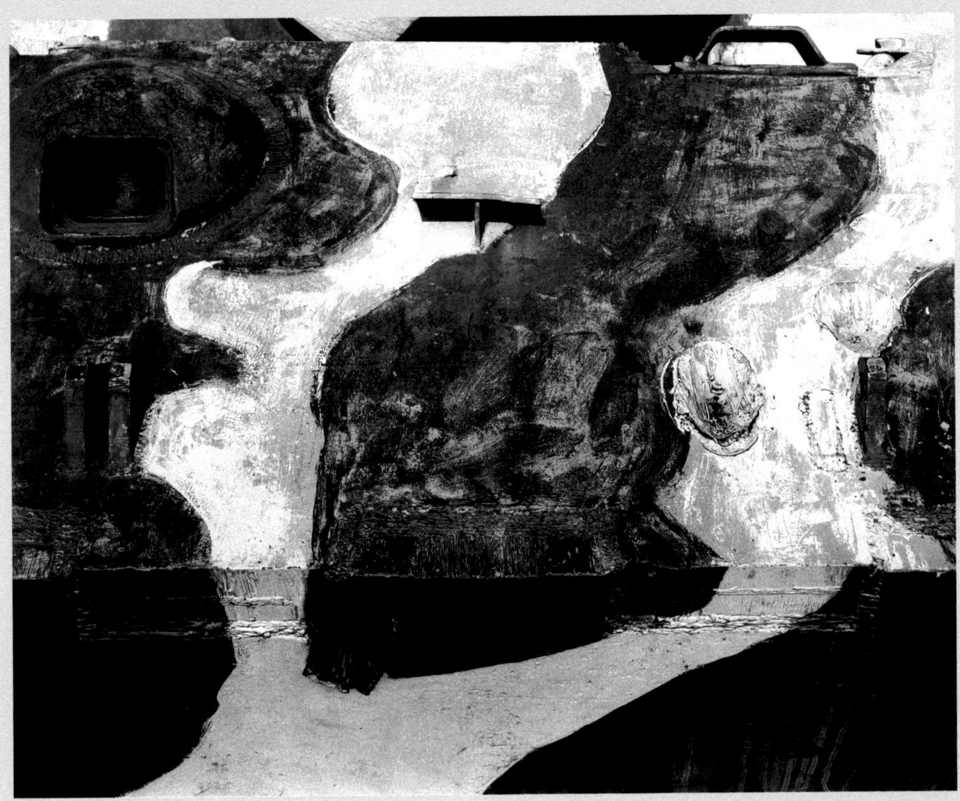

Camouflaged Glacis Detail, *Jagdtiger* The sloping frontal armor, or glacis plate, is nearly 10 inches thick on the Jagdtiger.<125> Note the evidence of a non-penetrating direct hit at center right of the image. *Former Collection of the US Army Ordnance Museum, Aberdeen Proving Ground, MD*

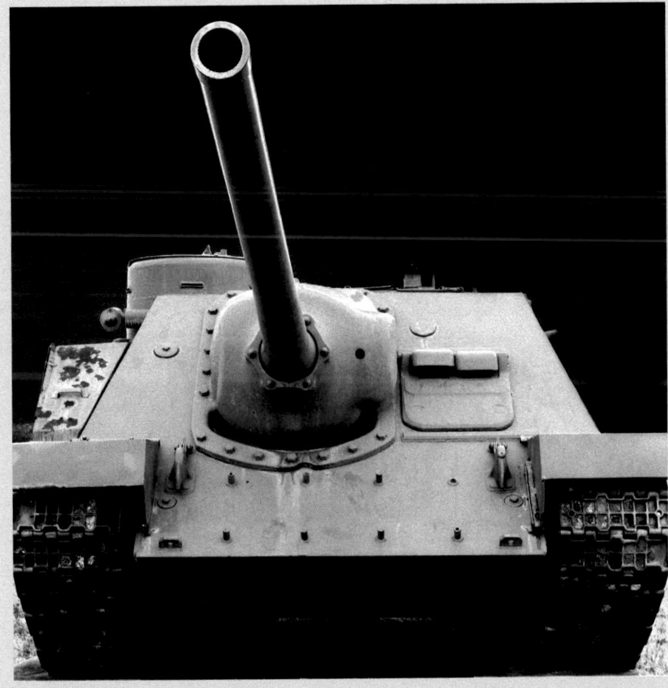

Soviet tank destroyer, SU-100 with 100-mm gun, 31.5 tons combat weight 1944 The introduction of the heavy German Panther and Tiger tanks necessitated a response from the Soviet Army that could counter them. The SU-100's powerful high-velocity gun filled the bill, enabling it to engage at great distance. It was built on the successful T-34 chassis. Over 1,500 were produced.<126> *Former Collection of the US Army Ordnance Museum, Aberdeen Proving Ground, MD*

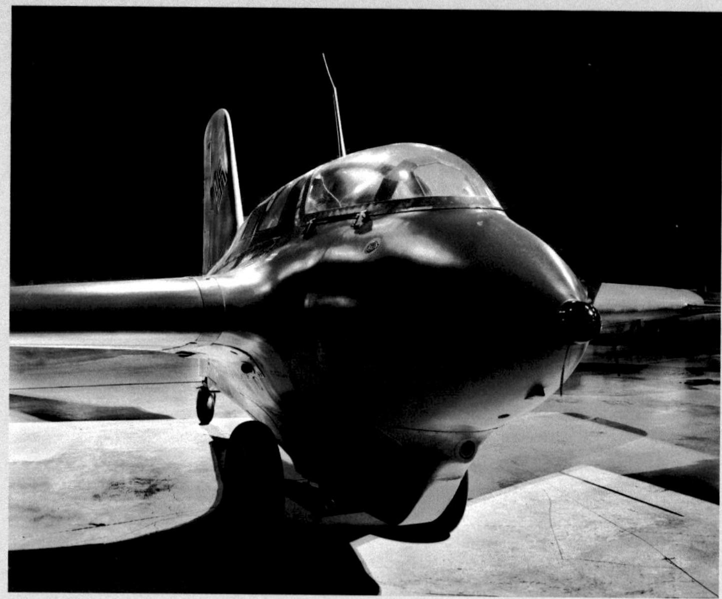

German rocket-powered fighter aircraft, Me 163B *Komet* 1944 The *Komet* was a quite extraordinary piece of wartime engineering. Its liquid-fueled rocket engine could carry just enough fuel to sustain 7.5 minutes of powered flight. With its extraordinary rate of climb, it would streak upward through Allied bomber formations firing its twin 15-mm MG 151 cannon, turn and glide back through the bomber formation firing at them for a second time. The plane was small and very agile, making a most difficult target for Allied guns. To save weight the plane took off with a two-wheeled dolly that was quickly jettisoned, and it was equipped with an extendable skid on which it would land on grassy fields. It was also a dangerous plane to fly. Fully one-third of planes taking off would never land again, with some 80% of these crashing on take-off or landing and 15% exploding midair or losing control.<127> *Collection of the National Museum of the US Air Force*

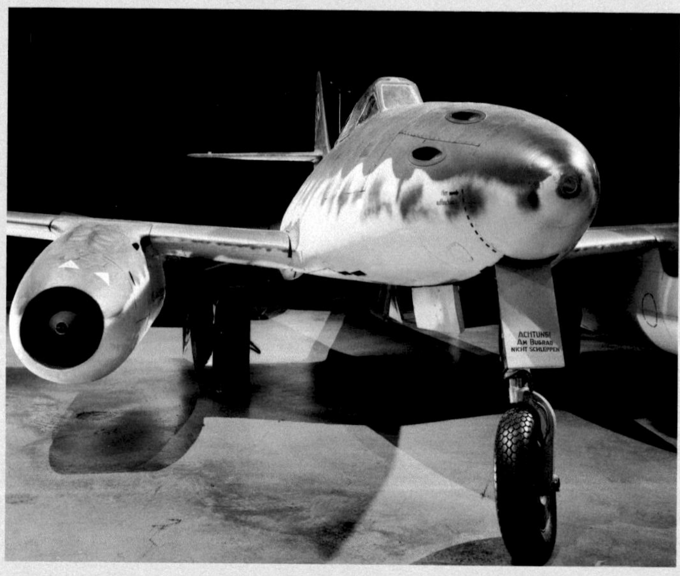

German jet fighter aircraft, Me-262A *Schwalbe* (swallow) 1944 The story of the Messerschmitt Me-262 is one of German technical genius pitted against Hitler's autocratic obstinacy. In his memoir Albert Speer, Nazi armaments minister, reported that Hitler ordered the Me-262 fighter production program be stopped in September 1943. Three months later he ordered that be restarted with the highest priority but to be fitted as a bomber. The bomber idea was opposed by Goering, all the air force experts, and quite a few army generals as well. Hitler would have none of it. As late as autumn 1944 he imperiously closed the subject.<128> By this time the P-51 Mustang was enabling fighter escort for long range bombing missions deep inside Germany. The Me-262, far superior to all Allied fighters, could have made a real difference to the growing success of Luftwaffe suppression. Finally, in November Hitler agreed to stop the bomber production and concentrate on the fighter. It was well into 1945, however, before enough pilots could be trained to become a serious threat, and by then the Allied bombing campaign was choking off the production of new Me-262s. Hitler's intransigence had delayed production of this vital weapon by at least six months at the most critical point in the war.<129> *Collection of the National Museum of the US Air Force*

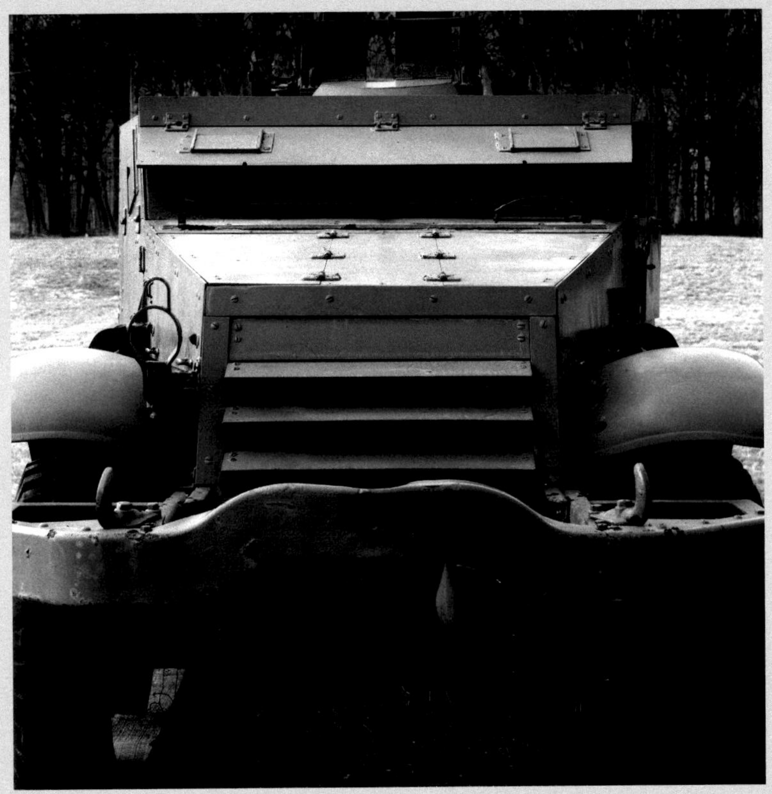

US M3 half-track armored personnel carrier 1944 These lightly armored vehicles were tried in many roles from anti-tank to howitzer platform to anti-aircraft platform. It was this last mission that suited it best. Some 3800 were produced.<130> *Former Collection of the US Army Ordnance Museum, Aberdeen Proving Ground, MD*

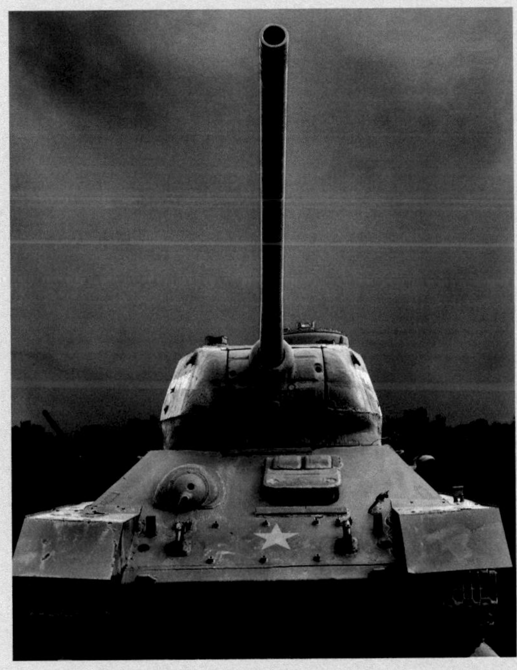

Soviet medium tank, T34 with 85-mm gun, 34.4 tons combat weight<131> 1944 The appearance of the German Panther and Tiger tank, themselves an answer to the Soviet T-34/76, required an upgrade in order to compete. The T-34/85 was that revision. With thicker armor and a more potent gun, it rivaled the new German tanks. However, with its simplicity and sheer numbers it ultimately overwhelmed the panzer forces. By the end of the war over 40,000 T34 tanks in all their variations had been built.<132> *Former Collection of the US Army Ordnance Museum, Aberdeen Proving Ground, MD*

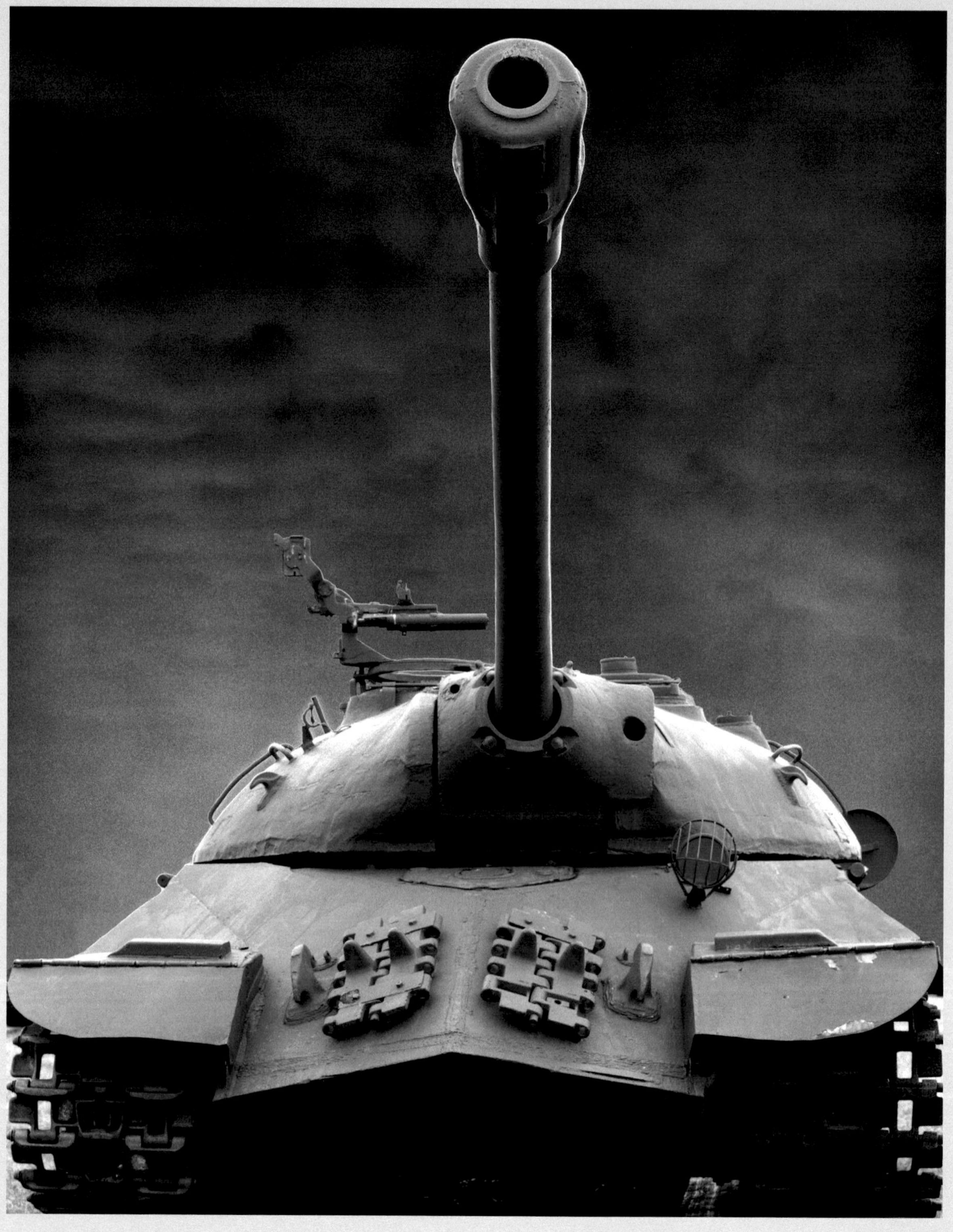

Russian heavy tank, IS-3 with 122-mm gun, 45.8 tons combat weight 1945 Not available until January 1945 and then in limited numbers, the IS-3 was a formidable war machine. Its low silhouette, well sloped armor, and powerful 122-mm gun made it the embodiment of the combat lessons of the war. Though not built in sufficient numbers to play a major role in the end game, it proved a great concern to the West in the early phases of the Cold War.<133> *Former Collection of the US Army Ordnance Museum, Aberdeen Proving Ground, MD*

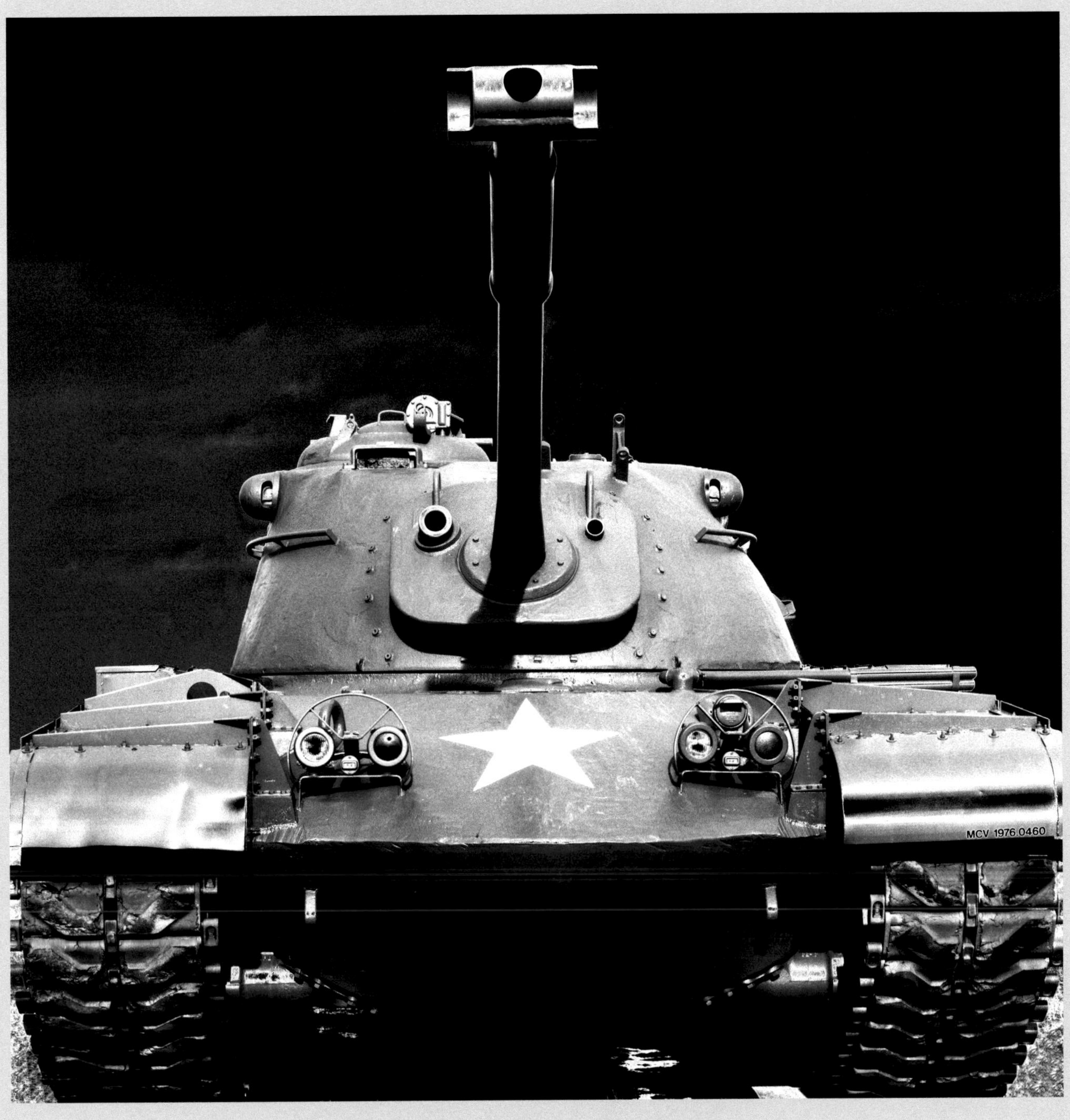

US medium tank, M48 Patton with 90-mm gun, 50 tons combat weight 1953 The M48 was the mainstay of the US tank force in the early postwar period. It was rushed into production for use in the Korean War and many found their way into the Israeli Army as the M60 replaced the US M48 in the early 1960s. Later versions saw the 90-mm gun replaced by a 105-mm gun.<134> *Former Collection of the US Army Ordnance Museum, Aberdeen Proving Ground, MD*

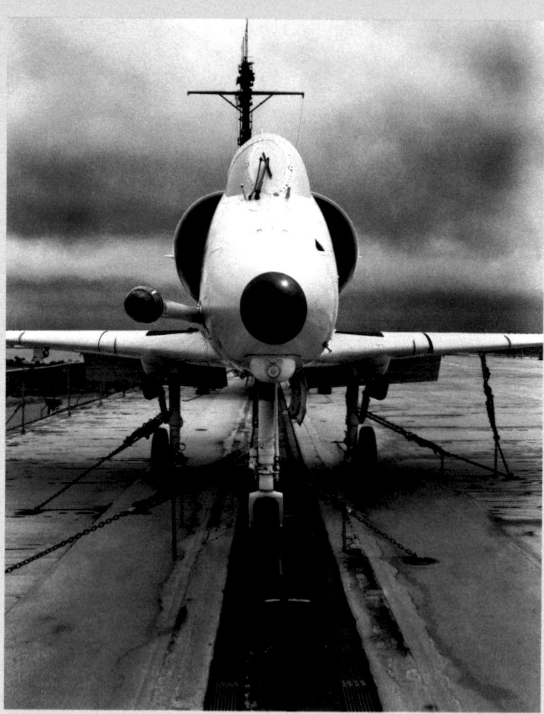

US carrier-based attack aircraft, A-4C Skyhawk 1956 The Skyhawk was a carrier-based fighter/bomber that stayed in production for an astonishing 25 years and in first-line deployment for 30 years. It was assigned an attack aircraft role using armament that included bombs, two 20-mm cannons, and missiles—also nuclear weapons. A-4s could be refueled in midair from either a tanker or from another A-4 via the fixed refueling probe on the right side of the forward fuselage. Skyhawks performed many bombing missions over Vietnam and quite a number of pilots were captured, prominent among them John McCain and James Stockdale. Some 3,000 A-4s were produced.<135> *At Patriots Point Naval & Maritime Museum, Mt. Pleasant, SC*

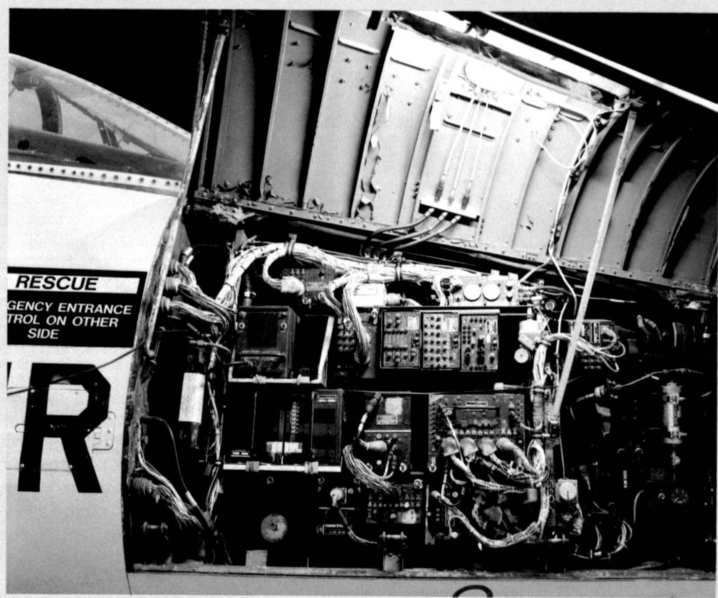

Avionics Systems, US supersonic jet fighter aircraft, F-101B Voodoo 1957 The Strategic Air Command, established in 1946, envisioned a new transoceanic fighter escort for the B-36 bomber analogous to the P-51 escorting B-17s deep into German territory during WWII. Existing fighters escorting B-29s in the Korean War were deemed unsatisfactory but funding issues and performance shortfalls led to years of an on-again, off-again pattern of support for the F-101. Models A, B, and C were all pursued somewhat concurrently, but the F-101B was the most satisfactory and 480 were built. They continued to be somewhat tricky to fly, and accidents cost some one third of the supersonic planes over their dozen years of operation. The complexity of their avionics systems, evident in the image here, was responsible in part.<136> *Collection of Hill Aerospace Museum, Hill Air Force Base, UT*

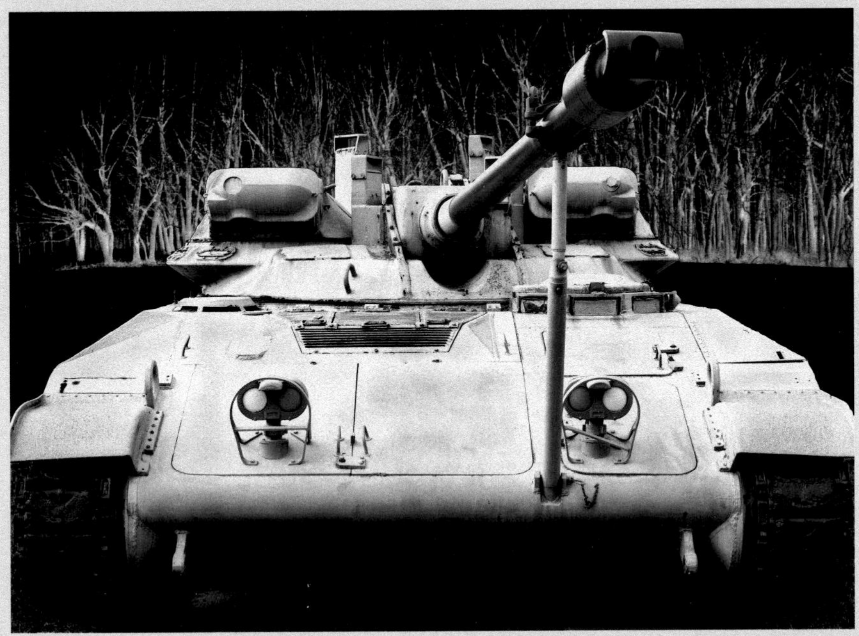

US experimental light tank, T92 with 76-mm gun, 18 tons combat weight 1958 This is one of two prototypes built by Aircraft Armaments, Inc. (AAI) competing to replace the M41 Walker Bulldog light tank. It was meant to be air-deployable by parachute. By January 1956 the T92 was the sole remaining candidate. The first prototype arrived at Aberdeen Proving Ground for testing at the end of 1956 and the second mid-1957. A number of improvements were recommended and production was expected to start by mid-1962. But in 1957 intelligence reports indicated that the Soviets were developing a light tank that was amphibious. The adequacy of the 76-mm gun was also being challenged. The T92 design did not lend itself to being made to swim, so the project was terminated in 1958. The T92 showed that obsolescence could occur before a weapon could even be fielded.<137> *Former Collection of the US Army Ordnance Museum, Aberdeen Proving Ground, MD*

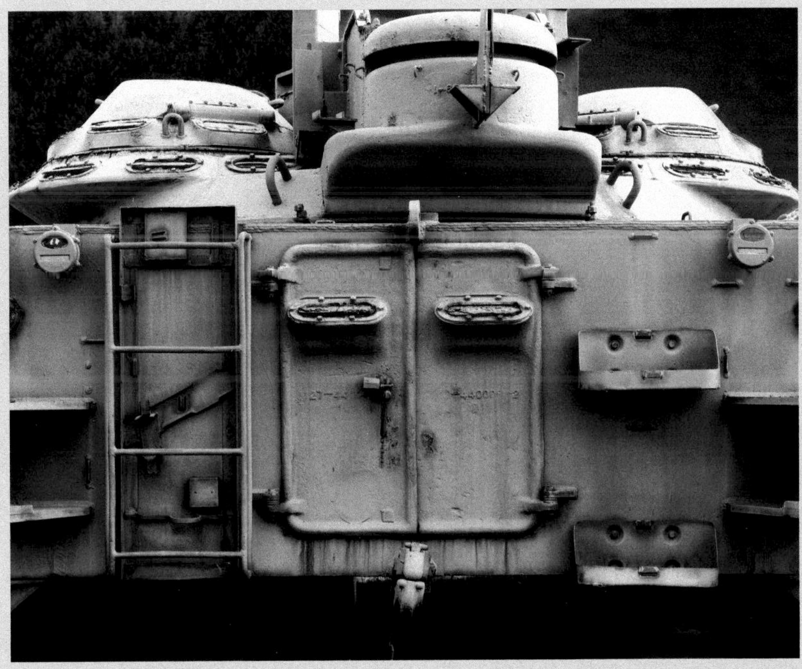

Rear view, US experimental light tank T92 1959 This tank was unusual in providing a rear escape hatch. The driver also had an escape hatch below his seat. On the previous image, one can see the driver's periscope vision blocks in the top of the hull to the right of the main gun. Many prototypes never make it to mass production either because they cannot meet performance or reliability requirements or because the competitive landscape has shifted. *Former Collection of the US Army Ordnance Museum, Aberdeen Proving Ground, MD*

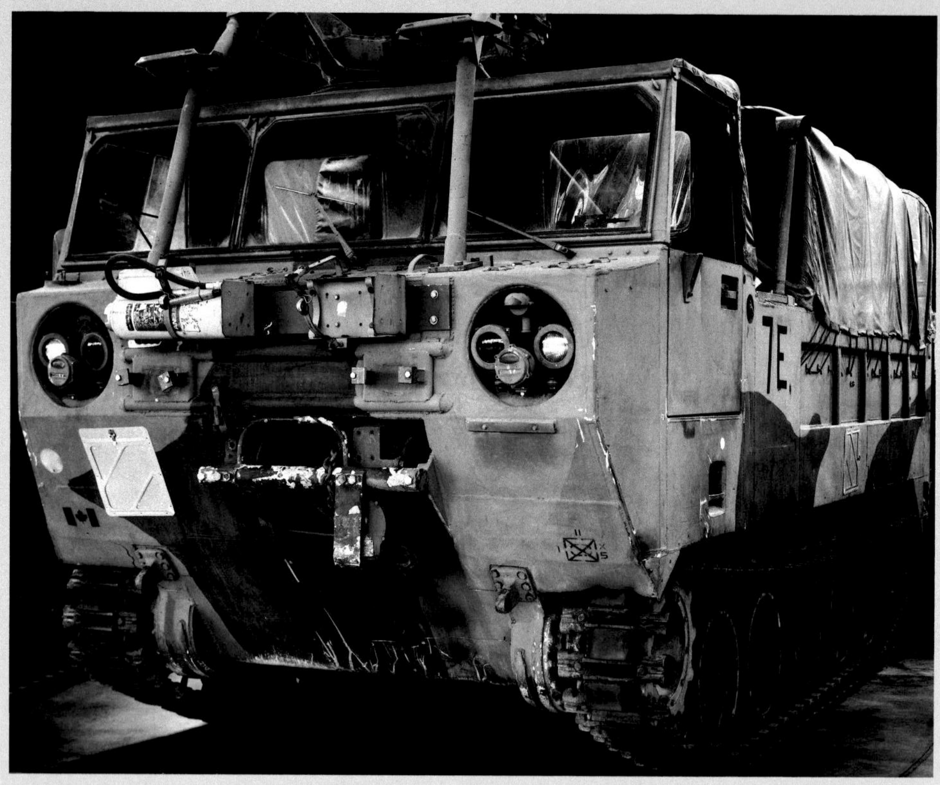

Canadian cargo carrier, M548 1960 Logistics is an unheralded but vital part of warfare. Conventional, wheeled trucks are more efficient where roadways exist but tracked vehicles may be required to supply the battlefield. This vehicle is an unarmored variant of the US M113A1 armored personnel carrier.<138> *Collection of the Canadian War Museum*

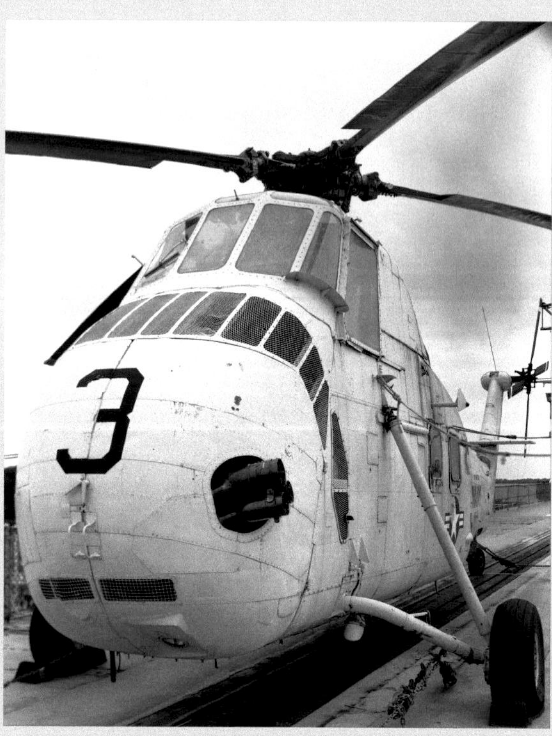

US search and rescue helicopter, UH-34D Seahorse 1957 Originally designated HUS-1 until 1962 when it became UH-34D, this Sikorsky helicopter appeared in a bewildering number of variants. The UH-34D was used by the Navy to transport Marines into combat in Vietnam and to rescue downed pilots.<139> *At Patriots Point Naval & Maritime Museum, Mt. Pleasant, SC*

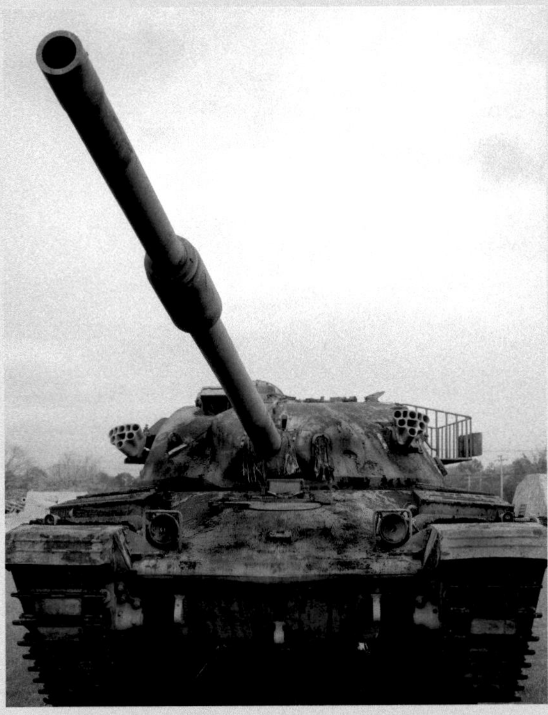

British main battle tank, Chieftain with 120-mm gun, 54 tons combat weight 1963 The Chieftain was the first among the NATO countries to move to the 120-mm caliber gun from the 105-mm, which had become the standard. Though production began in 1963, it was 1967 before the Chieftain was delivered to the British Army in numbers — 900 by 1967. It was partly phased out in the 1980s by the Challenger 1 and completely by the Challenger 2 in the 1990s. Some 700 found their way to Iran before the Shah was deposed and were used with great success in the Iran-Iraq War.<140> *Former Collection of the US Army Ordnance Museum, Aberdeen Proving Ground, MD*

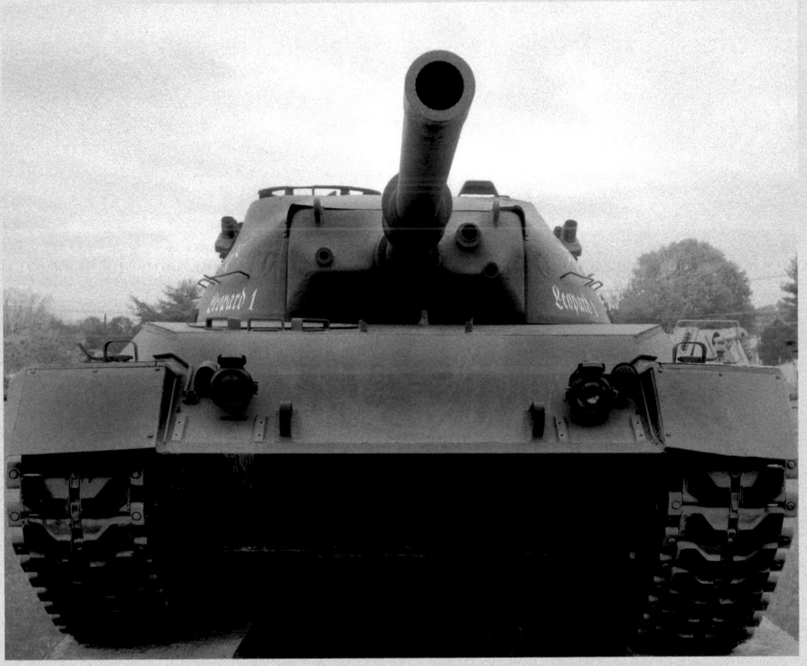

West German main battle tank, *Leopard* 1 with 105-mm gun, 40–45 tons combat weight 1965 Development begun initially as a joint program among West Germany, France, and Italy but eventually devolved to a German-only program. The German Army took delivery of 2,437 *Leopard* 1s, and they were gradually removed completely from service by the end of the century. As they were withdrawn, they were transferred to a host of countries: Denmark, Greece, Turkey, Brazil, Chile, Belgium, and the Netherlands.<141> *Former Collection of the US Army Ordnance Museum, Aberdeen Proving Ground, MD*

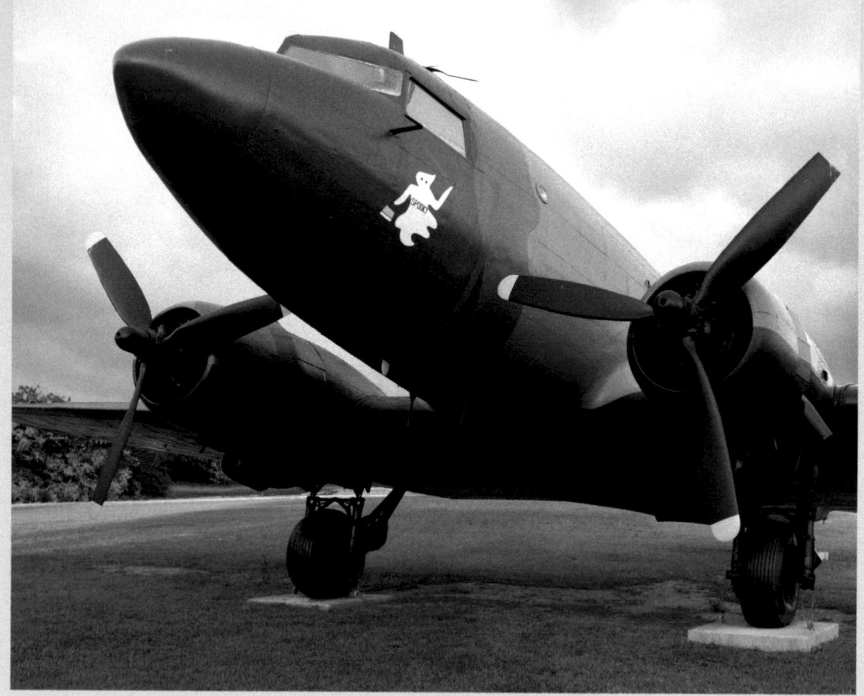

US close ground-support gunship, AC-47 Spooky "Puff the Magic Dragon" 1965 The AC-47 was developed to provide protection to outposts and friendly villages in the Vietnam War. C-47 cargo planes were modified to carry three 6-barrel Gatling guns that could fire at rates up to 6,000 7.62-mm NATO rounds per minute. In action the gunfire would directed continuously from the side of the plane placed a steeply banked turn, or "pylon turn," onto a target on the ground. With such high rates of fire from three guns directed to the same spot, the terminal effects were blistering. Grateful soldiers knew the plane as "Puff the Magic Dragon" after the then popular song by Peter, Paul, and Mary.<142> *Collection US Air Force Armaments Museum, Eglin, Air Force Base, FL*

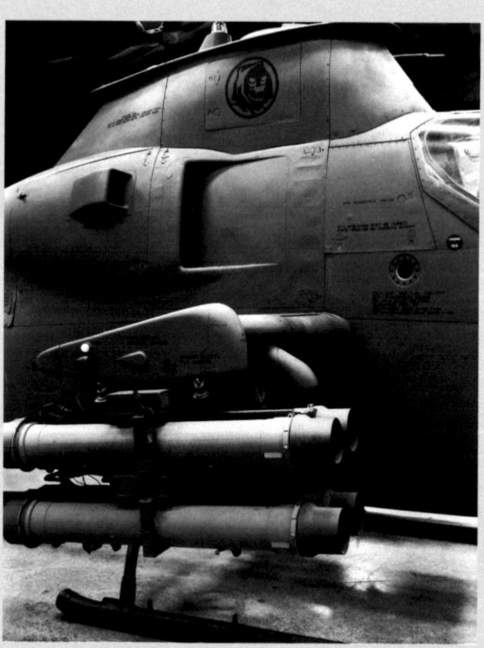

Weapons pod detail, US attack helicopter, AH-1S Cobra 1966 The Cobra was the result of converting the Bell UH-1 "Huey" helicopter to a more efficient weapons platform for escorting troop-carrying Hueys into battle. By placing the two pilots in-line instead of side-by-side, the fuselage could be made more streamlined and therefore faster, 138 knots to 170 knots maximum. It also improved the visibility of the co-pilot/weapons officer in front by giving him an unrestricted 230-degree field of view.<143> Visible here attached to the right weapons pod are four TOW missiles (Tube-launched, Optically tracked, Wire-guided). *Collection of Pima Air & Space Museum, Tucson, AZ*

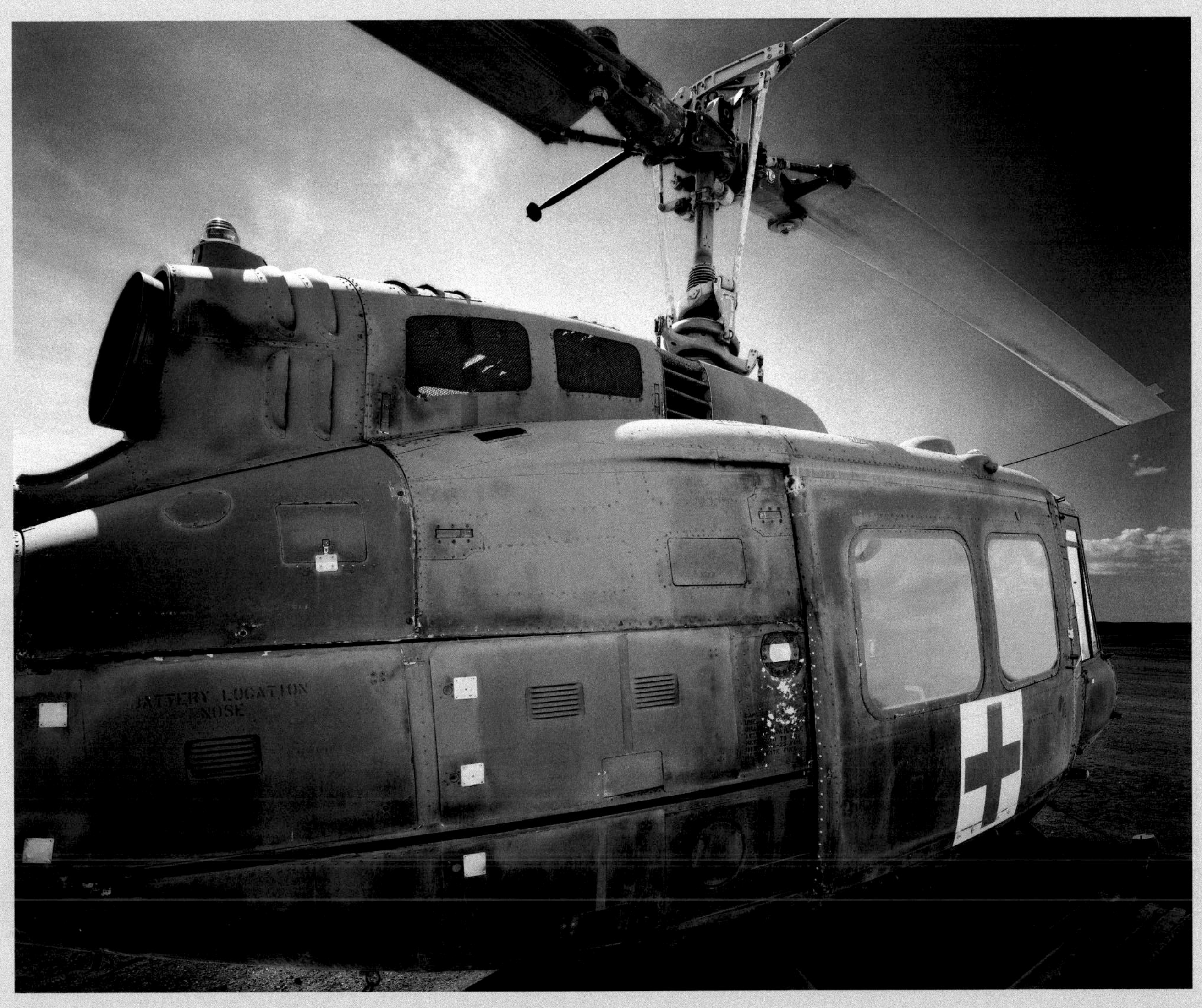

US medivac helicopter, UH1-1H Iroquois "Huey" 1967 The original designation of this helicopter was HU-1 and the nickname "Huey" derives from this period. The Huey and the M16 rifle became the defining symbols of the Vietnam War. Veterans describe its distinctive rotor beat as "the sound of our war." It was used in many roles, combat troop deployments/extractions and medical evacuation being uppermost. Some 90,000 medical evacuations were managed by Hueys, with the average time between wound and hospital being less than an hour. In WWII and Korea this time was typically measured in days.<144> *Collection of Pima Air & Space Museum, Tucson, AZ*

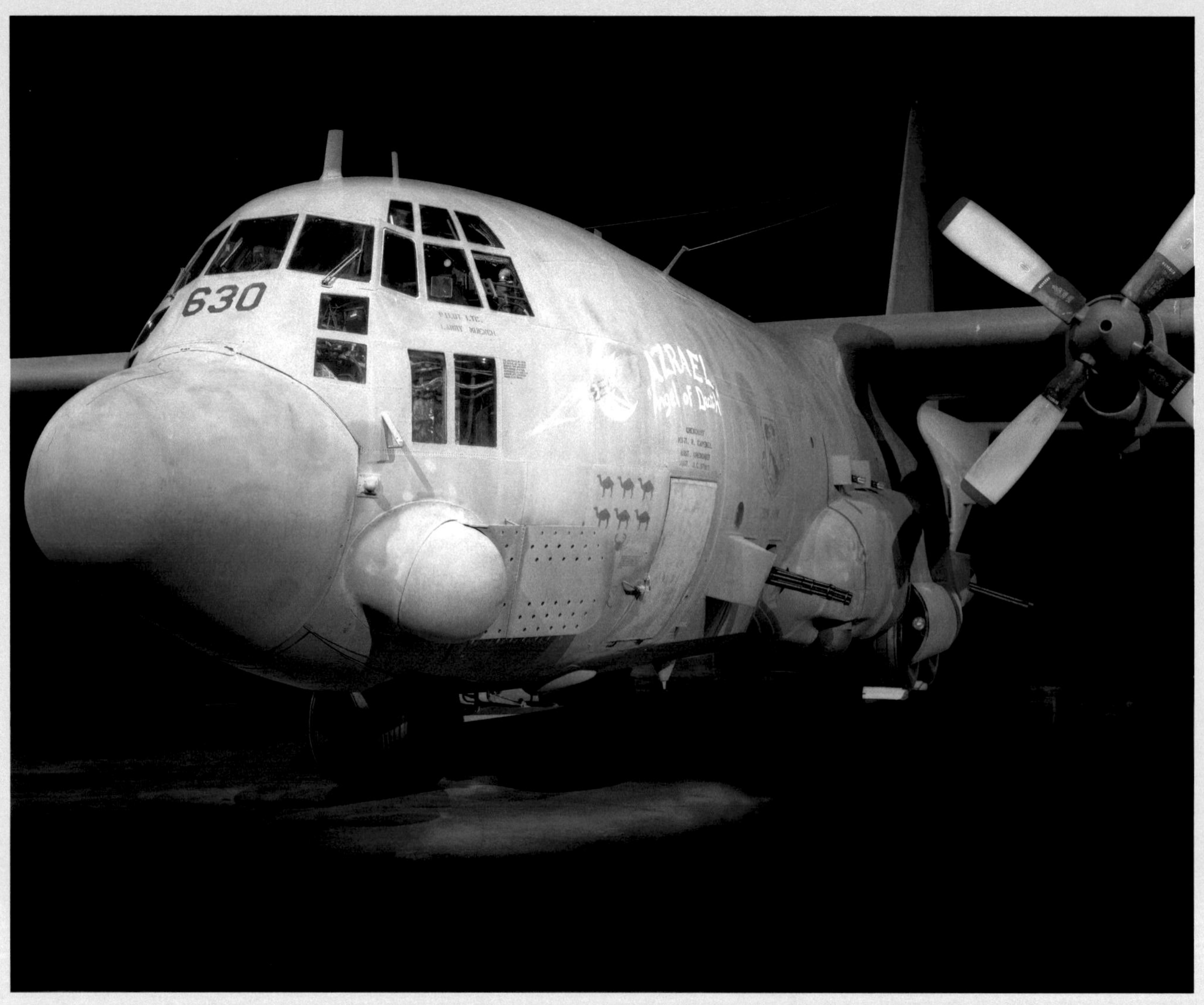

US ground-attack gunship, AC-130A Spectre 1968 The use of the much larger C-130 aircraft allowed a huge increase in the armaments carried by this successor to the AC-47. In this first version there are two 7.62 Gatling guns as in the AC-47, two 20-mm M-61 Vulcan cannons and two 40-mm Bofors cannons. The 20-mm cannons could fire 2,500 rounds per minute of high-explosive incendiary ammunition. In some versions one 40-mm gun is replaced by a 105-mm howitzer. AC-130s saw extensive use in Vietnam, especially in night operations, to disrupt supply convoys along the Ho Chi Minh Trail. During the closing act of Operation Desert Storm, this particular AC-130 attacked the Iraqi Army fleeing Kuwait City for Basrah, Iraq despite intense anti-aircraft fire from SA-6 and SA-8 surface-to-air missiles and radar-guided 37-mm and 57-mm anti-aircraft artillery.<145,146> *Collection of the National Museum of the US Air Force*

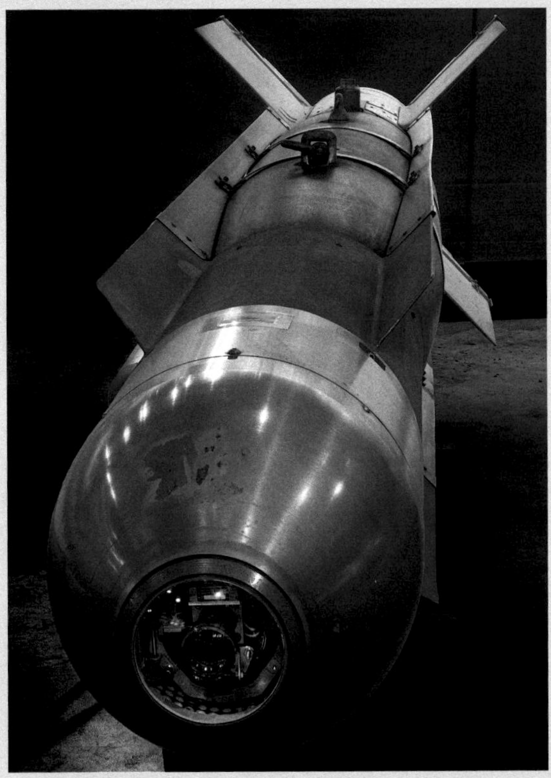

US GBU-8 Electro-Optical Guided Bomb 1969 An early "smart bomb" concept, the GBU-8 (Glide Bomb Unit) employed a television camera and image-contrast circuitry in the nose of an Mk-84, 2,000-pound bomb. The Weapons Systems Officer (WSO) aboard the attacking aircraft would observe the target on his monitor and, when the image contrast was adequate, allow the bomb to home in on the target on its own, using control fins attached to the aft end of the bomb. This was known as the Homing Bomb System or HOBOS. Darkness, cloud cover, or fog spoiled the effectiveness of HOBOS, but some 700 GBU-8s were deployed in Vietnam after its introduction in 1969.<147> *Collection of the National Museum of the US Air Force*

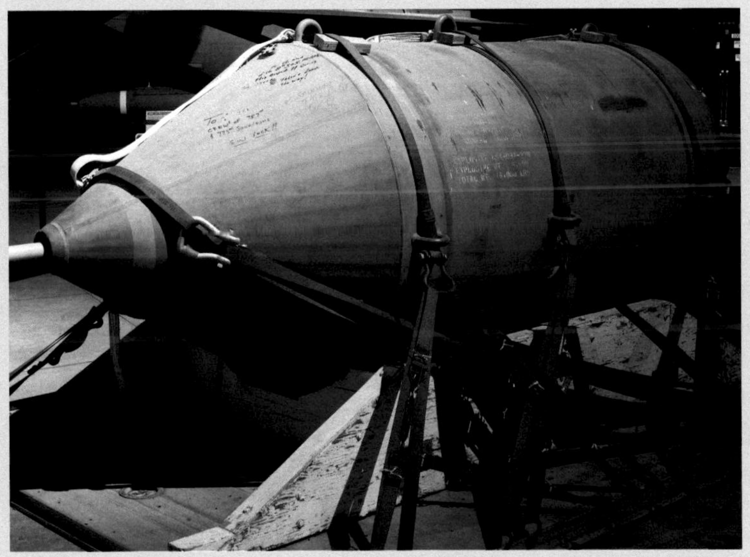

US Live Bomb Unit BLU-82/B 1970 This bomb was first used in Vietnam to clear helicopter landing zones and, as a result, acquired the name "Daisy Cutter." At 15,000 pounds it required a C-130 cargo plane for delivery. The bomb is basically a thin-walled (1/4 inch) tank holding an explosive slurry mixture of ammonium nitrate and aluminum powder along with a binder. The pipe protruding from its nose is a 38-inch fuze designed to detonate the bomb without making a crater. It was used again in the Persian Gulf War during Operation DESERT STORM and yet again in Afghanistan during Operation ENDURING FREEDOM following the Al Qaeda attack on the New York World Trade Center. About 225 of these bombs were made.<148> *Collection of the National Museum of the US Air Force*

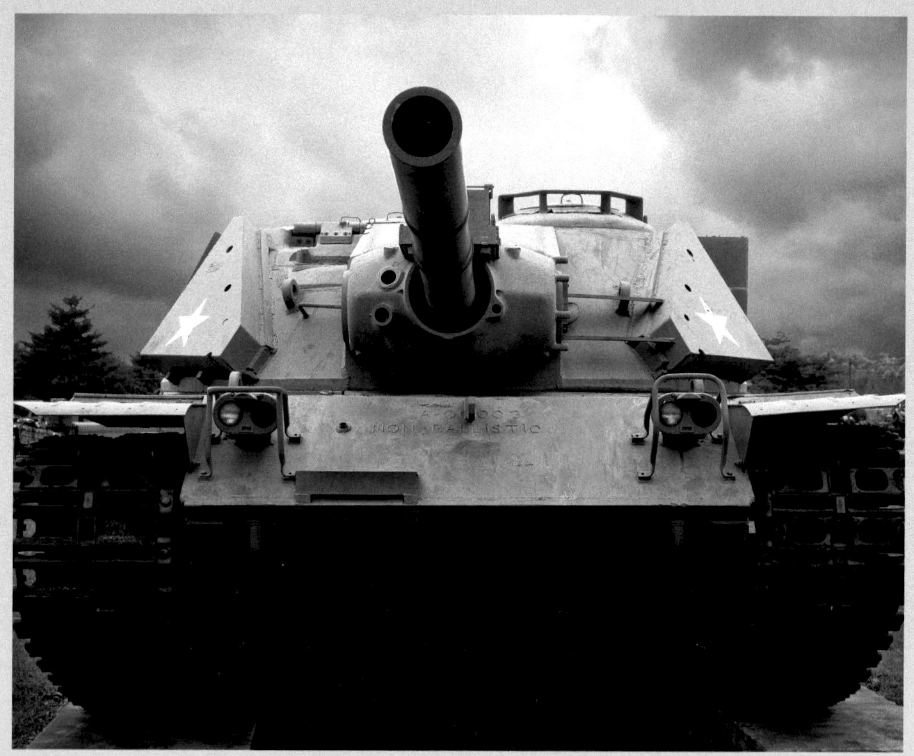

US main battle tank, MBT70 with 152-mm gun/missile launcher, 45 tons combat weight 1970 This tank was to be jointly developed by the US and West Germany, but ultimately they could not agree on a final design. So the tank shown existed in prototypes only. Germany went its own way and developed the Leopard 2 and the US the Abrams.<149> *Former Collection of the US Army Ordnance Museum, Aberdeen Proving Ground, MD*

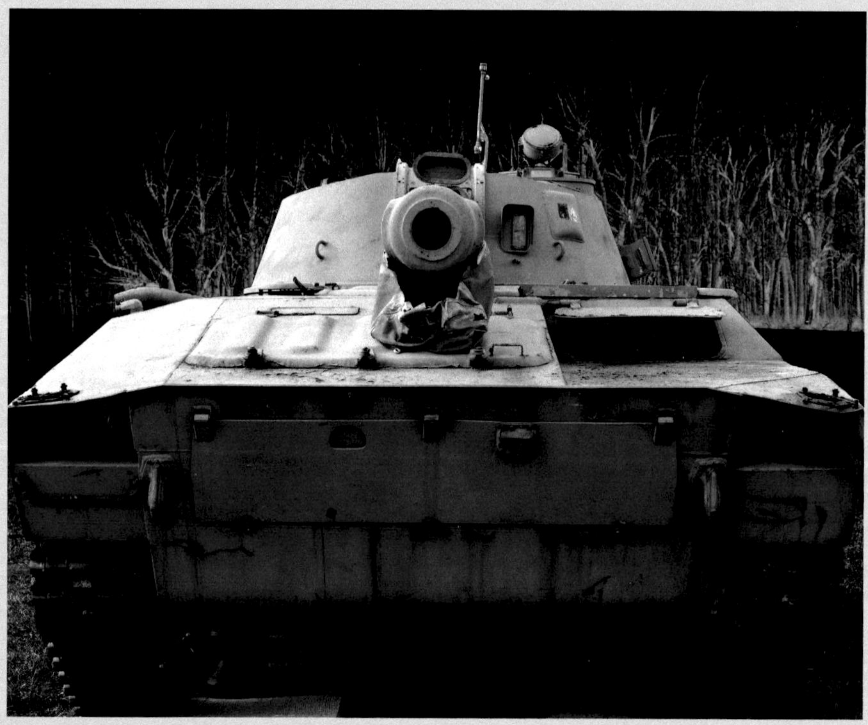

Soviet self-propelled howitzer, 2S1 *Gvozdika* (Carnation) with 122-mm gun, 15.7 tons combat weight 1971 Rivers had proved to be a serious obstacle in WWII, so the Soviets made a number of armored vehicles amphibious, including the *Gvozdika*. It had a crew of six and had night-vision capabilities and protection from nuclear, biological, and chemical threats.<150> *Former Collection of the US Army Ordnance Museum, Aberdeen Proving Ground, MD*

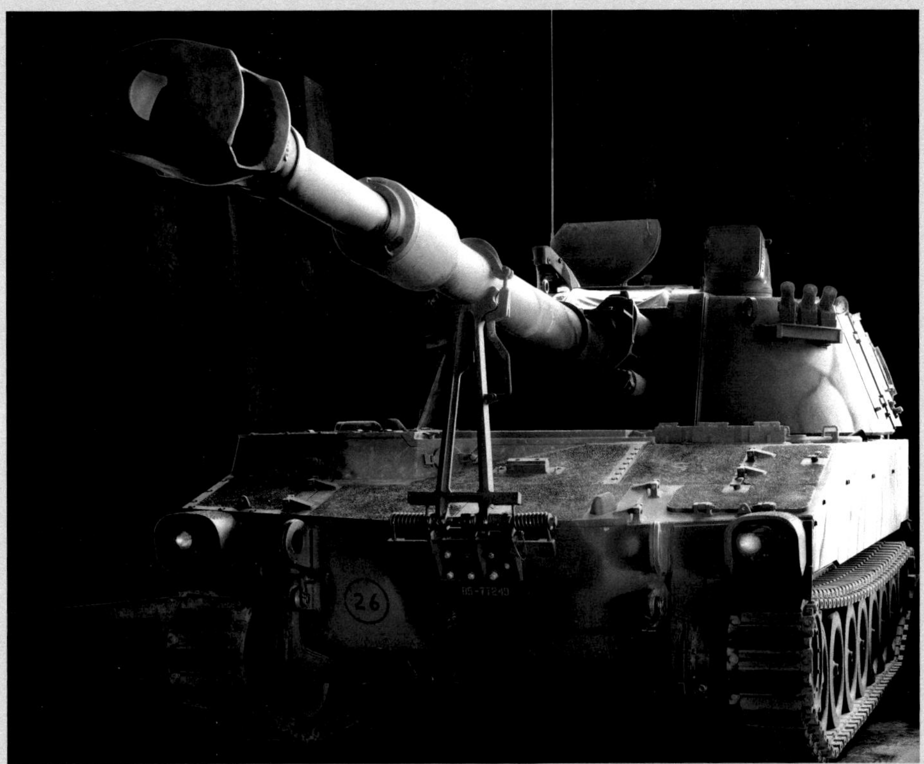

Canadian self-propelled howitzer, M109A4+ with 155-mm gun 1971 This is a Canadian version of the US M109, which has seen production of its various variants from 1963 to the present, currently the A7 model. It has nuclear, biological, and chemical protections and has been an extraordinarily successful export, having been adopted by 26 foreign countries besides Canada.<151> *Collection of the Canadian War Museum*

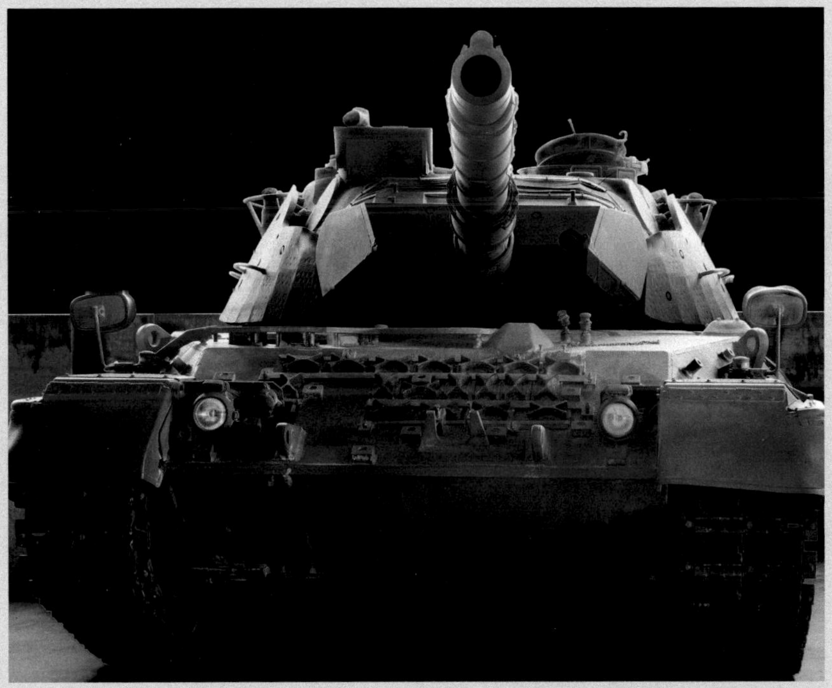

Canadian main battle tank, Leopard C2 1999 with 105-mm gun, 42.5 tons combat weight 1978 The Canadian Leopard C2 is an armor upgrade of their C1, which was a Leopard 1 variant. In 2008 Canada purchased 100 surplus Leopard 2 tanks from the Netherlands and expected the upgrading and deployment of these tanks by the end of 2017.<152> *Collection of the Canadian War Museum*

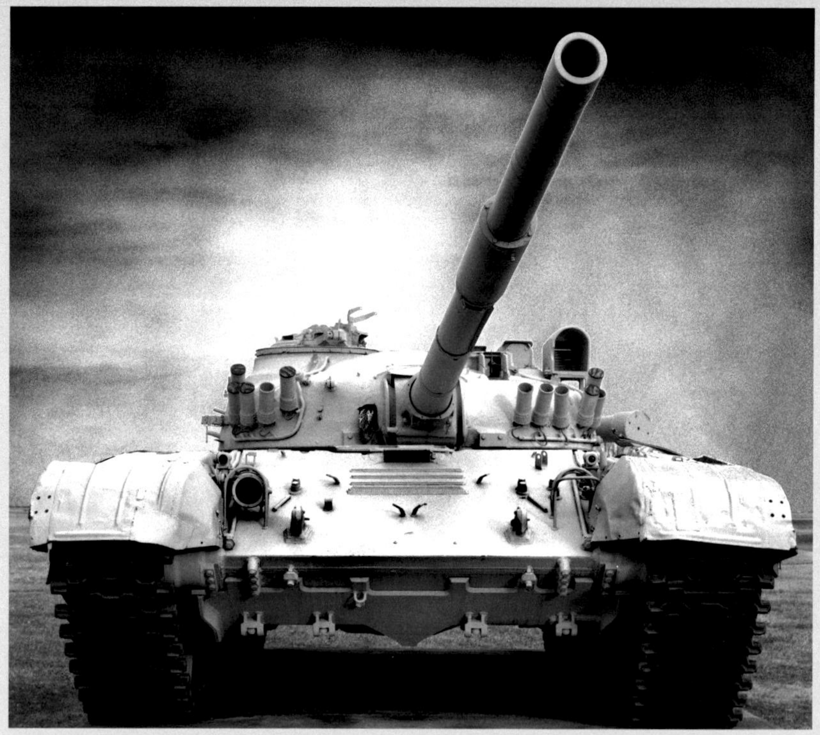

Russian main battle tank, T72M with 125-mm gun, 45 tons combat weight 1979 This tank, now out of date by first-world country standards, is still the world's most numerous tank. It was relatively cheap to produce and maintain, and its bulging armor at the front of the turret, called "Dolly Parton" up-armoring in some circles, was the most desirable of its time. During the Persian Gulf War, Iraq threw a thousand T-72s into the fight, but they were not even close to a match against the newer-generation US M1 Abrams tanks.<153> *Former Collection of the US Army Ordnance Museum, Aberdeen Proving Ground, MD*

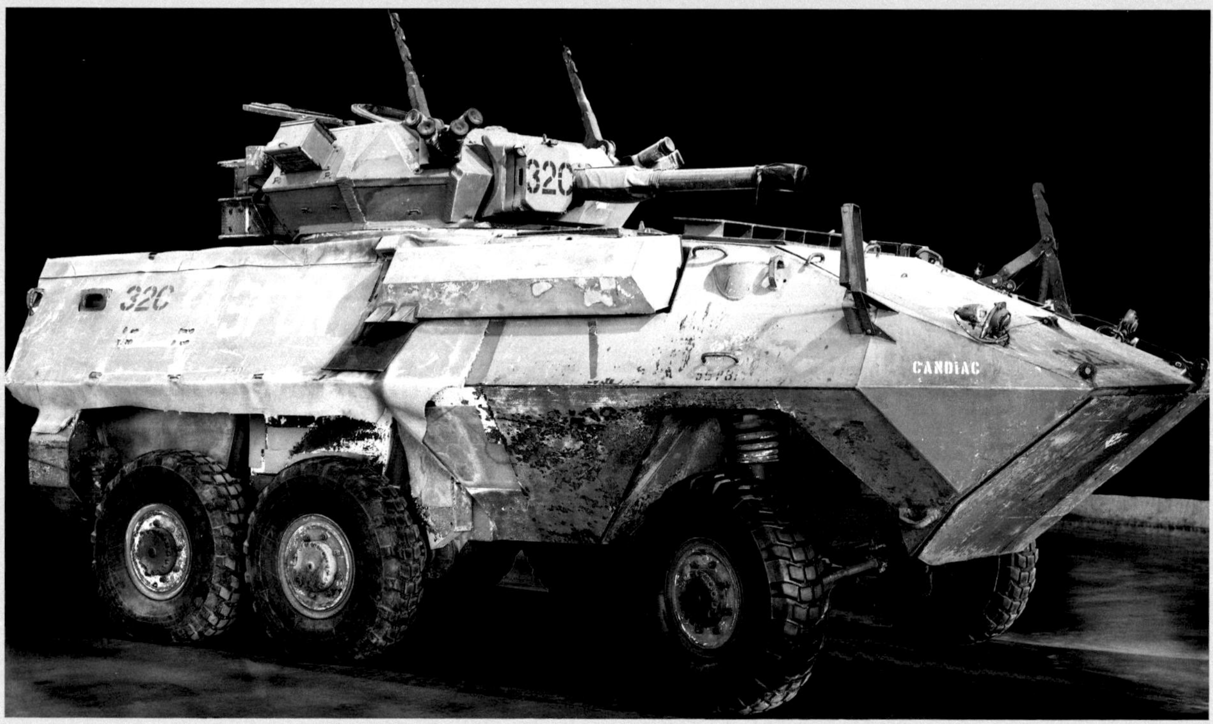

Canadian armored fire-support vehicle, Cougar with 76-mm gun, 10.7 tons combat weight 1977 A placard in the Canadian War Museum identified this vehicle as part of the Canadian component to the UN peacekeeping force in the Bosnian War from 1992–1995. The scars on its exterior were caused by exploding mortar shells. *Collection of the Canadian War Museum*

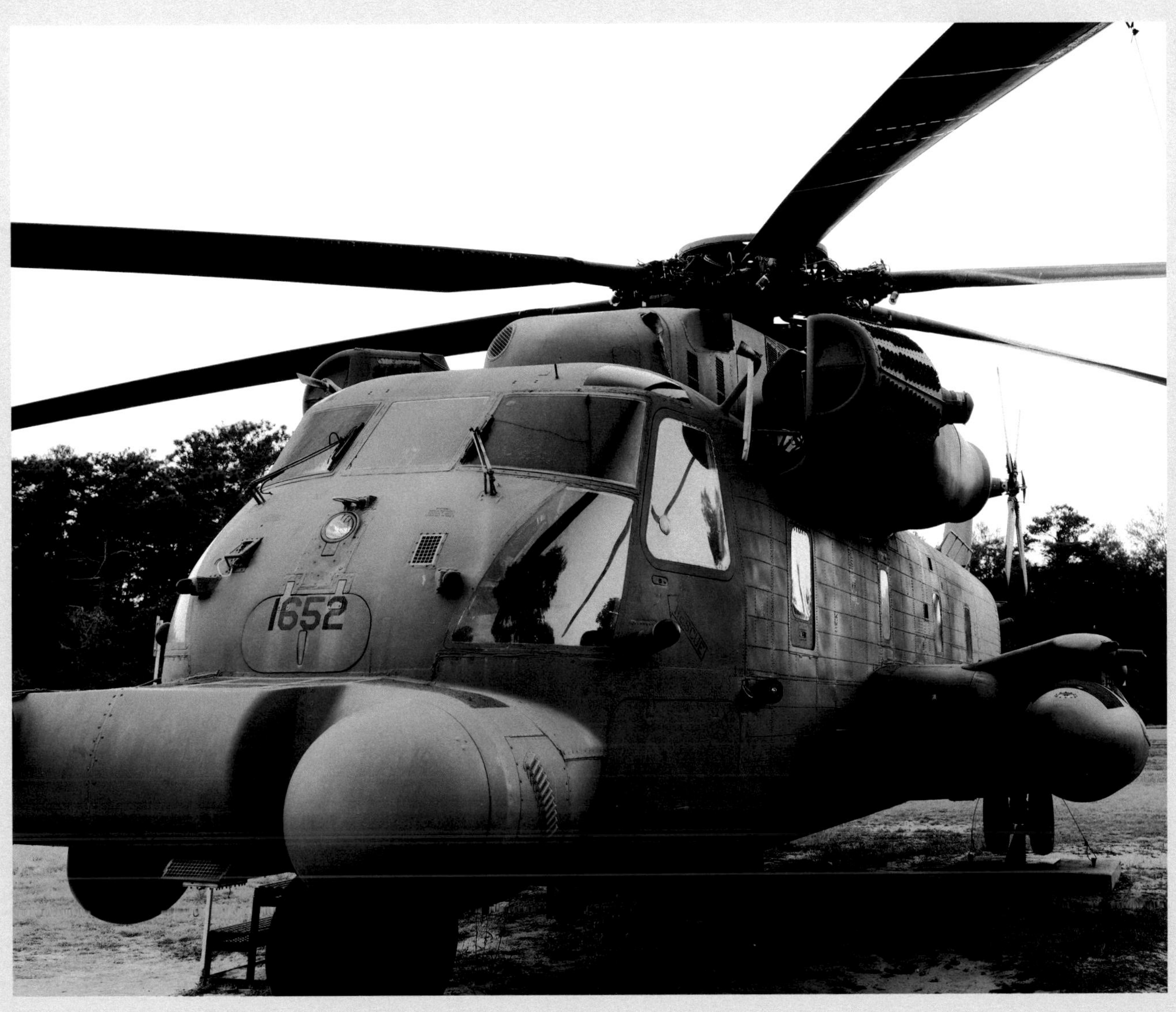

US multi-purpose combat helicopter, MH-53M Pave Low IV Helicopter 1979 (for Pave Low III) The Pave Low series of helicopters were originally developed for long-range combat insertions and rescue especially at night and/or in bad weather. This mission ultimately was refined and narrowed to special operations under high-threat conditions. In the aftermath of the failed attempt to rescue the Iranian hostages in April 1980, the Joint Special Operations Command (JSOC) was established for hostage rescue and similar specialized tasks. One of their key assets was the Pave Low helicopter. By 1999 a significant upgrade in communications and sensor capability created the new designation of Pave Low IV, which is the model shown here.<154> *Collection US Air Force Armaments Museum, Eglin, Air Force Base, FL*

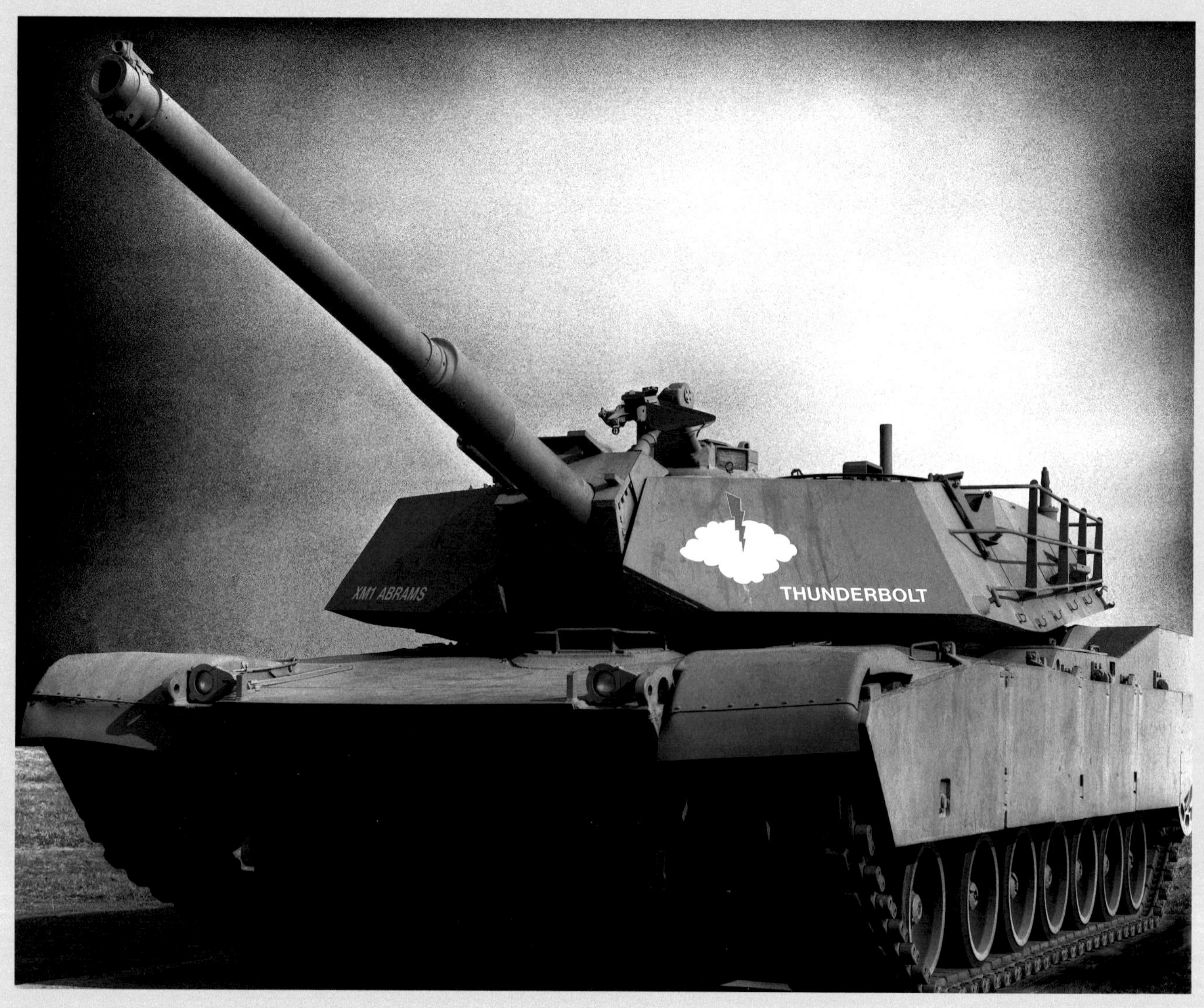

US main battle tank, XM1 Abrams with 105-mm gun (later increased to 120 mm), 60 tons combat weight (later increased to 72 tons) 1980 After the failure of the joint US-West German MBT70 project, a competition among US contractors to build the next generation tank was won by Chrysler. The development of the intended 120-mm smooth-bore gun was not yet complete so prototypes were built with a 105-mm gun. This is one of those prototypes. The version with 120-mm gun was out by 1984 as the M1A1. Though never needed for its intended role in Cold-War, it did distinguish itself in Desert Storm where its advanced steel-encased depleted uranium armor was not penetrated once even by a few friendly fire incidents with the latest American armor-piercing rounds. Its fire-control system almost guaranteed a first-round hit at targets up to nearly two miles. The last version, the M1A2, has a datalink system common to other M1A2s, the Bradley Fighting Vehicle, and the AH-64D Apache attack helicopter for integrated digital battlefield management.<155> *Former Collection of the US Army Ordnance Museum, Aberdeen Proving Ground, MD*

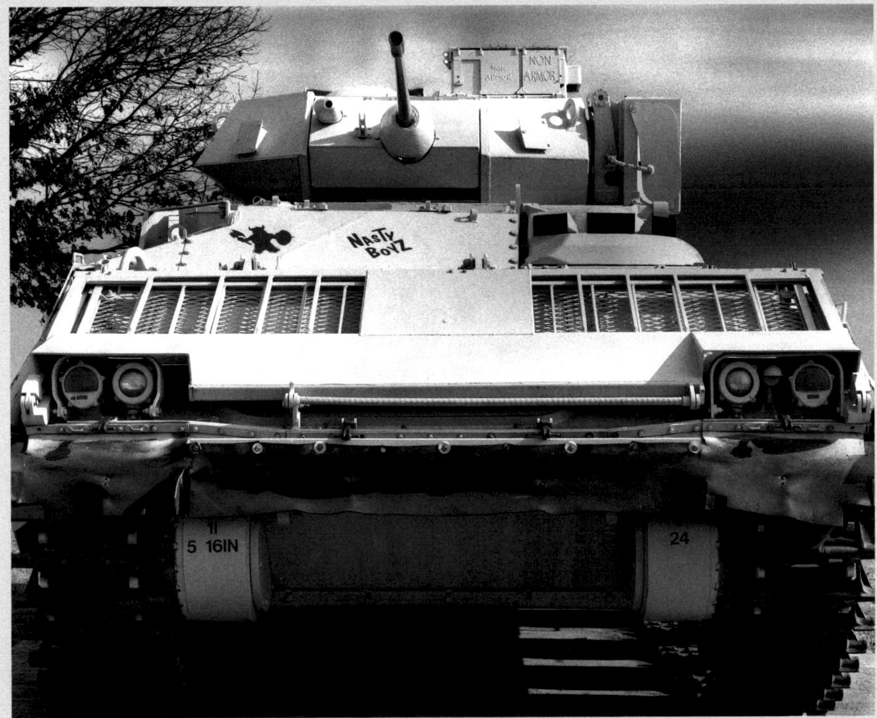

US armored infantry fighting vehicle, M2 Bradley with 25-mm gun, 22.6 tons combat weight 1981 Released about the same time as the M1 tank, the Bradley has room for a crew of three and six infantry personnel. Its main armament includes a 25-mm Bushmaster Chain Gun capable of 200 rounds per minute and dual TOW anti-tank missile launchers. The Bradleys survivability was a source of controversy during its development because of its aluminum armor, but its excellent performance in Desert Storm silenced its critics.<156> *Former Collection of the US Army Ordnance Museum, Aberdeen Proving Ground, MD*

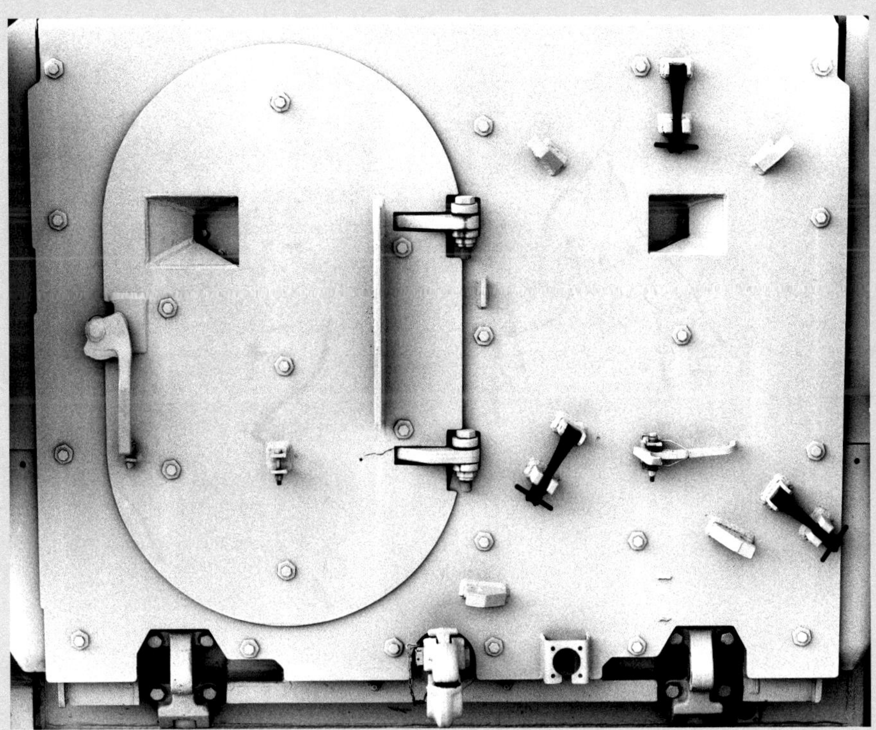

Rear Ramp, US M2 Bradley 1981 The rear ramp of the Bradley is lowered hydraulically for rapid troop ingress and egress, but a rounded hatch is also provided for single-soldier access. Visible here are the two firing ports in the ramp. *Former Collection of the US Army Ordnance Museum, Aberdeen Proving Ground, MD*

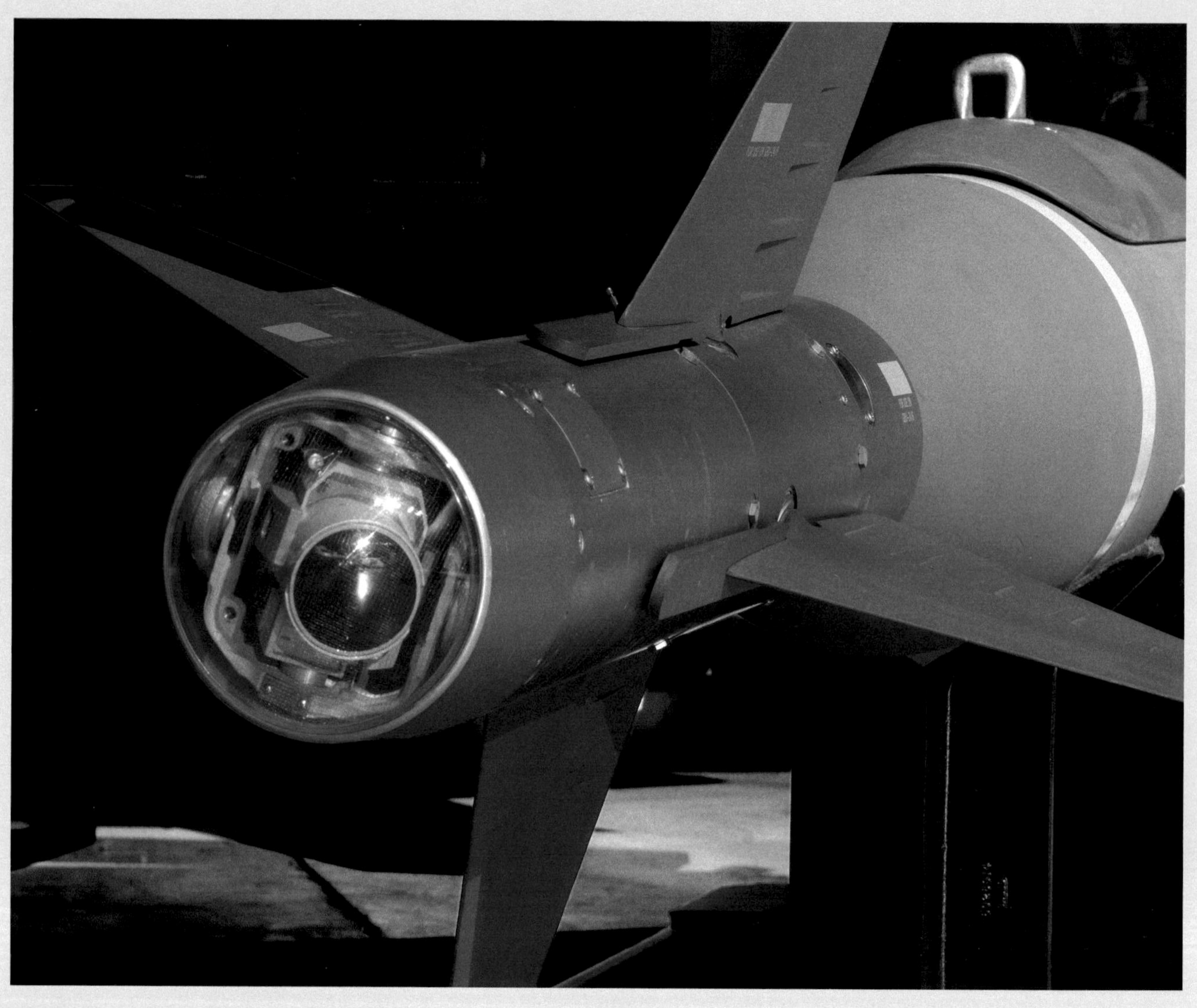

US GBU-24 Paveway III 1983 Originally developed during the Cold War for delivery by aircraft flying low to avoid Warsaw Pact radar, the laser-guided GBU-24 would have to wait until the Persian Gulf War to prove its combat worth. Attached payloads are 2,000-pound BLU-84 and BLU-109 bombs, the latter designed to penetrate hard targets and used to good effect on Iraqi hardened aircraft shelters by F-117 Night Hawk stealth fighters. More than 9,300 Paveway II and III laser-guided bombs were used in this war.<157> *Collection of the National Museum of the US Air Force*

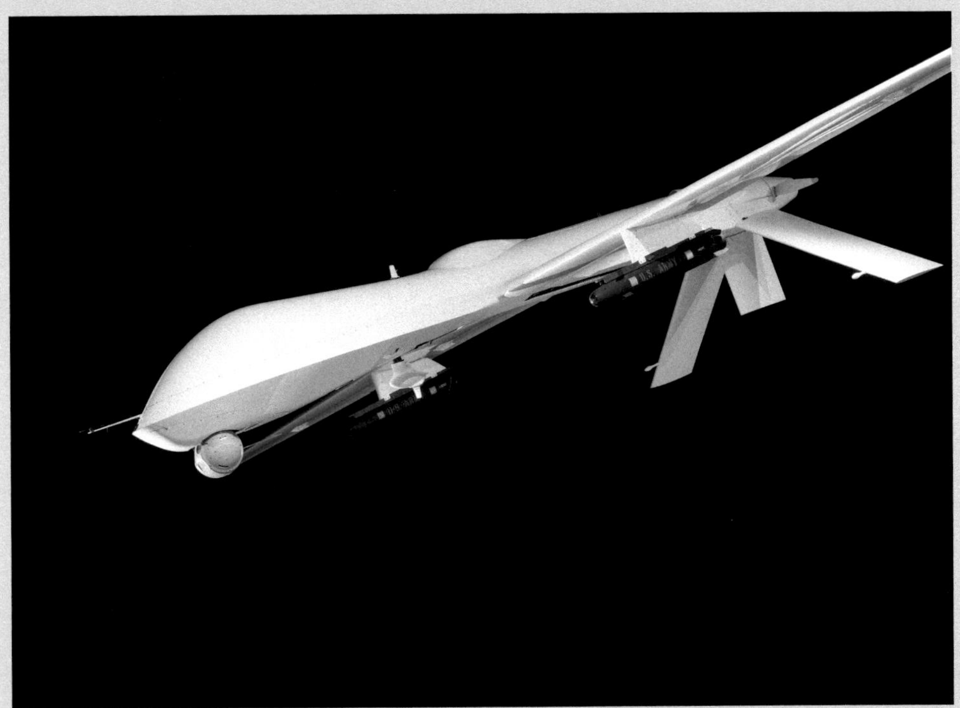

US RQ-1 Predator 1995 The Predator represented a major paradigm shift in weapons systems at the end of the century. When first deployed in the skies of Bosnia in 1995, it was used exclusively for reconnaissance and was unarmed. They went into full production in 1997. After 2001 their designation changed to MQ-1 and they could carry AGM-114 Hellfire laser-guided missiles. The RQ-1 shown here has been upgraded to display the Hellfire missiles. Their first armed deployment was in Afghanistan after the terrorist attacks of September 11, 2001. The Predator can be controlled by a local ground station or from across the globe by satellite links. Video and radar data can be provided in real time to commanders around the world.<158> *Collection of the National Museum of the US Air Force*

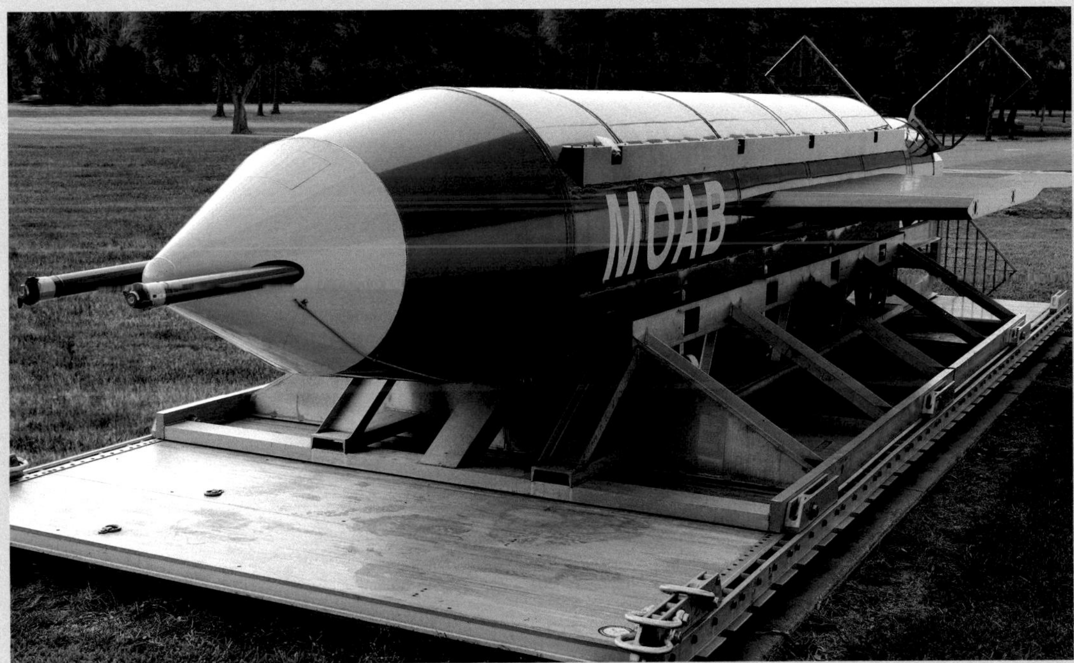

US large-yield conventional bomb, Massive Ordnance Air Blast bomb (MOAB), GBU-43/B "Mother of All Bombs" 2003 The GBU-43/B is the largest non-nuclear bomb in the US inventory. It has an explosive yield of about 11 tons — tiny by nuclear weapon standards — but it weighs some 21,600 pounds. It is GPS-guided. Development started in early 2002 so it is not strictly a 20th century weapon, but it does embody the precision-guided, late-century philosophy. It was used for the first time against an ISIS command center in the mountains of Afghanistan on 13 April 2017.<159,160> *Collection US Air Force Armaments Museum, Eglin, Air Force Base, FL*

ENDNOTES

PREFACE

1 Fussel, Doing Battle, 140.

CHAPTER 1. THE 20th CENTURY CALAMITY

1 Ferguson, The War of the World, 646.
2 Boot, War Made New, 299.
3 See for example: Eliot, Twentieth Century Book of the Dead, Appendix B; Tirman, The Deaths of Others, Chapter 10; White, "Source List and Detailed Death Tolls for the Primary Megadeaths of the Twentieth Century."
4 Bogart, Direct and Indirect Costs of the Great World War, 271, 282.
5 White, "Source List and Detailed Death Tolls for the Primary Megadeaths of the Twentieth Century."
6 Ibid.
7 Ibid.
8 Keegan, History of Warfare, 359.
9 Ibid., 305–6.
10 Wilbanks, Machine Guns, 31.
11 Haines, "The Population of Europe."
12 Roser, "Life Expectancy."
13 Keegan, History of Warfare, 361.
14 Boot, War Made New, 198.

CHAPTER 2. THE LONG PRELUDE

1 Wells, What are we to do with Our Lives?, 1.
2 Ergang, Europe from the Renaissance to Waterloo, 74.
3 Appleby, Relentless Revolution, 60.
4 Ibid., 70.
5 Ibid., 73.
6 Ibid., 80.
7 Ibid., 76.
8 Ergang, Europe from the Renaissance to Waterloo, 360.
9 Wootton, The Invention of Science, 286.
10 Caspar, Kepler, 138.
11 Bronowski & Mazlish, Western Intellectual Tradition, 122.
12 NASA Office of Space Science website.
13 Watson, Modern Mind, 2.
14 Forbes and Mahon, Faraday, Maxwell, and the Electromagnetic Field, 46.
15 Gribbin, Scientists, 246.
16 Ibid., 248-9.
17 Ibid., 271.
18 Ibid., 283.
19 Ibid., 250.
20 Trinder, Britain's Industrial Revolution, 54–7.
21 Camp and Francis, Making, Shaping, and Treating of Steel, 174.

22 Trinder, Britain's Industrial Revolution, 291–3.
23 Ricketts, "How a Blast Furnace Works."
24 Trinder, Britain's Industrial Revolution, 306.
25 Ibid., 296.
26 Ibid., 203.
27 Ibid., 297.
28 Camp and Francis. Making, Shaping, and Treating of Steel, 174.
29 Trinder, Britain's Industrial Revolution, 300.
30 Camp and Francis. Making, Shaping, and Treating of Steel, 188.
31 Ibid., 202.
32 Smiles, Life of George Stephenson, 16.
33 Ibid., 26.
34 Ibid., 89.
35 Wolmar, Iron Road, location 256.
36 Smiles, Life of George Stephenson, 139.
37 Wolmar, Iron Road, location 286.
38 Ibid., location 246-313
39 Brunel, Life of Isambard Kingdom Brunel, 86.
40 Ibid., 81.
41 Ibid., 232, 242–3.
42 Ibid., 245.
43 Ibid., 261.
44 Ibid., 282.
45 Ibid., 416.
46 Ibid., 382, 403.
47 Forbes and Mahon. Faraday, Maxwell, and the Electromagnetic Field, 165.
48 Ibid., 212–213.
49 Malanowski, Race for Wireless, 40–41.
50 Ferguson and Kirkpatrick. Internal Combustion Engines, 4–5.
51 Gribbin, Scientists, 435–439.
52 Boot, War Made New, 127.
53 Ibid., 151.
54 Ibid., 149.
55 Ibid., 176.
56 Hounshell, From the American System to Mass Production 1800–1932, 15–65.
57 Appleby, Relentless Revolution, 132.
58 Popkin, Columbia History of Western Philosophy, 554.
59 Malia, Soviet Tragedy, 37.
60 Kafka, Metamorphosis.
61 Eksteins, Rites of Spring, 10–12.
62 Canaday, Mainstreams of Modern Art, 431.

CHAPTER 3. THE GATHERING STORM

1 Manchester, The Arms of Krupp, 134.
2 Ergang, Europe from the Renaissance to Waterloo, 164.
3 Massie, Dreadnought, 58.
4 Ibid., 58.

5 Boot, War Made New, 130.
6 Ibid., 128–129.
7 Ergang, Europe from the Renaissance to Waterloo, 169.
8 Massie, Dreadnought, 64.
9 Maddison, Angus. Contours of the World Economy, 1— 230AD, 380.
10 Massie, Dreadnought, 135.
11 Ibid., 137.
12 Black, Great War and the Making of the Modern World, 10.
13 Hochschild, King Leopold's Ghost, 252.
14 Nichols, The Road to Trinity, 43.
15 Massie, Dreadnought, 178.
16 Massie, Castles of Steel, 9.
17 Massie, Dreadnought, 180.
18 Ibid., 180–181.
19 Ibid., 491.
20 Ibid., 184.
21 Hough, Dreadnought, location 286.
22 Brodie and Brodie. From Crossbow to H-Bomb, 189.
23 Boot, War Made New, 193.
24 Massie, Dreadnought, 478.
25 Ibid., 407.
26 Gardiner and Gray, Conway's All the World's Fighting Ships 1906–1921, 21–41.
27 Ibid., 145–155.
28 Koistinen, Mobilizing for Modern War, 30.
29 Simpson, Report of the Gun Foundry Board.
30 Koistinen, Mobilizing for Modern War, 48.
31 Ibid., 38–39.
32 Ibid., 39–40.
33 Engelbrecht and Hanighen, Merchants of Death.
34 Ibid., 272.
35 Gilbert, History of the Twentieth Century, Volume Two, 205.
36 Senate Historical Office, "Merchants of Death."
37 Herman, Freedom's Forge, location 129.
38 Camp and Francis, Making, Shaping, and Treating of Steel, 176.
39 Koistinen, Mobilizing for Modern War, 52.
40 Ibid., 34, 54.
41 Dougan, The Great Gun-Maker, 100.
42 Johnston and Buxton, Battleship Builders, 63.
43 Engelbrecht and Hanighen, Merchants of Death, 110.
44 Boot, War Made New, 187.
45 Macrosty, Trust Movement in British Industry, 53.
46 Ibid., 52.
47 Heald, William Armstrong, 332.
48 Jaques, "The Establishment of Steel Gun Factories in the United States," 694.
49 Ibid., 708.
50 Engelbrecht and Hanighen, Merchants of Death, 124–126.
51 Ibid., 73.
52 Ibid., 90.

53 Ibid., 93.
54 Manchester, Arms of Krupp, 102.
55 Manchester, Arms of Krupp, 98.
56 Ibid., 124.
57 Ibid., 131–132.
58 Menne, Blood and Steel, location 3930.
59 Warr, Sheffield's Great War and Beyond, 107.
60 Menne, Blood and Steel, location 4036.
61 Du Pont, E.I. Du Pont de Nemours and Company, 9–12.
62 Engelbrecht and Hanighen, Merchants of Death, 25.

CHAPTER 4. THE FIRST WORLD WAR

1 Fussell, The Great War and Modern Memory, 23–24.
2 Ergang, Europe from the Renaissance to Waterloo, 356.
3 Gilbert, The First World War, 28–29.
4 Black, Great War and the Making of the Modern World, 10–12.
5 Ibid., 14–15.
6 Massie, Dreadnought, 875.
7 Manchester, Arms of Krupp, 287.
8 Ibid., 261, 282.
9 Hogg, Twentieth Century Artillery, 92.
10 Black, Great War and the Making of the Modern World, 36.
11 Ibid., 21.
12 Cruttwell, A History of the Great War 1914–1918, 109–110.
13 Dyer, War, 250.
14 Koistinen, Mobilizing for Modern War, 110.
15 Ibid., 106.
16 Ibid., 109.
17 Ibid., 113.
18 Ibid., 114–138.
19 National Park Service, "John Pershing - Success and Tragedy."
20 Pershing, My Experiences in the World War, Vol. 1.
21 Ibid., location 5914.
22 Gilbert, First World War, 420.
23 Bidwell and Graham, Fire-Power, location 2045.
24 Ibid., location 1959.
25 Ibid., location 2083.
26 Beckett et al., British Army and the First World War, 230.
27 Ibid., 232.
28 Bidwell and Graham, Fire-Power, location 1840.
29 Bidwell and Graham, Fire-Power, location 1880.
30 Black, Great War and the Making of the Modern World, 113.
31 Gilbert, Somme: Heroism and Horror in the First World War, 37.
32 Strachan, First World War, 192.
33 Gilbert, Somme, 50.
34 Dyer, War, 270.
35 Dastrup, King of Battle, 171–174.

36 Strachan, First World War, 172.
37 Manchester, Arms of Krupp, 304–306.
38 Sumner, Kings of the Air, location 242–270.
39 Ibid., location 318.
40 Ibid., location 964.
41 Ibid., location 978.
42 NOAA, US Standard Atmosphere, 1976.
43 Ibid., location 1040.
44 Ibid., location 1070.
45 Ibid., location 1093.
46 Ibid., location 1238.
47 Gilbert, First World War, 42.
48 Ibid., 289, 448.
49 Ibid., 334–335.
50 Ibid., 396.
51 Fegan, Baby Killers, 63.
52 Cornish, Machine Guns and the Great War, location 739.
53 Ibid., location 1737.
54 Ibid., location 1843.
55 Ibid., location 2419.
56 Cornish, Machine Guns and the Great War, location 2505.
57 Ibid., location 191.
58 Gilbert, First World War, 259.
59 Cornish, Machine Guns and the Great War, location 191.
60 Ogorkiewicz, Tanks, 32.
61 Fuller, Tanks in the Great War, location 683.
62 Ibid., location 740.
63 Ibid., location 2477.
64 Ogorkiewicz, Tanks, 39–40.
65 Miller, Great Book of Tanks, 62–63.
66 Ibid., 48–49.
67 Ibid., 51–52.
68 Ogorkiewicz, Tanks, 45.
69 Ibid., 41.
70 Ibid., 42.
71 Massie, Castles of Steel, 681.
72 Strachan, First World War, 214.
73 Ibid., 214.
74 Ibid., 215.
75 Chickering, Imperial Germany and the Great War, 81.
76 Massie, Castles of Steel, 73.
77 Ibid., 126.
78 Black, Great War and the Making of the Modern World, 213.
79 Massie, Castles of Steel, 786–787.
80 Black, Great War and the Making of the Modern World, 36.
81 Kramer, Dynamic of Destruction, 6.
82 Massie, Castles of Steel, 140.
83 Thiel and Westerhoff, "Forced Labor," 7.

84 March et al., History of the World War, 226.
85 Gilbert, First World War, 358.
86 Black, Great War and the Making of the Modern World, 21.
87 Gilbert, First World War, 541.
88 Kershaw, To Hell and Back, 98.
89 McNutt, "Still No Man's Land."
90 Widdig, Culture and Inflation in Weimar Germany, 42.
91 Kaufmann et al., Maginot Line, locations 200-1255.

CHAPTER 5. THE SECOND WORLD WAR

1 Boot, War Made New, 212.
2 Herman, Freedom's Forge, location 17.
3 Ibid., location 2916.
4 Ibid., location 2223.
5 Ibid., location 3417.
6 Lane, Ships for Victory, 4, 204.
7 Herman, Freedom's Forge, locations 3971,4056.
8 Ibid., location 4115.
9 Ibid., location 4221.
10 Ibid., location 4238.
11 Ibid., location 4246.
12 Ibid., location 4322.
13 Ibid., location 4339.
14 Ibid., location 2973.
15 Ibid., location 2127.
16 Ibid., location 5385.
17 Koistinen, Arsenal of World War II, 498.
18 Herman, Freedom's Forge, location 2753.
19 Ogorkiewicz, Tanks, 40–45.
20 Macksey, Kenneth. Guderian, 27–40.
21 Ogorkiewicz, Tanks, 94.
22 Tucker, World War I Encyclopedia, Vol. I, 581.
23 Ogorkiewicz, Tanks, 94.
24 Ibid., 97.
25 Ibid., 96–97.
26 Macksey, Kenneth. Guderian, 103.
27 Ibid., 106.
28 Ibid., 124.
29 Keegan, The Second World War, 60.
30 Ogorkiewicz, Tanks, 119.
31 Ibid., 120.
32 Boot, War Made New, 238.
33 Ibid., 148.
34 Overy, Why the Allies Won, 5.
35 Manchester, Arms of Krupp, 356–357.
36 Ibid., 365.
37 Hogg, German Secret Weapons of World War II, location 2294.

38 Zaloga, Railway Guns of World War II, location 276.
39 Hogg, German Secret Weapons of World War II, location 2280–2287.
40 According to Manchester, Arms of Krupp, 438, only one was built; according to Hogg, German Secret Weapons of World War II, location 179, Gustav was built first then Dora; according to Zaloga, Railway Guns of World War II, location 179, Dora was built first then Gustav, but only one gun was assembled at any given time. Dora was the name of Gustav Krupp's chief designer's wife so it seems more probable that the first gun built would be named after Krupp himself.
41 Zaloga, Railway Guns of World War II, location 180.
42 Hogg, German Secret Weapons of World War II, location 2423.
43 Ibid., locations 783–937.
44 Ogorkiewicz, Tanks, 131.
45 Applied Physics Laboratory, 18.
46 Ibid., 18–19.
47 Ibid., 22.
48 Brown, 180.
49 Applied Physics Laboratory, 19.
50 Brown, 184.
51 Conant, location 4815.
52 Boot, War Made New, 262–263.
53 Neillands, Bomber War, 32.
54 Ibid., 36.
55 Overy, Bombing War, 70–83.
56 Brown, Radar History of World War II, 9.
57 Hijiya, Lee de Forest and the Fatherhood of Radio, 71.
58 Brown, Radar History of World War II, 35–36.
59 Ibid., 42–49.
60 Ibid., 45.
61 Ibid., 156.
62 Ibid., 155.
63 Ibid., 454.
64 Ibid., 173.
65 Bernstein, Hitler's Uranium Club.
66 Ibid., 43.

CHAPTER 6. IN THE SHADOW OF THE GREAT WARS

1 Ambrose, Rise to Globalism, 12.
2 Sarkees et al., Resort to War.
3 Kozak, LeMay, 305.
4 Ibid., 307.
5 Fehrenbach, This Kind of War, location 19.
6 Lewy, America in Vietnam, 4.
7 Ibid., 22, 24.
8 Ibid., 24.
9 Ibid., 10.
10 Ibid., 33.
11 Ibid., 453.
12 Ibid., 42, 74.

13 Ibid., vii.
14 Stevens and Ezell, Black Rifle, 57.
15 Ibid., 100–107.
16 Ibid., 107.
17 Ibid., 109.
18 Ibid., 121–123.
19 Ibid., 203.
20 Ibid., 209.
21 Ibid., 215.
22 Ibid., 216.
23 Pace et al., Global Positioning System, 245.
24 Atkinson, Crusade, 36.
25 Polmar, Ships and Aircraft of the US Fleet, 100-101.
26 Atkinson, Crusade, 15, 37.
27 Ibid., 226.
28 Ibid., 225.

CHAPTER 7. EPILOGUE

1 Bronowski, The Ascent of Man, 367.
2 Daub, "How Many People Have Ever Lived on Earth?"
3 Gilbert, Winston S. Churchill: Road to Victory, locations 11766-11768.
4 Gilbert, Churchill, 184.
5 Some of these ideas are discussed in Krauss, A Universe from Nothing.
6 Bohr, "The Structure of the Atom," 43.
7 Gleick, Genius, 244.
8 Rose, Harold Nicolson, 42.
9 Nicholson, Peacemaking, 5.

CHAPTER 8. GALLERY: PHOTOGRAPHS BY THE AUTHOR

1 Huyssen, Twilight Memories, 3.
2 Armstrong, Images of America: Maryland in World WAR I, 59.
3 Ibid., 59.
4 Ibid., 51.
5 Ibid., 59.
6 Cornish, Machine Guns and the Great War, location 1121.
7 Ibid., location 732.
8 Gardiner and Gray, eds. Conway's All the World's Fighting Ships 1906–1921, 112–115.
9 Ibid., 21-115.
10 Doyle, USS Texas: Squadron at Sea, 119.
11 Ibid., 31.
12 Power, Battleship Texas, 66–68.
13 Ibid., 34.
14 Power, Battleship Texas, 74.
15 Anonymous. Tank Data: Aberdeen Proving Grounds Series, 1.
16 Ogorkiewicz, Tanks, 40–41.
17 Anonymous. Tank Data: Aberdeen Proving Grounds Series, 31.

18 Ibid., 42.
19 Greenhalgh, The French Army and the First World War, 357.
20 Westermann, Flak: German Anti-Aircraft Defenses, 24.
21 Fredriksen, International Warbirds, 253.
22 National Museum of the US Air Force, "F-1 Camel."
23 Nijboer, Fighting Cockpits, 46–49.
24 Armstrong, Images of America: Maryland in World WAR I, 44.
25 Dastrup, King of Battle, 164–165.
26 Potts, J.R and Dan Alex, "M1919 16-inch Naval Gun."
27 "16"/50 (40.6 cm) Mark 2 and Mark 3," NavWeaps Naval Weapons, Naval Technology, and Naval Reunions website,
28 Potts, J.R. and Dan Alex, "M1919 16-inch Naval Gun."
29 Ibid.
30 "Model 1931 (B-4)," Military Factory website.
31 Ibid.
32 Anonymous. Tank Data: Aberdeen Proving Grounds Series, 37.
33 Miller, Tanks, 111.
34 Anonymous. Tank Data: Aberdeen Proving Grounds Series, 33.
35 Miller, Tanks, 130–133.
36 Anonymous. Tank Data: Aberdeen Proving Grounds Series, 127.
37 Miller, Tanks, 116–117.
38 Foss, Encyclopedia of Tanks and Armored Fighting Vehicles, 475.
39 Anonymous. Tank Data: Aberdeen Proving Grounds Series, 131.
40 Foss, Encyclopedia of Tanks and Armored Fighting Vehicles, 475–476.
41 "Model 1937 (ML-20)," Military Factory website.
42 Hogg, Twentieth Century Artillery, 55.
43 Ibid., 76.
44 Potts, J.R., "M59 (M2 Long Tom)."
45 Anonymous. Tank Data: Aberdeen Proving Grounds Series, 141.
46 Miller, Tanks, 202–203.
47 Gander, The Bofors Gun, 105.
48 Military Factory website, "21cm Mörser 18 (21cm Mrs 18)."
49 Anonymous. Tank Data: Aberdeen Proving Grounds Series, 117.
50 Miller, Tanks, 214–215.
51 Zaloga, Railway Guns of World War II, location 139.
52 Ibid., location 329.
53 Brown, Radar History of World War II, 35–36, 81.
54 Air Ministry (British), Introductory Survey of Radar Part II, Chapter 2, Section 44.
55 Dwyer, "Mitsubishi A6M Reisen (Zero-Sen)."
56 Ibid., "Focke-Wulf Fw 190."
57 Ibid., "Grumman F4F Wildcat."
58 Anonymous. Tank Data: Aberdeen Proving Grounds Series, 137.
59 Forty, Tanks of the World, 142.
60 Chamberlain and Ellis, Tanks of the World, 211.
61 Miller, Tanks, 234–237.
62 Anonymous. Tank Data: Aberdeen Proving Grounds Series, 15.
63 Miller, Tanks, 237–238.
64 Forty, Tanks of the World, 105.

65 Miller, Tanks, 235.
66 Juno Beach Centre, "Canadian-built Tanks."
67 Anonymous. Tank Data: Aberdeen Proving Grounds Series, 41.
68 Foss, Tanks and Armored Fighting Vehicles, 236–237.
69 Miller, Tanks, 172.
70 Anonymous. Tank Data: Aberdeen Proving Grounds Series, 50.
71 Foss, Tanks and Armored Fighting Vehicles, 237–239.
72 Miller, Tanks, 144–147.
73 Anonymous. Tank Data: Aberdeen Proving Grounds Series, 59.
74 Foss, Tanks and Armored Fighting Vehicles, 241–242.
75 Miller, Tanks, 292.
76 Anonymous. Tank Data: Aberdeen Proving Grounds Series, 61.
77 Foss, Tanks and Armored Fighting Vehicles, 241–242.
78 Anonymous. Tank Data: Aberdeen Proving Grounds Series, 67.
79 Foss, Tanks and Armored Fighting Vehicles, 268–269.
80 Anonymous. Tank Data: Aberdeen Proving Grounds Series, 71.
81 Foss, Tanks and Armored Fighting Vehicles, 268.
82 Anonymous. Tank Data: Aberdeen Proving Grounds Series, 79.
83 Foss, Tanks and Armored Fighting Vehicles, 273.
84 Potts, J.R., "12.8cm Flakzwilling 40."
85 Anonymous. Tank Data: Aberdeen Proving Grounds Series, 101.
86 Foss, Tanks and Armored Fighting Vehicles, 266.
87 Green, Weapons of Patton's Armies, 61–65.
88 Zaloga, Railway Guns of World War II, location 197.
89 Anonymous. Tank Data: Aberdeen Proving Grounds Series, 123.
90 Foss, Tanks and Armored Fighting Vehicles, 304.
91 Gardiner and Chesneau, eds., Conway's All the World's Fighting Ships 1922–1946, 98–99.
92 Thompson, Mark A., ed. Battleship USS Alabama, BB-60, 22–36.
93 Ibid., 12.
94 Ibid., 14.
95 Ibid., 14, 33–36.
96 Historic Naval Ships Organization, "Historic Naval Ships Guide," 27.
97 Ibid., "Historic Naval Ships Guide," 79.
98 Burr, Lawrence. US Fast Battleships 1936–47, 15.
99 Clymer, A. Ben. "The Mechanical Analog Computers of Hannibal Ford and William Newell," 23–27.
100 Lane, Ships for Victory, 257.
101 Herman, Freedom's Forge, locations 1846–1888.
102 Gardiner and Chesneau, eds., Conway's World's Fighting Ships 1922–1946, 102–104.
103 Stahura, USS Yorktown CV-10 CVA-10 CVS-10, 6–27.
104 Ibid., 22–23.
105 Anonymous. Tank Data: Aberdeen Proving Grounds Series, 143.
106 Foss, Tanks and Armored Fighting Vehicles, 423.
107 Anonymous. Tank Data: Aberdeen Proving Grounds Series, 83.
108 Foss, Tanks and Armored Fighting Vehicles, 275.
109 "Canadian-built Tanks," Juno Beach Centre.
110 Hyde. Arsenal of Democracy, location 2736.
111 Green and Brown. M4 Sherman at War, 8.
112 Anonymous. Tank Data: Aberdeen Proving Grounds Series, 85.

113 Doyle, German Military Vehicles, 194–195.
114 Reilly, Kamikazes, Corsairs, and Picket Ships,166–167.
115 Historic Naval Ships Organization, “USS Lionfish (SS-298).”
116 Green and Monroe-Jones. The Silent Service in World War II, 114–116.
117 Anonymous. Tank Data: Aberdeen Proving Grounds Series, 99.
118 Foss, Tanks and Armored Fighting Vehicles, 270.
119 Anonymous. Tank Data: Aberdeen Proving Grounds Series, 93.
120 Foss, Tanks and Armored Fighting Vehicles, 271–272.
121 Anonymous. Tank Data: Aberdeen Proving Grounds Series, 103.
122 Doyle, German Military Vehicles, 139–141.
123 Anonymous. Tank Data: Aberdeen Proving Grounds Series, 107.
124 Doyle, German Military Vehicles, 144–145.
125 Foss, Tanks and Armored Fighting Vehicles, 271.
126 Ibid., 425–426.
127 Pelt, Rocketing into the Future, 78–79.
128 Speer, Albert. Inside the Third Reich, 362–364.
129 Bradsher, Greg, “The German Jet Me-262 in 1944: A Failed Opportunity – Part II.”
130 Foss, Tanks and Armored Fighting Vehicles, 63–64.
131 Anonymous. Tank Data: Aberdeen Proving Grounds Series, 139.
132 Chamberlain and Ellis, Tanks of the World, 211.
133 Foss, Tanks and Armored Fighting Vehicles, 389.
134 Ibid., 44–46.
135 Petrescu and Petrescu, The Aviation History, 31–33.
136 Knaack, Encyclopedia of US Air Force Aircraft and Missile Systems, Vol.1, 135–157.
137 Nash, “76mm Gun Tank T92.”
138 Skaarup, “Ironsides.”
139 Camp, Assault from the Sky, 15–16.
140 Miller, Tanks, 430–433.
141 Foss, Tanks and Armored Fighting Vehicles, 247.
142 Ballard, Development and Employment of Fixed-Wing Gunships 1962–1972, 28–76.
143 Williams, A History of Army Aviation, 120.
144 Vietnam Helicopters Museum, “UH-1H Iroquois “Huey” Helicopter.”
145 National Museum of the US Air Force,“ AC-130A Spectre.”
146 Ballard, Development and Employment of Fixed-Wing Gunships 1962–1972, 28–129.
147 National Museum of the US Air Force, “International GBU-8 Electro-Optical Guided Bomb.”
148 Tucker, Persian Gulf War Encyclopedia, 59–60.
149 Forty, Tanks of the World, 206.
150 Foss, Tanks and Armored Fighting Vehicles, 427.
151 Ibid., 91.
152 National Defense and the Canadian Armed Forces, Government of Canada, “Tank Replacement Project,” 14 February 2017, http://www.forces.gc.ca/en/business-equipment/tank-replacement.page, accessed 18 March 2018.
153 Foss, Tanks and Armored Fighting Vehicles, 396–398.
154 Witcomb, On a Steel Horse I Ride, 101, 113, 165, 437.
155 Foss, Tanks and Armored Fighting Vehicles, 49–51.
156 Ibid., 73–74.
157 Cohen, Gulf War Air Power Survey, Vol. V, 553.

158 National Museum of the US Air Force, “Atomics Aeronautical Systems RQ-1 Predator.”
159 GlobalSecurity.org,“ GBU-43/B ‘Mother of All Bombs,’ MOAB — Massive Ordnance Air Blast Bomb.”
160 Wasserbly, “US forces in Afghanistan drop first GBU-43/B MOAB in combat.”

BIBLIOGRAPHY

"AC-130A Spectre," National Museum of the US Air Force, http://www.nationalmuseum.af.mil/Visit/Museum-Exhibits/Fact-Sheets/Display/Article/196341/lockheed-ac-130a-spectre/, accessed 18 March 2018.

Ambrose, Stephen E. *Rise to Globalism: American Foreign Policy Since 1938.* New York: Penguin Books, 1971.

Anonymous. *Tank Data: Aberdeen Proving Grounds Series*. Old Greenwich, CT: WE, Inc. undated c. 1969.

"APG History," Harford County Website, http://www.harfordcountymd.gov/1225/APG-History, accessed 6 March 2018.

Applied Physics Laboratory,"APL and the VT Fuze," APL Technical Digest, September–October 1962, 18–22.

Appleby, Joyce. *The Relentless Revolution: A History of Capitalism*. New York: W. W. Norton & Co., 2010, Kindle edition.

Armstrong, William M. *Images of America: Maryland in World WAR I*. Charleston, SC: Arcadia Publishing, 2017.

Atkinson, Rick. *Crusade: The Untold Story of the Persian Gulf War*. Boston Houghton Mifflin Company, 1993.

Ballard, Jack S. *Development and Employment of Fixed-Wing Gunships 1962–1972*. Washington, DC: Office of Air Force History (US Government Printing Office), 1982. https://media.defense.gov/2010/May/26/2001330293/-1/-1/0/AFD-100526-036.pdf, accessed 18 March 2018.

Beckett, Ian, Timothy Bowman, and Mark Connelly. *The British Army and the First World War*. Cambridge: The Cambridge University Press, 2017.

Bernstein, Jeremy. *Hitler's Uranium Club: The Secret Recordings at Farm Hall*. New York: Copernicus Books, 2001. First published 1996 by American Institute of Physics.

Bidwell, Shelford and Dominick Graham. *Fire-Power: British Army Weapons and Theories of War 1904–1945*. Yorkshire: Pen & Sword Books Limited, 2004, Kindle edition. First published by George Allen & Unwin 1982.

Black, Jeremy. *The Great War and the Making of the Modern World.* London: Continuum International Publishing Group, 2011.

Bogart, Ernest L. *Direct and Indirect Costs of the Great World War*. New York: Oxford University Press, 1920.

Bohr, Niels. "The Structure of the Atom." Nobel Prize Lecture, December 11, 1922. https://

www.nobelprize.org/nobel_prizes/physics/laureates/1922/bohr-lecture.html, accessed 31 July 2017.

Boot, Max. *War Made New: Weapons, Warriors, and the Making of the Modern World*. New York: Gotham Books, 2007, paperback edition. First published 2006.

Bradsher, Greg, "The German Jet Me-262 in 1944: A Failed Opportunity – Part II," The National Archives Text Message Blog, 18 December 2013. https://text-message.blogs.archives.gov/2014/12/18/the-german-jet-me-262-in-1944-a-failed-opportunity-part-ii/, accessed 16 March 2018.

Brodie, Bernard and Fawn M. Brodie. *From Crossbow to H-Bomb*. Bloomington, IN: Indiana University Press, 1973. First published 1962.

Bronowski, J. *The Ascent of Man*. Boston: Little, Brown and Co., 1973

Bronowski, J. and Bruce Mazlish. *The Western Intellectual Tradition: From Leonardo to Hegel*. New York: Harper & Row, 1974, paperback edition. First published 1960.

Brown, Louis. *A Radar History of World War II: Technical and Military Imperatives*. Bristol, UK: Institute of Physics Publishing, 1999.

Brunel, Isambard. *The Life of Isambard Kingdom Brunel*. London: Longmans, Green, and Co., 1870.

Burr, Lawrence. *US Fast Battleships 1936–47: The North Carolina and South Dakota Classes*. Oxford: Osprey Publishing Ltd., 2010.

Camp, Dick. *Assault from the Sky: U.S. Marine Corps Helicopter Operations in Vietnam*. Havertown, PS: Casemate Publishers, 2013.

Camp, J.M. and C.B. Francis. *The Making, Shaping, and Treating of Steel*. Pittsburgh: The Carnegie Steel Company, 1920, second edition.

Canaday, John. *Mainstreams of Modern Art*. New York: Holt, Reinhart and Winston, 1966. First published 1959.

"Canadian-built Tanks," Juno Beach Centre, https://www.junobeach.org/canada-in-wwii/articles/armoured-fighting-vehicles/canadian-built-tanks/, accessed 11 March 2018.

Caspar, Max. *Kepler*. Translated and edited by C. Doris Hellman. New York: Dover Publications, Inc., 1993. First published by Abelard-Schuman 1959.

Chamberlain, Peter and Chris Ellis. *Tanks of the World: 1915–1945*. London: Cassell & Co, 2002. First published by Arms and Armour Press 1972.

Chickering, Roger. *Imperial Germany and the Great War, 1914–1918*. Cambridge: Cambridge University Press, 2004. First published 1998.

Clymer, A. Ben. "The Mechanical Analog Computers of Hannibal Ford and William Newell." IEEE Annals of the History of Computing 15, no. 2 (1993): 19–34.

Conant, Jennet. Tuxedo Park: A Wall Street Tycoon and the Secret Palace of Science That Changed the Course of World War II. New York: Simon & Schuster Paperbacks, 2003, Kindle edition. First published 2002.

Cornish, Paul. *Machine Guns and the Great War*. South Yorkshire, UK: Pen & Sword Books Ltd, 2009, Kindle edition.

Cruttwell, C.R.M.F. *A History of the Great War 1914–1918*. Chicago: Academy Chicago Publishers, 1991. First published by Oxford Publishing Co. 1934.

Daub, Carl. "How Many People Have Ever Lived on Earth?" Population Reference Bureau, October, 2011, http://www.prb.org/Publications/Articles/2002/HowManyPeopleHaveEverLivedonEarth.aspx, accessed 20 JUL 2017.

Dougan, David. *The Great Gun-Maker: The Story of Lord Armstrong*. Newcastle upon Tyne, UK: Frank Graham, 1971.

Doyle, David. *Standard Catalog of German Military Vehicles*. Iola, WI: KP Books, 2005.

Doyle, David. *USS Texas: Squadron at Sea*. Carrollton, TX: Squadron/Signal Publications, 2012.

Du Pont, Bessie Gardner. *E.I. Du Pont de Nemours and Company: a History 1802–1902*. Boston: Houghton Mifflin Company, 1920.

Dwyer, Larry, "Focke-Wulf Fw 190," The Aviation History Online Museum, http://www.aviation-history.com/focke-wulf/fw190.html.

Dwyer, Larry, "Grumman F4F Wildcat," The Aviation History Online Museum, http://www.aviation-history.com/grumman/f4f.html, accessed 11 March 2018.

Dwyer, Larry, "Mitsubishi A6M Reisen (Zero-Sen)," The Aviation History Online Museum, http://www.aviation-history.com/mitsubishi/zero.html, accessed 11 March 2018.

Dyer, Gwynne. *War: The Lethal Custom*. New York: Carroll & Graf Publishers, 2005, revised edition. First published 1985 Random House.

Eksteins, Modris. *Rites of Spring: The Great War and the Birth of the Modern Age*. Boston: Houghton Mifflin Company, 1989.

Eliot, Gil. *Twentieth Century Book of the Dead*. New York: Charles Scribner's Sons, 1972.

Engelbrecht, H.C. and F.C. Hanighen. *Merchants of Death: A Study of the International Armament Industry*. New York: Dodd, Mead & Company, 1934.

Ergang, Robert. *Europe from the Renaissance to Waterloo*. Boston: D.C. Heath and Company, 1954 edition, 1966 printing. First published 1939.

Fegan, Thomas. *The Baby Killers: German Air Raids on Britain in the First World War*. South Yorkshire: Pen & Sword Ltd, 2012. First published by Leo Cooper 2002.

Fehrenbach, T.R. *This Kind of War: The Classic Military History of the Korean War*. New York: Open Road Media, 2014, Kindle edition.

Ferguson, Colin R. and Allan T. Kirkpatrick. *Internal Combustion Engines: Applied Thermosciences*. Chichester, UK: John Wiley & Sons Ltd, 2016, third edition. First edition 2014.

Ferguson, Niall. *The War of the World: Twentieth Century Conflict and the Descent of the West*. New York: Penguin Books, 2006.

Forbes, Nancy and Basil Mahon. *Faraday, Maxwell, and the Electromagnetic Field: How Two Men Revolutionized Physics*. New York: Prometheus Books, 2014.

Foss, Christopher F., ed. *The Encyclopedia of Tanks and Armored Fighting Vehicles: The Comprehensive Guide to Over 900 Armored Fighting Vehicles from 1915 to the Present Day*. San Diego: Thunder Bay Press, 2002.

Foss, Christopher F. *Jane's AFV Recognition Handbook*. 2nd ed. Coulsdon, UK: Jane's Information Group Limited, 1992. First published 1987.

Fussel, Paul. *Doing Battle: The Making of a Skeptic*. Boston: Little, Brown and Company, 1996.

Fussel, Paul. *The Great War and Modern Memory*. New York: Oxford University Press, paperback edition 1977. First published in 1975.

Gander, Terry. *The Bofors Gun.* Barnsley, UK: Pen and Sword Military, 2013.

Gardiner, Robert, Roger Cheseau, and Eugene M Kolesnik, eds. *Conway's All the World's Fighting Ships 1860–1905*. London: Conway Maritime Press Ltd, 1979.

Gardiner, Robert and Randall Gray, eds. *Conway's All the World's Fighting Ships 1906–1921*. Annapolis, MD: Naval Institute Press, 1986. First published by Conway Maritime Press Ltd 1985.

Gardiner, Robert and Roger Chesneau, eds. *Conway's All the World's Fighting Ships 1922–1946*. New York: Mayflower Books, 1980. First published by Conway Maritime Press Ltd.

"GBU-43/B 'Mother of All Bombs,' MOAB — Massive Ordnance Air Blast Bomb," GlobalSecurity.org, https://www.globalsecurity.org/military/systems/munitions/moab.htm, accessed 19 March 2018.

Gilbert, Martin. *Churchill: A Life.* New York: Rosetta Books LLC, 2014, Kindle edition. First published by William Heineman Ltd 1991.

Gilbert, Martin. *The First World War: A Complete History*. New York: Henry Holt and Company, 1994.

Gilbert, Martin. *A History of the Twentieth Century, Volume Two: 1933–1951.* New York: Avon Books, 2000, paperback edition. First published by William Morrow and Company, 1998.

Gilbert, Martin. *Winston S. Churchill: Road to Victory, 1941–1945 (Vol VII)*. Hillsdale, Michigan: Hillsdale College Press and Rosetta Books, eBook Edition, 2015. First published by William Heinemann Ltd. And Houghton Mifflin 1986.

Gleick, James. *Genius: The Life and Science of Richard Feynman*. New York: Open Road Integrated Media, 2011, Kindle edition. First published by Macmillan 1992.

Green, Michael, and Edward Monroe Jones. *The Silent Service in World War II: The Story of the US Navy Submarine Force in the Words of the Men Who Lived It*. Havertown, PA: Casemate Publishers, 2012.

Green, Michael and Gladys. *Weapons of Patton's Armies*. Osceola, WI: MBI Publishing Company, 2000.

Green, Michael and James D. Brown. *M4 Sherman at War*. St. Paul, MN: Zenith Press, 2007.

Greenhalgh, Elizabeth. *The French Army and the First World War*. Cambridge: Cambridge University Press, 2014, Kindle edition.

Gribbin, John. *The Scientists: A History of Science Told through the Lives of Its Greatest Inventors*. New York: Random House, 2002.

Haines, Michael R. "The Population of Europe: The Demographic Transition and After." http://www.encyclopedia.com/international/encyclopedias-almanacs-transcripts-and-maps/population-europe-demographic-transition-and-after, accessed 27 MAR 17.

Heald, Henrietta. *William Armstrong: Magician of the North*. Alnwick, UK: McNidder & Grace, 2012. First published by Northumbria Press 2010.

Herman, Arthur. *Freedom's Forge: How American Business Produced Victory in World War II*. New York: Random House, 2012, Kindle edition.

Hijiya, James A. *Lee de Forest and the Fatherhood of Radio*. Bethlehem, PA: Lehigh University Press, 1992.

Historic Naval Ships Organization, "Historic Naval Ships Guide," Chatsworth, CA: Challenge Publications, Inc., 2004.

Historic Naval Ships Organization, "USS Lionfish (SS-298)," http://www.hnsa.org/hnsa-ships/uss-lionfish-ss-298/, accessed 15 March 2018.

Hochschild, Adam. *King Leopold's Ghost: A Story of Greed, Terror, and Heroism in Colonial Africa*. Boston: Mariner Books, 1999, eBook edition. First published in 1998 by Houghton Mifflin.

Hogg, Ian V. *German Secret Weapons of World War II*. New York: Skyhorse Publishing, 2016. Kindle Edition.

Hogg, Ian. *Twentieth Century Artillery*. New York: Barnes & Noble Books, 2000.

Hough, Richard. *Dreadnought: A History of the Modern Battleship*. London: Endeavor Press Ltd., 2015, Kindle edition. First published by Michael Joseph Ltd. 1965.

Hounshell, David A. *From the American System to Mass Production 1800–1932: The Development of Manufacturing Technology in the United States*. Baltimore: The Johns Hopkins University Press. 1985. First published 1984.

Huyssen, Andreas. *Twilight Memories: Marking Time in a Culture of Amnesia*. New York: Routledge, 1995.

Hyde, Charles K. *Arsenal of Democracy: The American Automobile Industry in World War II*. Detroit: Wayne State University Press, 2013.

Jaques, Lieutenant W.H., U.S.N. "The Establishment of Steel Gun Factories in the United States," Proceedings of the United States Naval Institute vol. X, no. 4, 1884. https://books.google.com/books?id=U59DAAAAIAAJ&pg=PA633&dq=Joseph+Whitworth+of+Sheffield&hl=en&sa=X&ved=0ahUKEwi6zd-Kn_TTAhWlYiYKHdSGD2MQ6AEIWzAJ#v=onepage&q=Joseph%20Whitworth%20of%20Sheffield&f=false, accessed May 16, 2017.

Johnston, Ian and Ian Buxton. *The Battleship Builders: Constructing and Arming British Capital Ships*. Yorkshire: Seaforth Publishing, 2013.

Kafka, Franz. *The Metamorphosis*. Translated from German by David Wyllie. Project Gutenberg Ebook, 2005, https://www.gutenberg.org/files/5200/5200-h/5200-h.htm, accessed 10 APR 2017.

Kaufmann, J.E., H.W. Kaufmann, Aleksander Jankoviè-Potoènik, and Patrice Lang. *The Maginot Line: History and Guide*. South Yorkshire: Pen & Sword Military, 2001, Kindle edition.

Keegan, John. *A History of Warfare*. New York: Vintage Books, 1994. First published by Alfred A. Knopf, Inc. 1993.

Keegan, John. *The Second World War*. New York: Penguin Books, 1989.

Kershaw, Ian. To Hell and Back: Europe, 1914–1949. New York: Viking, 2015, Kindle edition. First published by Allen Lane.

Knaack, Marcelle Size. *Encyclopedia of US Air Force Aircraft and Missile Systems, Vol.1*. Washington, DC: Office of Air Force History (US Government Printing Office), 1978. https://ia802307.us.archive.org/15/items/EncyclopediaOfUSAirForceAircraftAndMissileSystemsVolumeOne/EncyclopediaOfUSAirForceAircraftAndMissileSystemsVolumeOne.pdf, accessed 16 March 2018.

Koistinen, Paul A. C. *Arsenal of World War II: The Political Economy of American Warfare, 1940–1945*. Lawrence, KS: University of Kansas Press, 2004.

Koistinen, Paul A. C. *Mobilizing for Modern War: The Political Economy of American Warfare, 1865–1919*. Lawrence, KS: University Press of Kansas, 1997.

Kozak, Warren. *LeMay: The Life and Wars of General Curtis LeMay*. Washington, D.C.: Regnery Publishing, Inc.,
2009.

Kramer, Allan. *Dynamic of Destruction: Culture and Mass Killing in the First World War*. New York: Oxford

University Press, 2008, paperback edition. First published 2007.

Krauss, Lawrence. *A Universe from Nothing: Why There Is Something Rather Than Nothing*. London: Simon and Schuster, 2012.

Lane, Frederic C. *Ships for Victory: A History of Shipbuilding under the US Maritime Commission in World War II*. Baltimore: The Johns Hopkins University Press, 2001. First published 1951.

Lewy, Guenter. *America in Vietnam*. Oxford: Oxford University Press, 1980, paperback edition. First published 1978.

Lukacs, John. *At the End of an Age*. New Haven: Yale University Press, 2002.

Maas, Frank. "Canada's Leopards on the Prowl," Laurier Centre for Military, Strategic, and Disarmament Studies, http://canadianmilitaryhistory.ca/canadas-leopards-on-the-prowl-by-frank-maas/, accessed 18 March 2018.

Macksey, Kenneth. *Guderian: Creator of the Blitzkrieg*. New York: Stein and Day, 1976.

Macrosty, Henry W. *The Trust Movement in British Industry: A Study of Business Organisation*. London: Routledge/Thoemmes Press, 1907.

Maddison, Angus. *Contours of the World Economy, 1— 230AD: Essays in Macro-Economic History*. Oxford: Oxford University Press, 2007.

Malanowski, Gregory. *The Race for Wireless: How Radio was Invented (or Discovered?)*. Bloomington, IN: AuthorHouse, 2011.

Malia, Martin. *The Soviet Tragedy: a History of Socialism in Russia, 1917–1991*. New York: The Free Press, 1994. Kindle Edition.

Manchester, William. *The Arms of Krupp 1587–1968*. Boston: Little, Brown and Company, 1968.

March, Francis A., Richard J. Beamish, and General Peyton C March. *History of the World War: An Authentic Narrative of the World's Greatest War*. Philadelphia: The United Publishers of the United States and Canada, 1919.

Massie, Robert K. *Dreadnought: Britain, Germany, and the Coming of the Great War*. New York: Random House, 1991, Kindle edition.

McNutt, Amelia. "Still No Man's Land," Normandy Research Foundation, 25 April 2017, https://www.normandyresearchfoundation.com/single-post/2017/04/25/Still-No-Mans-Land, accessed 18 June 2017.

Menne, Bernhard. *Blood and Steel: The Rise of the House of Krupp*. New York: Lee Furman Inc., 1938, Kindle editon.

Miller, David. *The Great Book of Tanks: The World's Most Important Tanks from World War I to the*

Present Day. St. Paul, MN: MBI Publishing Company, 2002.

Miller, Martin. *The Neutron's Long Shadow: Legacies of Nuclear Explosive Production of the Manhattan Project*. Atglen, PA: Schiffer Publishing, Ltd., 2017.

Miller, Martin. *Weapons of Mass Destruction: Specters of the Nuclear Age*. Atglen, PA: Schiffer Publishing, Ltd., 2017.

Mitchell, B.R. *International Historical Statistics: Europe 1750–1993*. New York: Stockton Press, 1998.

"Model 1931 (B-4)," Military Factory website, https://www.militaryfactory.com/armor/detail.asp?armor_id=771, accessed 9 March 2018.

"Model 1937 (ML-20)," Military Factory website, https://www.militaryfactory.com/armor/detail.asp?armor_id=474, accessed 10 March 2018.

NASA Office of Space Science website, "Tycho's Nova," https://www.nasa.gov/audience/forstudents/postsecondary/features/F_Tycho_Nova.html, accessed September 26, 2017.

Nash, Mark. "76mm Gun Tank T92," Tank Encyclopedia, http://www.tanks-encyclopedia.com/coldwar/US/76mm-gun-tank-t92, accessed 17 March 2018.

National Park Service, "John Pershing - Success and Tragedy." https://www.nps.gov/prsf/learn/historyculture/john-pershing.htm, accessed May 26, 2017.

Neillands, Robin. *The Bomber War*. New York: Barnes & Noble, 2005. First published 2001 The Overlook Press.

Nichols, Kenneth D. *The Road to Trinity: A Personal Account of How America's Nuclear Policies Were Made.* New York: William Morrow and Company, Inc., 1987.

Nicholson, Harold. *Peacemaking, 1919*. London: Faber and Faber Ltd, 2013, ebook edition. First published by Constable & Co Ltd 1945.

NOAA, NASA, and USAF, "US Standard Atmosphere, 1976," Report NASA-TM-X-74335, Washington, DC, 1976. https://ntrs.nasa.gov/archive/nasa/casi.ntrs.nasa.gov/19770009539.pdf, accessed October 14, 2017.

"#4 Sherman Builders: Just How Many Tank Factories Did the US Have Anyway?" Sherman Tank Site, http://www.theshermantank.com/tag/alco/, accessed 14 March 2018.

Ogorkiewicz, Richard. *Tanks: 100 Years of Evolution.*. Oxford: Osprey Publishing, 2015, Kindle edition.

Overy, Richard. *The Bombing War: Europe 1939–1945*. London: Penguin Books, 2014. First published by Allen Lane 2013.

Overy, Richard. *Why the Allies Won*. New York: W.W. Norton & Company, 1996. First published 1995.

Pace, Scott, Gerald Frost, Irving Lachow, David Frelinger, Donna Fossum, Donald K. Wassem, Monica Pinto.

The Global Positioning System: Assessing National Policies. Santa Monica, CA: RAND, 1995. https://www.rand.org/pubs/monograph_reports/MR614.html, accessed 21 JULY 2017.

Pelt, Michel van. *Rocketing into the Future: The History and Technology of Rocket Planes*. New York: Springer-Praxis, 2012.

Pershing, John J. *My Experiences in the World War, Vol. 1.* Pickle Partners Press, 2013, ebook edition. First published by Frederick A. Stokes 1931.

Petrescu, Relly Victoria and Florian Ion Petrescu. *The Aviation History*. Norderstedt: Books on Demand GmbH, 2013.

Polmar, Norman. *The Naval Institute Guide to the Ships and Aircraft of the US Fleet*. Annapolis, MD: Naval Institute Press, Fifteenth Edition, 1993.

Popkin, Richard H. The Columbia History of Western Philosophy. New York: MJF Books, 1999.

Potts, J.R., "M59 (M2 Long Tom)," Military Factory website, https://www.militaryfactory.com/armor/detail.asp?armor_id=318, accessed 10 March 2018.

Potts, J.R., "17cm Kanone 18 (17cm K18)," Military Factory website, https://www.militaryfactory.com/armor/detail.asp?armor_id=520, accessed 11 March 2018.

Potts, J.R., "12.8cm Flakzwilling 40," Military Factory website, https://www.militaryfactory.com/armor/detail.asp?armor_id=488

Potts, J.R. and Dan Alex, "M1919 16-inch Naval Gun," Military Factory website, https://www.militaryfactory.com/armor/detail.asp?armor_id=568, accessed 8 March 2018.

Power, Hugh. *Battleship Texas*. College Station, TX: Texas A&M University Press, 2007. First published 1993.

Reilly, Robin L. *Kamikazes, Corsairs, and Picket Ships: Okinawa 1945*.

Ricketts, John A. "How a Blast Furnace Works." http://www.steel.org/making-steel/how-its-made/processes/how-a-blast-furnace-works.aspx, accessed 3 April 2017.

Rose, Norman. *Harold Nicolson*. London: Pimlico, 2006. First published by Jonathan Cape, 2005.

Roser, Max. "Life Expectancy," *Published online at OurWorldInData.org*, https://ourworldindata.org/life-expectancy, accessed 3 August 2017.

Sarkees, Meredith Reid and Frank Wayman. *Resort to War: 1816–2007*. Washington DC: CQ Press, 2010. Database: COW War Data, 1816–2007 (v4.0), http://www.correlatesofwar.org/data-sets/COW-war, accessed 16 July 2017.

Skaarup, Harold A. *"Ironsides": Canadian Armoured Fighting Vehicle Museums and Monuments*. Bloomington, IN: iUniverse, Inc., 2011.

Schneider and Cie, "The Schneider Works in France," Special Supplement to *Motorship* 4, no. 6 (1919). https://books.google.com/books?id=eoFNAQAAMAAJ&pg=PA36&dq=th+schneider+works+in+france&hl=en&sa=X&ved=0ahUKEwi9n7et3O3TAhVl4yYKHTEqBnMQ6AEIJzAA#v=onepage&q=th%20schneider%20works%20in%20france&f=false, accessed May 13, 2017.

Senate Historical Office, "Merchants of Death." https://www.senate.gov/artandhistory/history/minute/merchants_of_death.htm, accessed May 12, 2017.

Simpson, E., board president. *Report of the Gun Foundry Board.* Washington: Government Printing Office, 1984. https://books.google.com/books?id=chssAAAAYAAJ&printsec=frontcover&dq=Report+of+the+Gun+Foundry+Board+1884&hl=en&sa=X&ved=0ahUKEwi_lfno0_TTAhVqwVQKHWBqBlsQ6AEIMzAC#v=onepage&q=Report%20of%20the%20Gun%20Foundry%20Board%201884&f=false, accessed May16, 2017.

"16"/50 (40.6 cm) Mark 2 and Mark 3," NavWeaps Naval Weapons, Naval Technology, and Naval Reunions website, http://www.navweaps.com/Weapons/WNUS_16-50_mk2.php, accessed 8 March 2018.

Smiles, Samuel. *The Life of George Stephenson*. London: John Murray, 1858.

Speer, Albert. *Inside the Third Reich.* New York: Simon & Schuster Paperbacks, 1970. Originally published in 1969 under the title *Erinnerungen* by Verlag Ullstein GmbH.

Stahura, Barbara. *USS Yorktown CV-10 CVA-10 CVS-10*. Paducah, KY: Turner Publishing Company, 1997.

Stevens, R. Blake and Edward C. Ezell. *The Black Rifle: M16 Retrospective*. Cobourg, Canada: Collector Grade Publications Inc., 2015. First published 1992.

Strachan, Hew. *The First World War*. London: Penguin Books, 2005, paperback edition. First published 2003.

Sumner, Ian. *Kings of the Air: French Aces and Airmen of the Great War*. South Yorkshire: Pen and Sword Aviation, 2015, Kindle edition.

"Tank Replacement Project," National Defense and the Canadian Armed Forces, Government of Canada, http://www.forces.gc.ca/en/business-equipment/tank-replacement.page, accessed 18 March 2018.

Thiel, Jens and Christian Westerhoff. "Forced Labor." International Encyclopedia of the First World War, ver. 1.0, last updated 8 October 2014, http://encyclopedia.1914–1918-online.net/pdf/1914-1918-Online-forced_labour-2014-10-08.pdf, accessed 12 June 2017.

Thompson, Mark A., ed. *Battleship USS Alabama, BB-60*. Paducah, KY: Turner Publishing Company, 1993.

Tirman, John. *The Deaths of Others: The Fate of Civilians in America's Wars*. New York: Oxford University Press, 2011, Kindle edition.

Trinder, Barrie. *Britain's Industrial Revolution: The Making of a Manufacturing People*. Lancaster, UK: Carnegie Publishing Ltd, 2013.

"21cm *Mörser* 18 (21cm Mrs 18)," Military Factory website, https://www.militaryfactory.com/armor/detail.asp?armor_id=761, accessed 10 March 2018.

"UH-1H Iroquois "Huey" Helicopter," Vietnam Helicopters Museum, http://www.vietnamhelicopters.org/uh-1h-huey/, accessed 18 March 2018

US Federal Reserve Bank, St. Louis. FRED Economic Data website, https://fred.stlouisfed.org/series/A824RE1A156NBEA, accessed 16 July 2017.

"U.S. 16-Inch Gun, Mark II Mod 1, No. 138," World War 2 Headquarters, http://worldwar2headquarters.com/HTML/museums/Aberdeen/16-inch-gun.html, accessed 8 March 2018.

Warr, Peter. *Sheffield's Great War and Beyond*. South Yorkshire, UK: Pen & Sword Military, 2015.

Wasserbly, Daniel. "US forces in Afganistan drop first GBU-43/B MOAB in combat," HIS Jane's Defence Weekly, http://www.janes.com/article/69586/us-forces-in-afghanistan-drop-first-gbu-43-b-moab-in-combat, accessed 19 March 2018.

Watson, Peter. *The Modern Mind: An Intellectual History of the 20th Century*. New York: HarperCollins Publishers, 2001. First published by Weidenfeld & Nicolson 2000.

Webster, Donovan. *Aftermath: The Remnants of War*. New York: Vintage Books, 1998. First published by Pantheon books 1996.

Wells, Herbert George. *What are we to do with Our Lives?* Garden City, NY: Doubleday, Doran & Company, Inc., 1931.

Westermann, Edward. Flak: *German Anti-Aircraft Defenses, 1914–1945*. Lawrence, KS: University Press of Kansas, 2001.

White, Matthew. "National Death Tolls for the Second World War." http://necrometrics.com/ww2stats.htm#ww2chart, accessed March 19, 2017.

White, Matthew. "Source List and Detailed Death Tolls for the Primary Megadeaths of the Twentieth Century." http://necrometrics.com/20c5m.htm, accessed March 19, 2017.

Widdig, Bernd. *Culture and Inflation in Weimar Germany*. Berkeley, CA: University of California Press, 2001.

Wilbanks, James H. *Machine Guns: An Illustrated History of their Impact*. Santa Barbara, CA: ABC-CLIO Inc., 2004.

Williams, James W. *A History of Army Aviation: From Its Beginnings to the War on Terror*. New York: iUniverse, Inc, 2005.

Wiper, Steve and Tom Flowers. *USS Texas BB–35*. Revised ed. Tucson, AZ: Classic Warships Publishing, 2006. First published 1999.

Witcomb, Darrel D. *On a Steel Horse I Ride: A History of the MH-53 Pave Low Hicopters in War and Peace*. Maxwell Air Force Base, AL: Air University Press, 2012. http://www.afsoc.af.mil/Portals/86/documents/history/AFD-131112-025.pdf, accessed 18 March 2018.

Wolmar, Christian. *The Iron Road: An Illustrated History of the Railroad*. New York: DK Publishing, 2014, Kindle edition.

Woodward, David R. *The American Army and the First World War*. Cambridge: Cambridge University Press, 2014.

Wootton, David. *The Invention of Science: A New History of the Scientific Revolution*. New York: HarperCollins Publishers, 2015.

Zaloga, Steven J. *Railway Guns of World War II*. Oxford: Osprey Publishing Ltd., 2016, Kindle edition.

20th CENTURY WAR

Triumph and Tragedy, the keystone volume in the series, describes the five-hundred-year rise of science and relates and illustrates the uncertain development of the atomic bomb in WWII. *Weapons of Mass Destruction* c

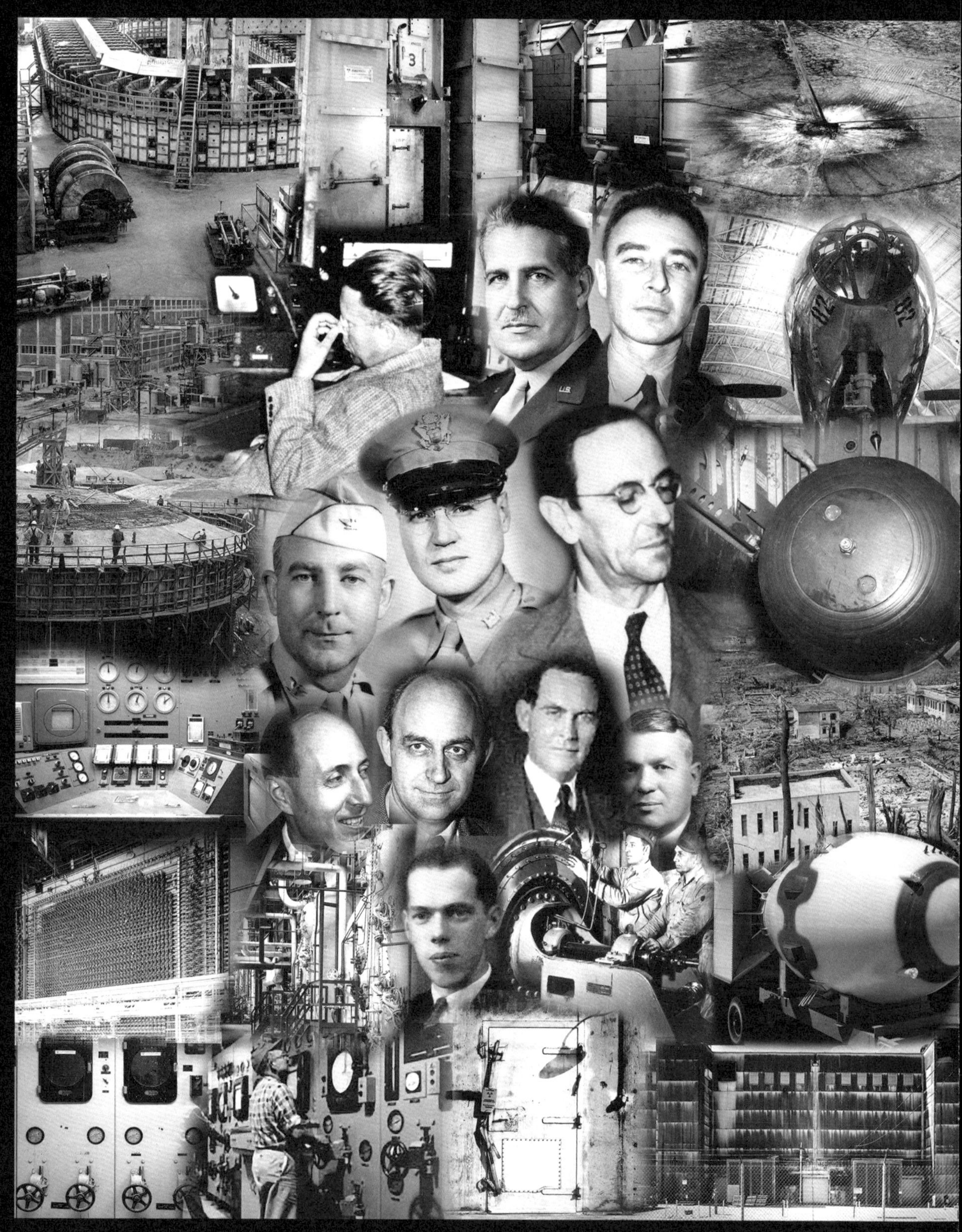

The Neutron's Long Shadow: Legacies of Nuclear Explosives Production in the Ma